In Situ Metal Matrix Composites: Theory and Applications—Celebration on the 40th Teaching Anniversary of Prof. Chunxiang Cui

In Situ Metal Matrix Composites: Theory and Applications—Celebration on the 40th Teaching Anniversary of Prof. Chunxiang Cui

Guest Editors

Xin Wang
Lichen Zhao

MDPI

Basel • Beijing • Wuhan • Barcelona • Belgrade • Novi Sad • Cluj • Manchester

Guest Editors
Xin Wang
School of Materials Science and Engineering
Hebei University of Technology
Tianjin
China

Lichen Zhao
School of Materials Science and Engineering
Hebei University of Technology
Tianjin
China

Editorial Office
MDPI AG
Grosspeteranlage 5
4052 Basel, Switzerland

This is a reprint of the Special Issue, published open access by the journal *Materials* (ISSN 1996-1944), freely accessible at: https://www.mdpi.com/journal/materials/special_issues/657K21AEAV.

For citation purposes, cite each article independently as indicated on the article page online and as indicated below:

Lastname, A.A.; Lastname, B.B. Article Title. *Journal Name* **Year**, *Volume Number*, Page Range.

ISBN 978-3-7258-5197-3 (Hbk)
ISBN 978-3-7258-5198-0 (PDF)
https://doi.org/10.3390/books978-3-7258-5198-0

Contents

About the Editors

Xin Wang

Xin Wang, Ph.D., is an Associate Professor of Metal Matrix Composites at the Hebei University of Technology, Tianjin, China. His research interests focus on the pressure infiltration molding of amorphous composite materials, functional amorphous alloys for wastewater treatment, and inoculation treatment techniques for aluminum, magnesium, and steel materials. He is the Vice Dean of the School of Materials Science and Engineering, the Deputy Director of the Hebei Key Laboratory of New Functional Materials and the Director of the Department of Metal Materials Engineering. He teaches professional courses such as Fundamentals of Materials Science, Mechanical Engineering Materials, and Material Surface and Interface at the School of Materials Science and Engineering. He obtained his PhD in Materials Science and Engineering (2013) from Tsinghua University, China.

Lichen Zhao

Lichen Zhao, Ph.D, is an Associate Professor of Metal Matrix Composites at the Hebei University of Technology, Tianjin, China. He received his Ph.D. in Materials Science from the School of Materials Science and Engineering at the Hebei University of Technology in 2011. He serves as the Deputy Director of the Department of Metal Materials Engineering and lectures on core courses including "Metal Matrix Composites", "Polymer Matrix Composites", and "Materials Synthesis and Preparation". His research interests include biodegradable metals for biomedical applications and metal matrix composites, structurally functional integrated porous metallic materials, magnesium alloys and their micro-arc oxidation, and the controlled synthesis and strengthening–toughening of aluminum–magnesium matrix composites.

Preface

Metal matrix composites (MMCs) have garnered significant attention in aerospace, automotive, electronics, and biomedical applications due to their design flexibility, lightweight nature, and superior performance. This Special Issue focuses on in situ-synthesized MMCs, compiling cutting-edge research on synthesis techniques, interface characterization, microstructure control, and performance optimization. Prof. Chunxiang Cui from the Hebei University of Technology has dedicated four decades to pioneering research and talent development in MMCs. His distinguished contributions include over 500 peer-reviewed publications and the mentorship of 400+ graduates, among whom 158 have earned master's degrees and 28 have obtained PhDs. To honor his 40-year academic career, we present this Special Issue to foster innovation in the field. The Special Issue features 14 high-quality papers covering aluminum alloys, metallic glass matrix composites, alloy steels, shape memory alloys, Al-Mg-based composites, Ti-Al-based superalloys, and magnetic materials, demonstrating the versatility and promising potential of MMCs. We hope that this Special Issue will inspire researchers of metallic and composite materials, offer fresh perspectives and drive advancements in structurally functionalized composites.

Xin Wang and Lichen Zhao

Guest Editors

materials

Article

Grain Refinement Mechanisms of $TiC_{0.5}N_{0.5}$ Nanoparticles in Aluminum

Kui Wang [1,2,*], Haiyan Jiang [1,2], Qudong Wang [1,2] and Yingxin Wang [1,2,*]

1 National Engineering Research Center of Light Alloys, Net Forming, Shanghai Jiao Tong University, Shanghai 200240, China
2 State Key Laboratory of Metal Matrix Composite, Shanghai Jiao Tong University, Shanghai 200240, China
* Correspondence: fateratory@sjtu.edu.cn (K.W.); wangyx@sjtu.edu.cn (Y.W.)

Abstract: In this study, $TiC_{0.5}N_{0.5}$ nanoparticles (NPs) are shown to induce a remarkable grain refinement of aluminum at various cooling rates. The grain refinement mechanisms are systematically investigated by microstructure observation, edge-to-edge matching (E2EM) model prediction, and first-principles calculations. The experimental results suggest that as the cooling rates increase from 10 K/s to 70 K/s, a transition from intergranular to intragranular distribution of NPs occurs and the Al/$TiC_{0.5}N_{0.5}$ interface varies from incoherent to coherent. Based on the E2EM analysis combined with first-principles calculation, it is found that $TiC_{0.5}N_{0.5}$ can act as a potent nucleant for the heterogeneous nucleation of α-Al. By analyzing the NP effects on the nucleation and growth of α-Al, the grain growth restriction and nucleation promotion mechanisms are proposed to elucidate the refinement phenomena at low and high cooling conditions, respectively.

Keywords: nanoparticle; grain refinement; cooling rate; first-principles calculation

1. Introduction

Microstructure control is of particular importance in the metal casting industry, and it is often the case that a uniformly fined, equiaxed grain structure is favorable for both cast and wrought alloys, as it benefits the casting process and improves the mechanical properties of the cast metal, facilitating subsequent mechanical working [1–3]. Grain refinement by inoculation is a well-established practice, which is usually accompanied by the addition of a grain refiner to the metallic melt [4,5]. For effective grain refinement, not only do the nucleants in the grain refiner need to provide potent heterogeneous sites for nucleation of the primary solid phase, but also some solute elements are required in the melt to restrict the grain growth and facilitate nucleation, either at a columnar front competing with equiaxed solidification or from particles where nucleation has already occurred [6–8].

While grain refiners are widely used in the foundry, their performance is not always satisfactory. Research shows that the refiner effectiveness is influenced by many factors, including the size and size distribution of nucleating particles, the alloy composition, as well as the cooling rate [9–11]. Firstly, according to the free-growth model [9], the critical undercooling at which the particle can be activated for grain initiation is inversely proportional to particle size, and because of the particle size distribution, only a minority of the particles can reach the critical undercooling where they are active. Therefore, grain refinement often has low efficiency, with <1% of the particles acting as effective nucleants. Furthermore, the refiner effectiveness is dramatically compromised in the presence of certain elements. For example, the grain refinement via Al-Ti-B inoculation is achievable for a majority of Al alloys. However, it has a poor refining effect in Zr/Si/Cr-containing Al alloys. This is attributable mainly to the loss of nucleation potency by the formation of intermetallic phases with Zr/Si/Cr that possess a poor crystallographic matching with α-Al [12–14]. In addition, recent research has demonstrated that a deterioration in the grain refining effect occurs at increased cooling conditions, including high-pressure die casting [15]. Ali and Liang et al. [16,17] suggested that such a grain coarsening effect at high cooling rate results from the small constitutional undercooling zone ahead of the solid/liquid interface due to the high temperature gradient.

Most recently, the practice of incorporating nanoparticles (NPs) into the metallic melt has presented exciting grain refinement results for Al and Mg alloys [18–20]. For instance, J. Xu et al. achieved the control of architecture and microstructure of an Al composite by the assembly of NPs with liquid Al in molten salt [20]. Unlike the inoculation, the added NPs are capable of restricting the grain growth through their accumulation at the solid/liquid interface. The physical nature of NP-induced refinement makes it become a potential avenue to address the intrinsic limitations of inoculation. In previous studies [21–23], the NPs are taken to serve as growth restrictors rather than nucleants. However, if possessing high nucleation potency, NPs are likely to promote nucleation of the solid phase. Although the growth-restricting effect of NPs has been experimentally observed by synchrotron X-ray tomography [19,20], the exact mechanisms underlying the grain refinement by NPs remain incompletely understood, especially from the perspective of their effects as heterogeneous nucleants. First-principles calculation based on density functional theory (DFT) is taken as a powerful and efficient tool for the prediction and estimation of the nucleation potency of nucleants. Up to now, little experimental work has been conducted to clarify the nucleation potency of NPs for α-Al, especially by the DFT calculation. Moreover, the cooling rate is a crucial factor affecting the grain size of cast metals. It is well accepted that a high imposed cooling rate can produce a large thermal undercooling at which NPs may be activated for heterogeneous nucleation [9,16,17]. Hence, the purpose of the present study is to understand the combined effects of cooling rate and NP addition on the grain refinement of aluminum. Density functional theory (DFT) was employed to quantitatively analyze the nucleation potency of TiCN as the nucleant for α-Al. The research results may be conducive to further revealing the NP-induced refinement mechanisms.

2. Experimental Procedures

In this work, commercially pure (CP) Al ingots at 99.7% purity were firstly melted using an electrical resistance furnace. After complete melting and mixing of the raw metals, 0, 0.5, 1.0, 1.5, and 2.0 vol.% $TiC_{0.5}N_{0.5}$ NPs with the mean size of ~100 nm were added into the melt and dispersed ultrasonically at 720 °C for 15 min under Ar gas protection. Then, a sonotrode was taken away from the melt and it was reheated to 750 °C before the pour was completed in a preheated cast iron mold. The average cooling rate is determined to be around 3 K/s. To achieve various cooling rates, the melt was also poured into a V-shaped copper mold where the cooling rate changes from 10 to 250 K/s. For comparison with the conventional inoculation, 0.1, 0.3, and 0.5 wt.% Al-3B and Al-5Ti-1B (wt.%) grain refiners were also added to the Al melt, respectively, followed by stirring and 5 min holding, before casting in the preheated cast iron mold.

The specimens for microstructural observation were sectioned from the bottom of each casting and grounded and polished mechanically. The etched specimens were characterized by an optical microscope with polarized light. The grain size was determined via the linear intercept method, as described by ASTM E112-10. The microstructure was examined by JEOL 78600F SEM, JEOL 2100 TEM, and JEOL 2100F FEGTEM.

In order to analyze the atomic structure of the interface, first-principles density functional theory (DFT) calculations were conducted by Vienna Ab initio Simulation Package (VASP) [24] with the Projector Augmented Wave (PAW) [25] method, and the Perdew–Burke–Ernzerhof (PBE) variant of generalized gradient approximation (GGA) [26] was used. The plane wave cutoff energy was equal to 520 eV. The energy tolerance for the electronic relaxation was 10^{-6} eV. The Hellmann–Feynman force tolerance for the ion relaxation was 0.01 eV·Å^{-1}. The Brillouin zone was sampled with a Gamma-centered *k*-point mesh. For the slab, a 12 × 12 × 1 *k*-point mesh was used and for the bulk a 12 × 12 × 12 *k*-point mesh was used. The Monte Carlo Special Quasi-random Structure (MCSQS) method [27] was employed for modeling the bulk $TiC_{0.5}N_{0.5}$ with the N-site in L1$_2$ TiN randomly substituted by C. In order to avoid surface–surface interactions, a vacuum space of at least 10 Å was imposed in the supercell [28].

3. Results and Discussion

The typical microstructures of the castings are illustrated in Figure 1. It is evident that the coarse columnar grains dominate in the microstructure of the CP Al. After an initial addition of 0.5 vol.% $TiC_{0.5}N_{0.5}$ NPs, the coarse columnar grains show a pronounced shift towards the refined equiaxed grain structure, and the average grain size is decreased from 1350 to 285 μm. Further addition of NPs to 2.0 vol.% triggers a significant refinement of equiaxed grains, and the average grain size is further reduced to 138 μm. It is also notable that the grain size shows a sensitivity to the NP addition level, and no obvious saturation of refining effect is observed even at high additions, implying the high refinement efficiency of $TiC_{0.5}N_{0.5}$ nanoparticles.

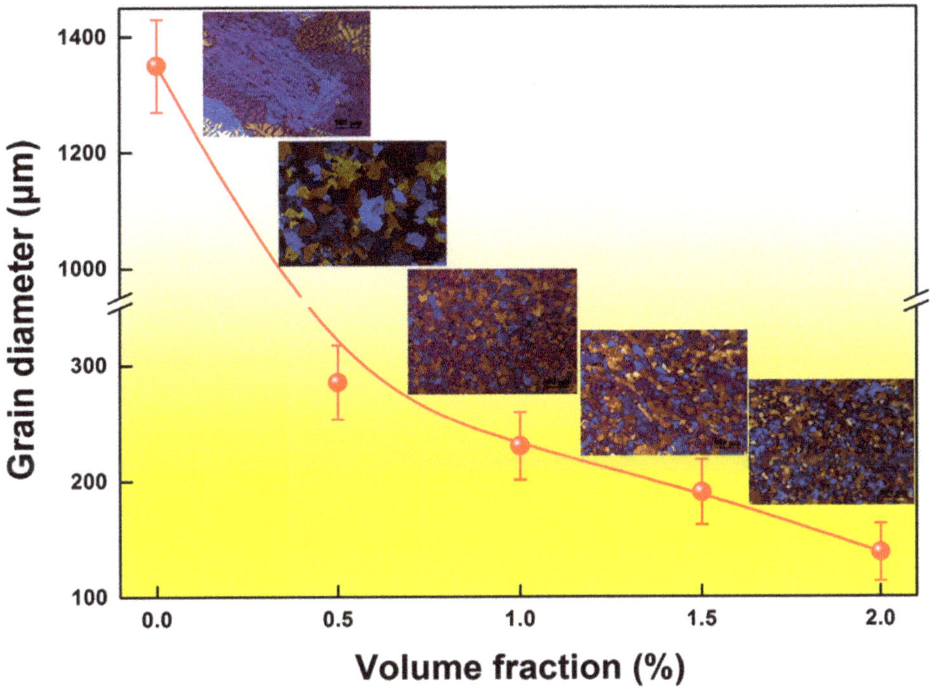

Figure 1. Variation in α-Al grain sizes with different NP addition levels. The insets are the anodized micrographs of the CP Al with various NP addition levels.

Figure 2 shows the comparison of NP-induced grain refinement with the conventional inoculation. The macrostructure of the CP Al castings is characterized by coarse columnar grain structure. When the addition level is raised, a large quantity of equiaxed grains is formed at the center of the sample inoculated by the Al-3B mater alloy. However, as no titanium element can be used as the growth restrictor, the grain refinement effect is far from satisfactory. In stark contrast, a pronounced reduction in grain size and transition to equiaxed grain structure are observed throughout the sample inoculated by the Al-5Ti-1B master alloy. In addition, an increase in the addition level results in fine equiaxed grain structure. As compared to inoculation, a stronger grain refinement occurs when $TiC_{0.5}N_{0.5}$ nanoparticles are incorporated into the melt. The finer and more uniform equiaxed grain structure shown in Figure 2 demonstrates that $TiC_{0.5}N_{0.5}$ nanoparticles can induce a more effective grain refinement than conventional inoculation.

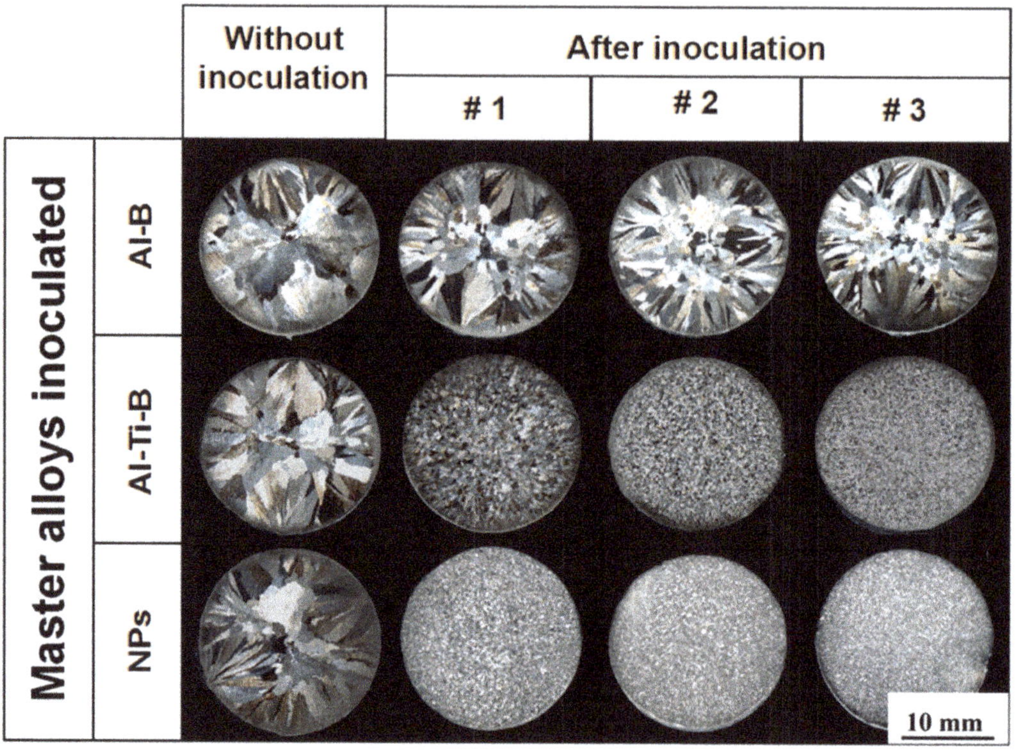

Figure 2. Comparison of NP-induced grain refinement with the conventional inoculation. For Al-B and Al-Ti-B master alloys, #1, #2, and #3 represent the addition level of 0.1 wt.%, 0.3 wt.%, and 0.5 wt.%, respectively. For NPs, #1, #2, and #3 represent the addition level of 0.5 vol.%, 1.0 vol.%, and 2.0 vol.%, respectively.

To unveil the influence of cooling rate on the NP-induced microstructural evolution, a wedge chill casting technique [29] was employed to produce the rapidly solidified specimens, which were subjected to various cooling rates. The anodized micrographs from the different positions on the V-shaped specimens with different NP contents are presented in Figure 3. Additionally, included in Figure 3 are the variations in grain size with cooling rate. It can be seen that the grains of the CP Al maintain the columnar structure, although an increased cooling rate leads to the finer α-Al grains. With the addition of 2 vol.% NPs, the coarse columnar grains are transformed to refined equiaxed ones, and the enhanced cooling rate leads to the further gain refinement. When the cooling rate reaches 70 K/s, the mean grain size is reduced up to 78 μm.

It is obvious that an increased cooling rate can lead to a decreased grain size by producing an increased melt undercooling. However, because no effective nucleants in the melt can induce the heterogeneous nucleation of α-Al, the bulky grains with columnar structure prevail in the matrix. By contrast, the samples with NP additions evince much more refined and homogeneous grain structure. This discrepancy can mainly be attributed to the NP influence on the nucleation and growth of α-Al. The microstructure evolution substantiates that the NPs might outweigh the cooling rate in terms of grain refinement.

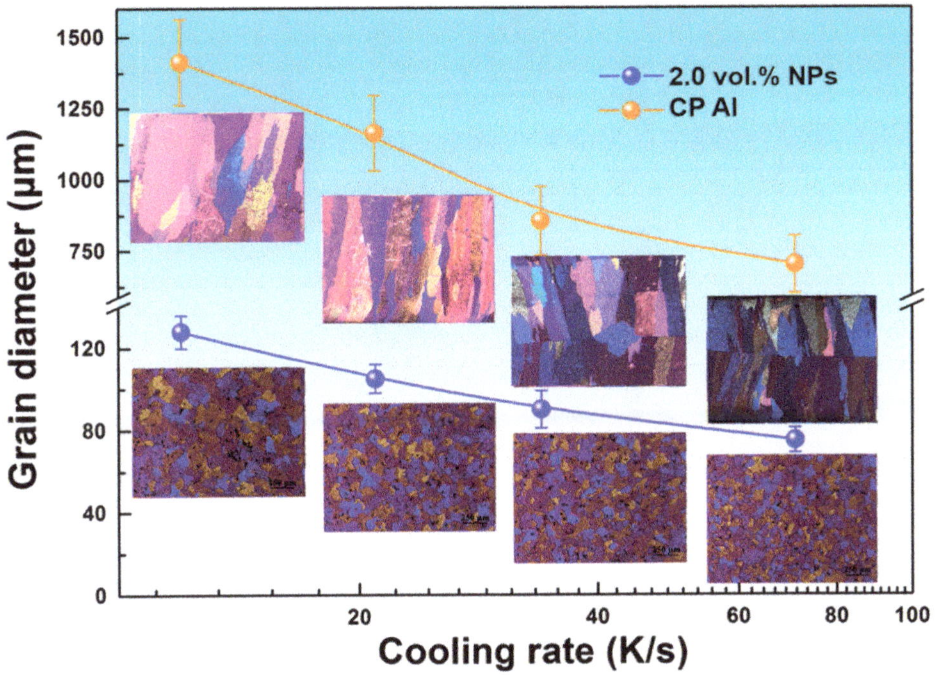

Figure 3. Variations in CP Al grain sizes with and without NP addition as functions of cooling rates. The insets are the anodized micrographs of the CP Al with and without NP addition solidified at different cooling conditions.

Figure 4 illustrates the distribution of NPs in the matrix of the V-shaped sample with the local cooling rate of 12 K/s. It is evident from Figure 4a that a myriad of NPs with the mean size of ~100 nm assemble onto the grain boundary. The EDS analysis in Figure 4b confirms the presence of TiCN nanoparticles. In addition to discrete nanoparticles, many nanoparticles aggregate together to form nanoclusters at the grain boundary. As depicted in Figure 4b, the accumulation of these NPs along the grain boundary results in the formation of an NP layer in between the α-Al grains. Figure 4c shows a nanoparticle located at the grain boundary. From the crystallographic information provided in Figure 4d, this nanoparticle has a face-centered cubic structure with the unit cell of NaCl-type, which corresponds to $TiC_{0.5}N_{0.5}$ [30]. Figure 4d also highlights the interfacial structure between the α-Al and the intergranular NP. It is indicated that no fixed or preferential crystallographic orientation relationship (OR) can be observed and thus an incoherent interface is formed between them.

From the analysis above, it may be speculated that the grain refinement which occurs at relatively low cooling conditions is predominantly ascribed to the growth restriction caused by NPs. According to the nanoparticle capture model proposed by Xu et al. [31], NPs tend to be pushed by the solidification fronts because the solidification front velocity is usually much smaller than the critical velocity for nanoparticle capture during regular solidification processes for which the cooling rate is generally very low. As crystallites grow, the NP pushing by the solid/liquid interface occurs, and then the assembly of NPs onto the grain boundary gives rise to an NP layer to significantly suppress the solute atom migration, thus restricting the grain growth.

On the contrary, once the solidification cooling rate exceeds 70 K/s, a great majority of NPs are distributed inside α-Al grains, as illustrated in Figure 5a. Figure 5b reveals the intragranular distribution of NPs. Instead of accumulating at the growing interface, they are evenly distributed in the Al matrix. Figure 5c presents a nanoparticle embedded inside the α-Al grain. From the HRTEM analysis in Figure 5d, it can be seen that the $TiC_{0.5}N_{0.5}$/Al interface with little formed reactant is clear and smooth. Furthermore, $TiC_{0.5}N_{0.5}$ forms a coherent interface with α-Al. Hence, a fixed OR can be achieved and is $(11\bar{1})_{Al}$ parallel to $(11\bar{1})_{NP}$, and $[011]_{Al}$ parallel to $[011]_{NP}$. Figure 5e exhibits the other nanoparticle entrapped in the matrix. Likewise, it possesses a coherent interface with the α-Al as shown in Figure 5f, and the OR is given as $(1\bar{1}1)_{Al}$ $[\bar{1}12]_{Al}//(1\bar{1}1)_{NP}$ $[\bar{1}12]_{NP}$.

To better understand the effect of $TiC_{0.5}N_{0.5}$ NPs on the grain refinement of α-Al at enhanced cooling conditions, their nucleation potency was examined in terms of the crystallographic matching, the work of adhesion, and the interfacial energies of Al/$TiC_{0.5}N_{0.5}$ interfaces.

As far as the crystallite nucleation is concerned, the edge-to-edge matching (E2EM) model proposed by Zhang et al. [32–34] is well documented to be effective in evaluating potential ORs between a nucleant substrate and solid phase. The interatomic spacing misfit (f_r) and the interplanar spacing mismatch (f_d) are the two pivotal factors in this model. If they are less than 10%, an OR between two phases can be predicted.

The possible orientation relationships between TiCN and Al are illustrated in Figure 6. Four low-energy and low-index interfaces between TiCN and Al are considered, including $(111)_{Al}//(111)_{TiCN}$ with Ti and C/N termination and $(200)_{Al}//(200)_{TiCN}$ and $(220)_{Al}//(220)_{TiCN}$ interfaces, in which the close-packed planes are matched across the interface. By calculation, seven potential ORs are given in Table 1. The determined values of f_r and f_d are less than 6.0% for almost all the potential Ors, implying a desirable crystallographic match. As commented by Zhang [32], when $f_d \leq 6\%$, the matching planes are parallel or nearly parallel. The predicted Ors involving a plane pair of $\{1\ 1\ 1\}_{Al}//\{1\ 1\ 1\}_{TiCN}$ are consistent with the HRTEM analysis in Figure 5. The combination of experimental results and the E2EM analysis indicates that pairs of Ors and interfaces may exist between Al and $TiC_{0.5}N_{0.5}$. Consequently, α-Al may be able to nucleate and grow on the surface of $TiC_{0.5}N_{0.5}$.

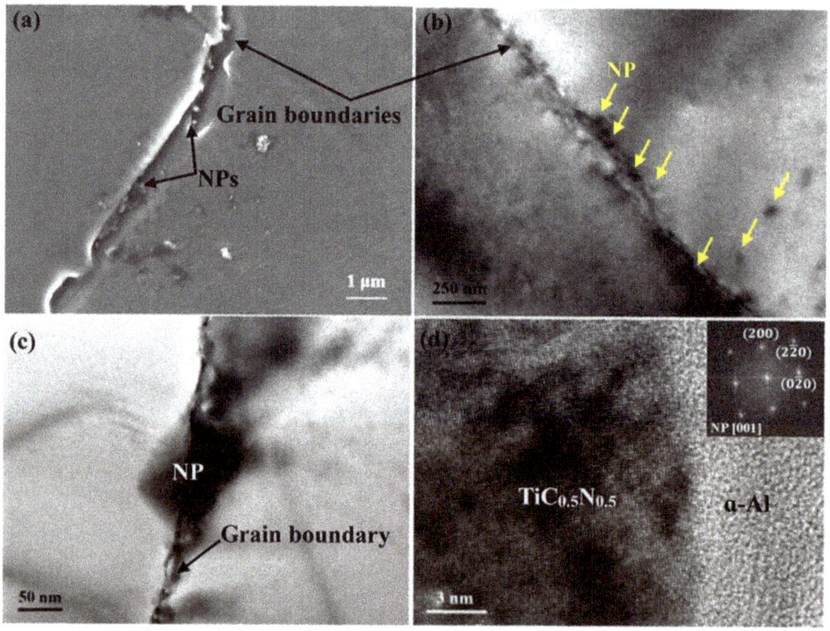

Figure 4. (**a**) SEM micrograph of CP Al with 2.0 vol.% NP addition solidified at the cooling rate of 10 K/s; (**b**,**c**) TEM bright field images showing the NP distribution along the grain boundary; (**d**) HRTEM image of the Al/$TiC_{0.5}N_{0.5}$ interface. The inset in (**d**) is the fast Fourier transformation (FFT) of the $TiC_{0.5}N_{0.5}$ nanoparticle.

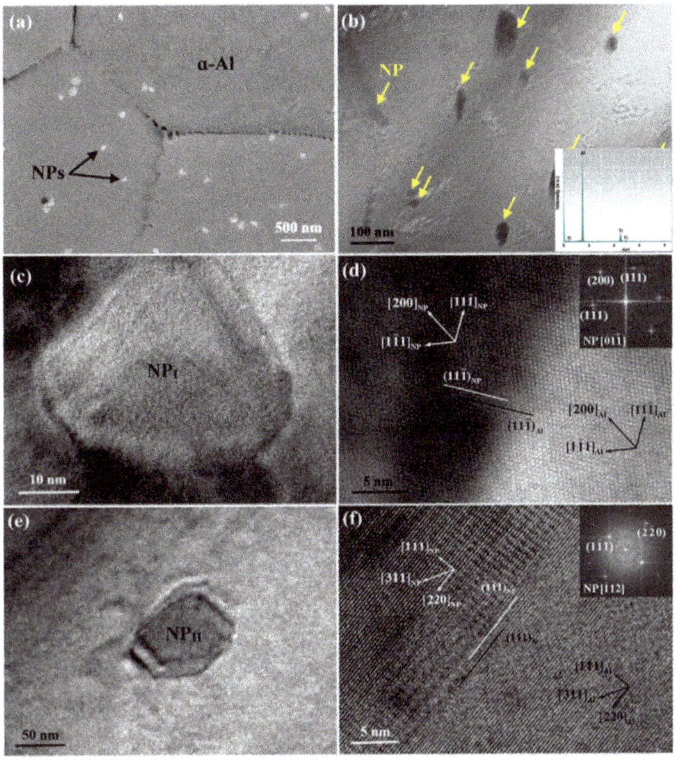

Figure 5. (**a**) SEM micrograph of CP Al with 2.0 vol.% NP addition solidified at the cooling rate of 70 K/s; (**b**) TEM bright field image showing the NP distribution within the interior of grain; (**c**,**e**) magnified view of the nanoparticle I (NP_I) and nanoparticle II (NP_{II}) in (**b**), respectively; (**d**,**f**) HRTEM images of the Al/NP_I and Al/NP_{II} interfaces, respectively. The insets in (**b**,**d**,**f**) are the EDS spectra of NPs and the fast Fourier transformation (FFT) of NP_I and NP_{II}, respectively.

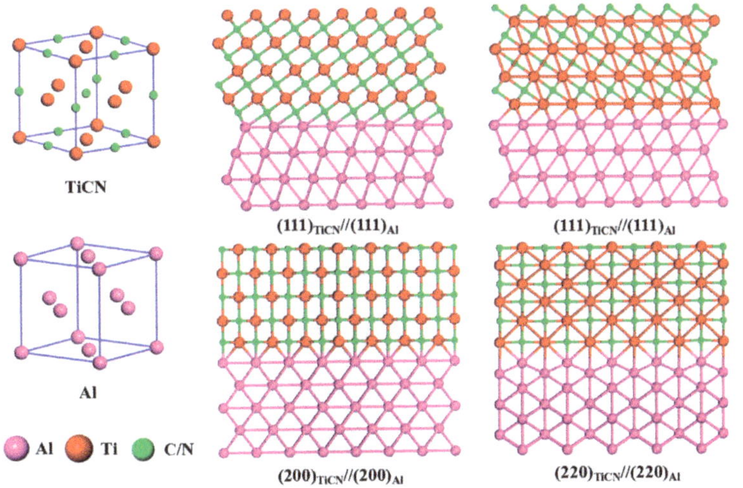

Figure 6. Interface structures of the Al matrix and TiCN nanoparticles.

Table 1. Possible orientation relationships between $TiC_{0.5}N_{0.5}$ and Al.

Crystal Structure and Lattice Parameters (nm)	Matching Planes	f_d (%)	Matching Direction	f_r (%)	OR
N: Al FCC a = 0.4049 S: $TiC_{0.5}N_{0.5}$ FCC a = 0.4286	$\{111\}_N//\{111\}_S$	5.85	$<110>_N//<110>_S$ $<112>_N//<112>_S$	5.86	$\{111\}_N//\{111\}_S$, $<110>_N//<110>_S$ $\{111\}_N//\{111\}_S$, $<112>_N//<112>_S$
	$\{200\}_N//\{200\}_S$	5.84	$<100>_N//<100>_S$ $<110>_N//<110>_S$	5.85	$\{200\}_N//\{200\}_S$, $<100>_N//<100>_S$ $\{200\}_N//\{200\}_S$, $<110>_N//<110>_S$
	$\{220\}_N//\{220\}_S$	5.87	$<100>_N//<100>_S$ $<110>_N//<110>_S$ $<112>_N//<112>_S$	5.83	$\{220\}_N//\{220\}_S$, $<100>_N//<100>_S$ $\{220\}_N//\{220\}_S$, $<110>_N//<110>_S$ $\{220\}_N//\{220\}_S$, $<112>_N//<112>_S$

Based on the Ors predicted by E2EM, three supercells containing $(111)_{Al}//(111)_{TiCN}$, $(200)_{Al}//(200)_{TiCN}$, and $(220)_{Al}//(220)_{TiCN}$ interfaces, respectively, were constructed to further reveal the nature of $Al/TiC_{0.5}N_{0.5}$ interfaces through density functional theory (DFT). The supercell models of (100), (110), and (111) $Al/TiC_{0.5}N_{0.5}$ are displayed in Figure 7a–c, respectively. The models of the polar Al_2Ti_2NC interface are the $(1\,1\,1)_{Ti2NC}$ | | $(1\,1\,1)_{Al}$, $(1\,1\,0)_{Ti2NC}$ | | $(1\,1\,0)_{Al}$, and $(1\,0\,0)_{Ti2NC}$ | | $(1\,0\,0)_{Al}$, in which the close-packed planes are matched across the interface with N/C-terminated Ti_2NC. To accommodate the periodic boundary condition inherent in a supercell calculation, we invoke the coherent interface approximation in which the softer Al matches the dimensions of the Ti_2NC. To make the results comparable and reasonable, the ratio between the content of C and that of N is adopted as 1 on the interface for all Al/Ti_2NC, and all the models are strictly stoichiometric, Al_2Ti_2NC.

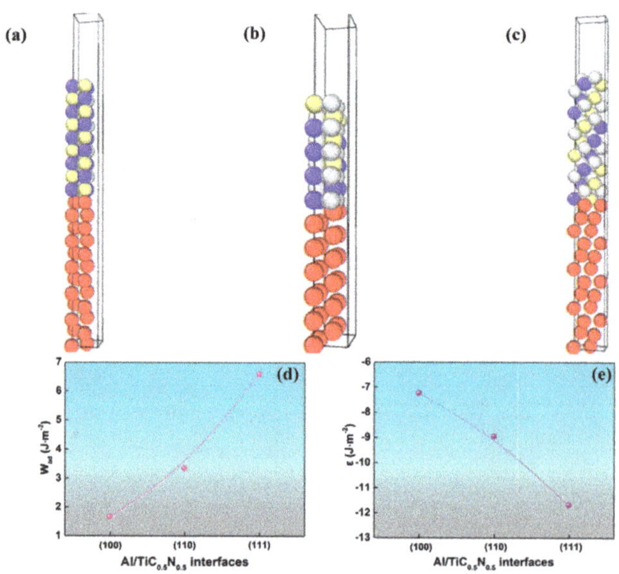

Figure 7. (**a**–**c**) Supercell models of (100), (110), and (111) $Al/TiC_{0.5}N_{0.5}$, respectively; (**d**) work of adhesion of three interface structures; (**e**) interfacial energy of three interface structures. Note that the red, blue, yellow, and white balls in (**a**–**c**) represent the Al, Ti, C, and N atoms, respectively.

The ideal work of adhesion (W_{ad}), which is defined as the reversible work against the adhesion strength of interface atoms, is commonly used to estimate the interfacial quality [35]. It can be determined by the difference in total energy between the interface and its isolated slabs [35,36]:

$$W_{ad} = (E_{Al}^{Slab} + E_{TiCN}^{Slab} - E_{Al/TiCN}^{Slab})/S \quad (1)$$

where E_{Al}^{Slab} and E_{TiCN}^{Slab} are the total energy of the fully relaxed, isolated Al and TiCN slabs, respectively, and $E_{Al/TiCN}^{Slab}$ is the total energy of the Al/TiCN interface system. S is the total interface area of the unit cell.

Figure 7d shows the calculated W_{ad} for the three interface structures. There is an obvious trend of increasing W_{ad} with the index of the lattice plane. The Al/$TiC_{0.5}N_{0.5}$ (111) interface exhibits the largest W_{ad} value of 6.6 J/m^2, whereas the (100) interface shows the smallest W_{ad} value of 1.2 J/m^2. Increased W_{ad} is correlated with increased thermodynamic stability. Therefore, the Al/$TiC_{0.5}N_{0.5}$ (111) interface is expected to be the most thermodynamically stable.

The interfacial energy is an important measure of interfacial stability, which can be defined as [35,36]

$$\varepsilon = (E_{Al/TiCN} - \frac{n_1}{N_1}E_{Al}^{Bulk} - \frac{n - n_1}{N_2}E_{TiCN}^{Bulk})/S \quad (2)$$

where ε is the interfacial energy, n_1 and n are the number of Al and all atoms in the interfacial structure, respectively, E_{Al}^{Bulk} and E_{TiCN}^{Bulk} are the per-atom energy of the relaxed bulk Al and TiCN in the same supercell consisting of N_1 and N_2 atoms without vacuum, respectively, $E_{Al/TiCN}$ is the total energy of the Al/TiCN interface system, and S is the total interface area of the unit cell. Generally, the lower the interfacial energy, the more stable the interface structure.

Figure 7e shows the interfacial energies of three Al/$TiC_{0.5}N_{0.5}$ interfaces. Although the matching plane pairs predicted by E2EM have the similar f_d values, there is a great difference in interfacial energy. From Figure 7d, it can be seen that the interfacial energy is decreased with increasing the index of the lattice plane, and the Al/$TiC_{0.5}N_{0.5}$ (111) interface has the lowest interfacial energy of −11.7 J/m^2. For all three Al/$TiC_{0.5}N_{0.5}$ interfaces, the calculated interfacial energies are far less than the solid–liquid interfacial energy of Al (0.158 J/m^2) [9], which means that the α-Al grain is able to nucleate on the $TiC_{0.5}N_{0.5}$ surface.

Based on the DFT analysis, it can be inferred that of the three Al/$TiC_{0.5}N_{0.5}$ interfaces, the (111) interface with the largest work of adhesion and smallest interfacial energy is the most thermodynamically stable and energetically favored nucleation site for the heterogeneous nucleation of α-Al. As a result, it is highly possible that α-Al preferentially nucleates on the most close-packed plane of $TiC_{0.5}N_{0.5}$, i.e., $(111)_{TiCN}$. The experimental results together with the DFT analysis provide the robust evidence that $TiC_{0.5}N_{0.5}$ can act as a potent nucleant for the α-Al grain to promote the grain refinement of α-Al.

It is noted that although NPs can be captured by the advancing solid/liquid interface at enhanced cooling conditions and thus distributed intragranularly throughout the matrix, they cannot be responsible for α-Al nucleation. In light of the classic nucleation theory, each nucleation site corresponds to one grain. As a result, merely a minority of NPs can participate in the nucleation events. Furthermore, the size of NPs and their size distribution have a significant influence on the nucleation efficiency. Based on the free growth theory [9], for a given particle size, the critical undercooling for grain initiation on the particle is fixed, above which the nuclei can continue to grow and below which the free growth of the crystal is stifled. At enhanced cooling conditions, the melt undercooling may be large enough to activate NPs as potent nucleants for the heterogeneous nucleation of α-Al, generating a high nucleation rate. However, a large undercooling can also promote the grain growth, which in turn gives a coarse grain size [17]. In fact, the nucleation of new crystals competes with the growth of nucleated grains [7,9,37]. The results of this work demonstrate that the effect of an enhanced nucleation rate outweighs that of fast grain growth. In addition, compared with the particles in grain refines, the added NPs have a relatively narrow size distribution, which favors intrinsically a fine grain size. As a result, it is inferred that the grain refinement at a high cooling rate is principally due to the NP-induced nucleation promotion.

On the basis of the analysis above, the grain refinement at different cooling rates can be qualitatively elucidated as follows. At relatively low cooling rates, the assembly of intergranular NPs onto the surface of growing crystals could stifle their growth. Meanwhile, a majority of NPs could not be activated as potent nucleants at very low undercoolings. As a consequence, the grain refinement at relatively low cooling rates may be due to the growth restriction caused by NPs. With increasing the cooling rates, the NP capture by SF proceeds and the growth restricting the effect of NPs is decreased. Additionally, the increased melt undercooling may reach the free-growth undercooling to activate NPs for grain initiation. Therefore, NP-induced heterogeneous nucleation could dominate in the grain refinement at relatively high cooling rates.

Apart from the microstructural refinement, according to our previous work [21–23], the TiCN NPs can impart high strength and ductility to the Al alloy. For one thing, the grain refinement induced by TiCN NPs can lead to the performance enhancement of materials, which is well described by classic Hall–Petch equation. For another, the ceramic nanoparticles distributed in the matrix, especially the intragranular ones at high cooling rates, could function as desirable reinforcements to immobilize dislocations via the Orowan strengthening and the coefficient of thermal expansion mismatch strengthening mechanisms, contributing to the improvement of material properties.

4. Conclusions

The addition of $TiC_{0.5}N_{0.5}$ nanoparticles can result in the grain refinement of α-Al at various cooling conditions. The microstructure analysis shows that the nanoparticles exhibit intergranular distribution and form an incoherent interface with α-Al at low cooling rates (<15 K/s), while they exhibit intragranular distribution and form the coherent interface with α-Al at high cooling rates (>70 K/s). The first-principles calculations reveal that among $(111)_{Al}//(111)_{TiCN}$, $(200)_{Al}//(200)_{TiCN}$, and $(220)_{Al}//(220)_{TiCN}$ interfaces, the (111) interface with the largest work of adhesion and smallest interfacial energy is the most thermodynamically stable and energetically favored nucleation site for the heterogeneous nucleation of α-Al. The experimental and theoretical results demonstrate that the NP-induced growth restriction and NP-induced nucleation promotion are responsible for the grain refinement at low and high cooling rates, respectively.

Author Contributions: K.W.: conceived the research idea, designed the research plan, conducted the experiments, and wrote the paper. H.J., Q.W. and Y.W.: participated in oral discussion. All authors have read and agreed to the published version of the manuscript.

Funding: The present study was sponsored by the National Key Research and Development Program of China under grant no. 2022YFC2905204 and the National Natural Science Foundation of China no. 51821001.

Institutional Review Board Statement: Not applicable.

Data Availability Statement: All data included in this study are available upon request by contact with the corresponding author.

Conflicts of Interest: The authors declare that they have no known competing financial interest or personal relationship that could have appeared to influence the work reported in this paper.

References

1. Wang, L.; Liu, D.; Chen, T.; Wang, S.; Cao, Y. Grain refinement and high thermal stability in laser surface remelted Mg-4Y-3Nd-1.5Al alloy. *Scr. Mater.* **2023**, *222*, 115000. [CrossRef]
2. Rastogi, A.; Neelakantan, S. Enhanced thermo-mechanical processing through stress-induced martensitic transformation and its effect on grain refinement in metastable β titanium alloys. *Scr. Mater.* **2023**, *226*, 115222. [CrossRef]

3. Li, H.; Wang, K.; Xu, G.; Jiang, H.; Wang, Q.; Wang, Y. Effective inhibition of anomalous grain coarsening in cast AZ91 alloys during fast cooling via nanoparticle addition. *J. Magnes. Alloy.* **2021**, in press. [CrossRef]
4. Luo, Q.; Li, X.; Li, Q.; Yuan, L.; Peng, L.; Pan, F.; Ding, W. Achieving grain refinement of α-Al and Si modification simultaneously by La-B-Sr addition in Al-10Si alloys. *J. Mater. Sci. Technol.* **2023**, *135*, 97–110. [CrossRef]
5. Fan, Z.; Gao, F.; Wang, Y.; Wang, S.; Patel, J. Grain refinement of Mg-alloys by native MgO particles: An overview. *J. Magnes. Alloy.* 2022, in press. [CrossRef]
6. Easton, M.; StJohn, D. An Analysis of the Relationship between Grain Size, Solute Content, and the Potency and Number Density of Nucleant Particles. *Metall. Mater. Trans. A* **2005**, *36*, 1911–1920. [CrossRef]
7. Men, H.; Fan, Z. Effects of solute content on grain refinement in an isothermal melt. *Acta Mater.* **2011**, *59*, 2702–2704. [CrossRef]
8. Easton, M.; Qian, M.; Prasad, A.; StJohn, D. Recent advances in grain refinement of light metals and alloys. *Curr. Opin. Solid State Mater. Sci.* **2016**, *20*, 13–24. [CrossRef]
9. Greer, A.; Bunn, A.; Tronche, A.; Evans, P.; Bristow, D. Modelling of inoculation of metallic melts application to grain refinement of aluminum by Al-Ti-B. *Acta Mater.* **2000**, *48*, 2823–2835. [CrossRef]
10. Zhao, C.; Li, Y.; Xu, J.; Luo, Q.; Jiang, Y.; Xiao, Q.; Li, Q. Enhanced grain refinement of Al-Si alloys by novel Al-V-B refiners. *J. Mater. Sci. Technol.* **2021**, *94*, 104–112. [CrossRef]
11. Fan, Z.; Gao, F.; Wang, Y.; Men, H.; Zhou, L. Effect of solutes on grain refinement. *Prog. Mater. Sci.* **2022**, *123*, 100809. [CrossRef]
12. Wang, Y.; Fang, C.; Zhou, L.; Hashimoto, T.; Zhou, X.; Ramasse, Q.; Fan, Z. Mechanism for Zr poisoning of Al-Ti-B based grain refiners. *Acta Mater.* **2019**, *164*, 428–439. [CrossRef]
13. Qiu, D.; Taylor, J.; Zhang, M.-X. Understanding the co-poisoning effect of Zr and Ti on the grain refinement of cast aluminium alloys. *Metall. Mater. Trans. A* **2010**, *41A*, 3412–3421. [CrossRef]
14. Qiu, D.; Taylor, J.; Zhang, M.-X.; Kelly, P. A mechanism for the poisoning effect of silicon on the grain refinement of Al-Si alloys. *Acta Mater.* **2007**, *55*, 1447–1456. [CrossRef]
15. Yang, W.; Ji, S.; Zhang, R.; Zhang, J.; Liu, L. Abnormal Grain Refinement Behaviour in High-Pressure Die Casting of Pure Mg with Addition of Zr as Grain Refiner. *JOM* **2018**, *70*, 2555–2560. [CrossRef]
16. Ali, Y.; You, G.; Pan, F.; Zhang, M. Grain coarsening of cast magnesium alloys at high cooling rate: A new observation. *Metall. Mater. Trans. A* **2017**, *48*, 474–481. [CrossRef]
17. Liang, G.; Ali, Y.; You, G.; Zhang, M. Effect of cooling rate on grain refinement of cast aluminium alloys. *Materialia* **2018**, *3*, 113–121. [CrossRef]
18. Chen, L.-Y.; Xu, J.-Q.; Choi, H.; Konishi, H.; Jin, S.; Li, X.-C. Rapid control of phase growth by nanoparticles. *Nat. Commun.* **2014**, *5*, 3879. [CrossRef]
19. Guo, E.; Shuai, S.; Kazantsev, D.; Karagadde, S.; Phillion, A.; Jing, T.; Li, W.; Lee, P.D. The influence of nanoparticles on dendritic grain growth in Mg alloys. *Acta Mater.* **2018**, *152*, 127–137. [CrossRef]
20. Xu, J.; Chen, L.; Choi, H.; Konish, H.; Li, X. Assembly of metals and nanoparticles into novel nanocomposite superstructures. *Sci. Rep.* **2013**, *3*, 1730–1735. [CrossRef]
21. Wang, K.; Jiang, H.; Jia, Y.; Zhou, H.; Wang, Q.; Ye, B.; Ding, W. Nanoparticle-inhibited growth of primary aluminum in Al-10Si alloys. *Acta Mater.* **2016**, *103*, 252–263. [CrossRef]
22. Wang, K.; Jiang, H.; Wang, Y.; Wang, Q.; Ye, B.; Ding, W. Microstructure and mechanical properties of hypoeutectic Al-Si composite reinforced with TiCN nanoparticles. *Mater. Des.* **2016**, *95*, 545–554. [CrossRef]
23. Wang, K.; Jiang, H.; Wang, Q.; Ding, W. Influence of nanoparticles on microstructural evolution and mechanical properties of Sr-modified Al-10Si alloys. *Mater. Sci. Eng. A* **2016**, *666*, 264–268. [CrossRef]
24. Kresse, G.; Furthmüller, J. Efficient iterative schemes for ab initio total-energy calculations using a plane-wave basis set. *Phys. Rev. B* **1996**, *54*, 11169–11186. [CrossRef]
25. Blöchl, P. Projector augmented-wave method. *Phys. Rev. B* **1994**, *50*, 17953–17979. [CrossRef]
26. Perdew, J.; Burke, K.; Ernzerhof, M. Generalized gradient approximation made simple. *Phys. Rev. Lett.* **1996**, *77*, 3865–3868. [CrossRef]
27. van de Walle, A.; Tiwary, P.; de Jong, M.; Olmsted, D.; Asta, M.; Dick, A.; Shin, D.; Wang, Y.; Chen, L.-Q.; Liu, Z.-K. Efficient stochastic generation of special quasirandom structures. *Calphad* **2013**, *42*, 13–18. [CrossRef]
28. Christensen, A.; Carter, E. Adhesion of ultrathinZrO_2(111)films on Ni(111) from first principles. *J. Chem. Phys.* **2001**, *114*, 5816–5831. [CrossRef]
29. Pryds, N.; Huang, X. The effect of cooling rate on the microstructures formed during solidification of ferritic Steel. *Metall. Mater Trans. A* **2000**, *31*, 3155–3166. [CrossRef]
30. Villars, P.; Calvert, L. *Pearson's Handbook of Crystallographic Data for Intermetallic Phases*; ASM, Metals Park: Ohio, OH, USA, 1985.
31. Xu, J.; Chen, L.; Choi, H.; Li, X. Theoretical study and pathways for nanoparticle capture during solidification of metal melt. *J. Phys. Condens. Matter* **2012**, *24*, 255304. [CrossRef]
32. Zhang, M.; Kelly, P.; Easton, M.; Taylor, J. Crystallographic study of grain refinement in aluminum alloys using the edge-to-edge matching model. *Acta Mater.* **2005**, *53*, 1427–1438. [CrossRef]
33. Zhang, M.; Kelly, P. Edge-to-edge matching and its applications Part II. Application to Mg-Al, Mg-Y and Mg-Mn alloys. *Acta Mater.* **2005**, *53*, 1085–1096. [CrossRef]

34. Zhang, M.; Kelly, P.; Qian, M.; Taylor, J. Crystallography of grain refinement in Mg-Al based alloys. *Acta Mater.* **2005**, *53*, 3261–3270. [CrossRef]
35. Liu, L.; Wang, S.; Ye, H. First-principles study of polar Al/TiN (1 1 1) interfaces. *Acta Mater.* **2004**, *52*, 3681–3688. [CrossRef]
36. Liu, L.; Wang, S.; Ye, H. Adhesion of metal carbide/nitride interfaces: Al/TiC and Al/TiN. *J. Phys. Condens. Matter* **2003**, *15*, 8103–8114. [CrossRef]
37. Kumar, M.; Sasikumar, R.; Nair, P. Competition between nucleation and early growth of ferrite from austenite-studies using cellular automaton simulations. *Acta Mater.* **1998**, *46*, 6291–6303. [CrossRef]

materials

MDPI

Article

Fast Degradation of Azo Dyes by In Situ Mg-Zn-Ca-Sr Metallic Glass Matrix Composite

Rui Jin, Gaojiong Wang, Xin Wang *, Wei Yang and Yumin Qi *

Hebei Key Laboratory of New Functional Materials, School of Material Science and Engineering, Hebei University of Technology, No. 5340, Xiping Road, Beichen District, Tianjin 300401, China

* Correspondence: ahaxin@hebut.edu.cn (X.W.); ymqi@hebut.edu.cn (Y.Q.); Tel.: +86-022-6020-4125 (X.W.)

Abstract: Mg-based metallic glass (MG) has attracted extensive attention in the field of wastewater treatment due to its high decolorization rate in degrading azo dyes. However, the azo dye degradation rate of Mg-based MGs is strongly dependent on the particle size. Improving the intrinsic degradation efficiency using large particles is of great interest for future applications. In this work, in-situ metallic glass matrix composites (MGMCs) with high Mg content were successfully prepared by melt spinning. It is found that when the Mg content is 79–82%, the as-spun sample shows typical glassy characteristics. The SEM and XRD tests confirm that the as-spun sample is composed of α-Mg dendrite, multiple Mg-Zn intermetallic particles and an MG matrix. The degradation experiment using Direct Blue 6 and a 500 μm particle sample demonstrate that the $Mg_{82}Zn_{14}Ca_3Sr_1$ MGMC sample degrades azo dyes faster than typical Mg-Zn-Ca MG alloy. It can be attributed to the galvanic cell effect on the α-Mg/MG interface, which reduces the waste of active Mg atoms in the MG matrix according to the corrosion protection mechanism by the α-Mg anode sacrifice. This result provides a new perspective and insight into the design of azo dye degradation alloys and the understanding of degradation mechanisms.

Keywords: metallic glass matrix composite; magnesium; in situ composite; azo dye degradation

1. Introduction

Mg-based metallic glass (MG) has attracted new research interests in the field of wastewater treatment due to its excellent performance in removing azo dyes and heavy metal ions from aqueous solutions [1,2]. It is well known that azo dyes are toxic to aquatic life and mutagenic for humans, and they are difficult to degrade completely in natural water because of the presence of a stable chemical bond -N=N- [3]. Therefore, azo dye pollution has been regarded as a stubborn environmental pollution factor that has threatened human health for a long time. At present, the effective treatment of azo dye wastewater depends on complex conditions, expensive equipment or high energy consumption [3,4]. It is difficult to obtain the ideal treatment effect (decolorization) at low costs, which forces printing and dyeing enterprises to stop using azo dyes. However, the threat of azo dye molecules accumulated in natural water to the ecological environment cannot be ignored. In recent years, it has been found that the environment-friendly Mg-Zn-Ca MG alloy composed of non-toxic metallic elements can rapidly degrade azo dyes without adding other chemicals, such as hydrogen peroxide [5], providing a new choice for treating azo dye pollution existing in natural water bodies.

Mg-based bulk metallic glasses (BMGs) are usually strong, corrosion-resistant and very brittle due to their homogeneous single-phase structure and disordered atomic arrangement [6,7]. Due to the aim of improving the application potential of Mg-based BMGs as structural materials, it has been a hot research topic to design metallic glass matrix composites (MGMCs) to adjust deformation behavior and improve plasticity [8–10]. On the other hand, Mg-based MG materials also show good potential for functional applications without regard to the mechanical properties. In 2012, Wang et al. reported for the first time that the azo dye degradation performance of Mg-Zn-Ca MG alloys was much higher than that of commercial crystalline Fe powders [5]. The high resistance of the glassy structure to water corrosion reduces the waste of zero-valent Mg (Mg^0), thus increasing the proportion of active metal actually participating in the degradation reaction. Subsequently, the research of Mg-Zn-Ca MG alloys in the treatment of azo dye wastewater has attracted more attention, and relevant reports mainly focus on the following aspects: (1) low-cost preparation methods; (2) optimization of alloy composition; (3) the influence of second-phases or partial crystallized precipitates; (4) surface treatment.

As with most azo dye degradation catalysts [11–13], the particle size of Mg-Zn-Ca MGs is an important factor affecting the degradation efficiency of azo dyes [14]. For example, when the particle size of Mg-Zn-Ca MG after ball milling is smaller than 20 μm, the larger specific surface area enhances the physical adsorption and increases the reaction site, resulting in complete decolorization of dye wastewater within 10 min [5]. However, Iqbal et al. used ribbon samples of similar Mg-Zn-Ca alloy to degrade azo dyes and found that it took several hours for the fully glassy ribbon sample to achieve full decolorization [15]. Likewise, Zhao et al. prepared Mg-Zn-Ca MG powder by gas atomization and found that the azo dye degradation performance of MG powder was significantly enhanced with the decrease in particle size (size effect) [14]. As a result, they found that the complete decolorization time of $Mg_{66}Zn_{29}Ca_5$ MG was less than 2 min when the particle size was smaller than 25 μm [14]. In our opinion, the size effect should be sufficiently considered in the degradation performance test of Mg-Zn-Ca MG alloys, which may be the main reason for the difference in the performance results of similar material systems.

The phase composition of Mg-based MG alloys containing crystalline inclusions, including the crystal structure, volume fraction and size of the second phase, is also an important factor affecting the efficiency of sewage treatment. It is well known that alloy composition, cooling rate, mold surface, etc., will affect the precipitation behavior of the second phase from the glass-forming liquid before it freezes into a solid state [16]. For as-spun Mg-Zn-Ca MGs, it is reported that a sample with a partly amorphous structure containing some crystalline phases had a worse azo dye degradation performance than the fully glassy sample [15]. However, Ramya et al. synthesized partially glassy Mg-Zn-Ca alloy powder by one-step ball milling and proved that Mg-Zn-Ca amorphous/crystalline composite powder showed excellent azo dye degradation efficiency [17]. Chen et al. prepared MgZn alloys with full crystalline, semi-amorphous and full amorphous structures and found that the second phase had a significant impact on the dye degradation behavior of the ribbon [18]. It can be seen that whether the formation of the crystalline second phase in Mg-Zn-Ca systems is conducive to the rapid degradation of azo dyes needs further study.

By controlling the corrosion process of Mg-based MG alloy, especially through surface treatment, the azo dye degradation efficiency of MG alloy can be effectively improved. Luo et al. prepared nano-porous copper on the surface of ball-milled $Mg_{65}Cu_{25}Y_{10}$ MG powder by dealloying, which not only increased the specific surface area, but also accelerated the dissolution of active atoms through a local Galvani cell reaction, thus improving the azo dye degradation efficiency [19]. Chen et al. found that the addition of NaCl can improve the degradation rate of the Mg-Zn-Ca MG ribbon by accelerating the surface corrosion coarsening and the formation of ZnO-like nanosheets in the later stage [20]. Recently, Chen et al. treated Mg-Zn-Ca MG ribbons using a citric acid solution to form a porous surface, together with the Zn/ZnO crystalline phase and found that the azo dye degradation property was significantly enhanced by delivering more primary hydrogen [H] and $O_2^{\bullet-}$ radicals to Mg [21]. Similarly, the Mg-Cu-Ag-Y MG ribbons after dealloying treatment also improved the catalytic degradation efficiency of the pesticide cis-cyanthrin through the nano-porous network [22]. Zhang et al. studied the azo dye degradation behavior of the $Mg_{63}Cu_{16.8}Ag_{11.2}Er_9$ MG ribbon and pointed out that the degradation process of dye molecules includes physical adsorption and a reduction/degradation reaction ($Mg^0 - 2e^- = Mg^{2+}$), and the efficient physical adsorption on the surface of an MG alloy plays a dominant role in the early stage of azo dye degradation [23].

Mg-Zn-Ca BMG is a lightweight alloy composed of all non-toxic elements [24], which is very commendable for environmentally friendly sewage treatment. Therefore, it is important to adjust the composition of the alloy without changing the main elements to improve the intrinsic degradation performance for azo dyes. Ren et al. investigated the effect of Mg content in Mg-Zn-Ca MGs on azo dye degradation performance and found that high Mg content is beneficial for improving the degradation efficiency because samples with high Mg concentration can provide more Mg^0 required for the degradation reaction of azo dyes [25]. However, when the Mg content exceeds 75% (in at.), the Mg-Zn-Ca alloy cannot maintain a fully amorphous structure [25]. This is not consistent with the early strategy of obtaining a fully glassy structure for designing an azo dye degradation alloy, so the azo dye degradation behavior of the high Mg content alloy has not been further studied. In this work, we designed several in-situ MgZn-based MGMCs with high Mg content of more than 79% (in at.) based on an $Mg_{66}Zn_{30}Ca_3Sr_1$ alloy, which has the highest glass-forming ability (GFA) up to date [26]. We analyzed the glassy structure characteristics using XRD and DSC and investigated the azo dye degradation behavior and decolorization mechanism of the sample with a high Mg content of up to 82%.

2. Materials and Methods

2.1. Materials Preparation

Raw materials with purities of Mg 99.9%, Zn 99.99%, Ca 99% and Sr 99.9% (in wt.) were used to prepare the master alloys with the nominal composition $Mg_{66+x}Zn_{30-x}Ca_3Sr_1$ (x = 13, 14, 15 and 16, in at.%). All alloys were weighed by using a precision balance (Mettler, ME204E, Columbus, OH, USA) with an accuracy of 0.0001 g. The weighed raw materials were put into quartz cups in a vacuum furnace with an ultimate vacuum of about 1×10^{-3} Pa, and induction heating was used to melt the alloys with Ar protection. The melted liquid was poured into the copper mold to obtain master alloy ingots, which were cut into pieces with pliers. The master alloy pieces were put into a quartz tube with a round hole (1 mm diameter) at the bottom, which was put in a spinning furnace (Shanghai MTINST, SDM-0.02, Shanghai, China) with Ar protection for re-melting. The melted liquid was blown onto a running copper rod with a rotating speed of 2000–3000 rpm through Ar to prepare ribbons with a width of 5–10 mm. For the confirmation of composition, an energy dispersive spectrometer (EDS) was performed, and the results are shown in Table 1. It shows that the actual composition of the sample prepared in this paper is similar to the nominal composition, which can be used as a model material to study the effect of Mg content on the azo dye degradation behavior. The as-spun ribbons were ground into a powder with scissors and mortar. Stainless steel sieves with 80-mesh and 100-mesh holes were used to screen all the powder samples to avoid the influence of particle size difference on the subsequent experimental results. For the convenience of description, $Mg_{66+x}Zn_{30-x}Ca_3Sr_1$ (x = 13, 14, 15 and 16) alloy powder samples are simplified as Mg79, Mg80, Mg81 and Mg82 according to the content of Mg.

Table 1. Compositions of as-spun samples with different Mg content determined by EDS.

Sample	Nominal Composition (at.%)	Tested Composition (at.%)				
		Mg	Zn	Ca	Sr	O
M79	$Mg_{79}Zn_{17}Ca_3Sr_1$	78.14	16.98	2.66	0.89	1.33
M80	$Mg_{80}Zn_{16}Ca_3Sr_1$	79.45	15.90	2.55	0.99	1.11
M81	$Mg_{81}Zn_{15}Ca_3Sr_1$	80.50	15.12	2.76	0.78	0.84
M82	$Mg_{82}Zn_{14}Ca_3Sr_1$	81.53	13.84	2.65	0.78	1.20

2.2. Material Characterization

An X-ray diffractometer (XRD, Bruker D8 Advance, Cu Kα radiation, Billerica, MA, USA) was used to examine the phase composition of powder samples with a scanning speed of 8°/min and a scanning angle range of 10–90°. A differential scanning calorimeter

(DSC, Setaram Themys ONE, Cranbury, NJ, USA) was used to measure the glass transition characteristics and thermal stability of the as-prepared samples under N_2 protection at a heating rate of 20 °C/min. A scanning electron microscope (SEM, Hitachi S-4800, Tokyo, Japan) was used to examine the microstructure of the ribbon samples and the surface micromorphology of the particles after azo dye degradation experiments. An energy dispersive spectrometer (EDS) was used to examine the micro-area chemical composition. Image Pro Plus software was used to carry out particle size statistical analysis based on the optical photos taken by a metallurgical microscope (Zeiss, Axio Imager M2m, Jena, Germany). X-ray photoelectron spectroscopy (XPS, Thermo Scientific, Waltham, MA, USA, K-Alpha, with Al Kα X-ray monochromator) was used to analyze the electronic states of various elements on the surface of samples after azo dye degradation.

2.3. Azo Dye Degradation Test

Commercial direct blue 6 powder (DB6, $C_{32}H_{20}N_6Na_4O_{14}S_4$) and deionized water were mixed at a ratio of 0.2 g/L. In order to fully dissolve DB6 molecules, the mixture was stirred for 1 h with a magnetic stirrer at room temperature. For each degradation experiment, 0.2 g of alloy powder was put into 50 mL of azo dye simulation solution, and the mixture was kept in a water bath at 25 °C with magnetic stirring. After stirring for a certain time (0.5, 1, 2, 5, 10, 20, 40, 60 min), a syringe was used to extract about 0.5 mL of the treated solution and pass it through the membrane filter (nylon 66, 0.45 μm diameter pore) for filtering. The degraded liquid was injected into the quartz cuvette after different times, and a UV-visible spectrophotometer (UV, Shanghai Yuanxi UV-6000, Shanghai, China, wavelength scanning mode, tungsten lamp light source, wavelength range of 400–800 nm, scanning speed interval of 1 nm, scanning accuracy of 10°) was used to test the absorbance of different solutions. In order to recover the powder sample after each degradation test, the mix solution was filtered with a medium-speed filter paper with a pore diameter of 30–50 μm, and then the obtained wet powder was dried in a vacuum drying oven (Tianjin Taist, Tianjin, China, DZ-1BC II) at 102 °C for 1 h.

3. Results

3.1. Microstructure

Figure 1 shows the XRD patterns of the as-prepared samples. All patterns are composed of a diffuse peak and several sharp crystalline phase diffraction peaks. The diffuse peak comes from the amorphous structure, indicating that all samples contain a certain amount of amorphous phase. The crystalline diffraction peaks can be divided into two groups: Mg and Mg-Zn intermetallic. For the Mg groups, the diffraction peak angle is highly consistent with the primary α-Mg dendrite phase in reported Mg-based MGMCs [27,28]. It is suggested that the as-prepared samples with different Mg content all contain α-Mg dendrite. The intensity of α-Mg peaks in the XRD pattern increases with the increase in Mg content, indicating that the addition amount of Mg significantly affects the volume fraction of the primary α-Mg crystalline phase. For the Mg-Zn intermetallic compounds, the diffraction peaks are much lower, which are generally hidden and interfered with by background noise. Furthermore, the common Zn-rich intermetallic compounds in the Mg-Zn alloy system, such as Mg_7Zn_3, Mg_4Zn_7 and $MgZn_2$, have similar diffraction peak angles, so it is difficult to accurately distinguish them through the XRD patterns in this work. According to the reported XRD analysis of Mg-Zn phases in the literature [29–32], it can be inferred that the intermetallic phase in the Mg79-Mg82 sample may be a mixture of Mg_7Zn_3, Mg_4Zn_7 and $MgZn_2$ phases. In addition, the diffraction peak of α-Mg in sample M82 is significantly higher than that of Mg_Zn compounds, which implies that the amount of crystalline phase α-Mg is significantly higher than the other crystalline phases.

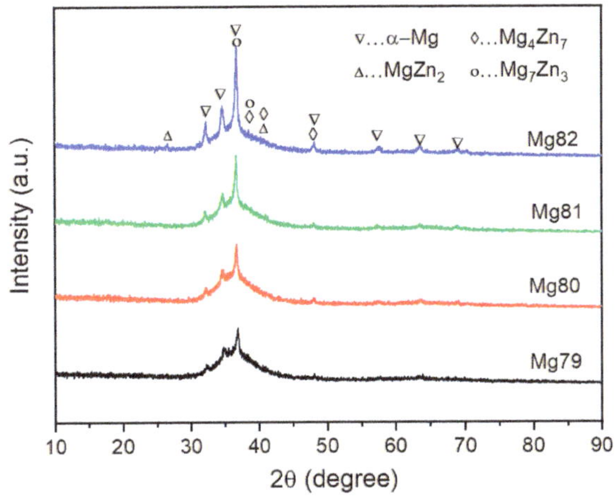

Figure 1. XRD patterns of Mg-Zn-Ca-Sr MGMC samples with different Mg content.

In order to further confirm the presence of the MG phase, the DSC test is performed on the powder samples. The inset of Figure 2a shows a partially enlarged view of the low temperature range marked by the dotted box, which indicates that all samples have the characteristics of a glass transition at about 350 K. It is suggested that there is a certain amount of glassy phase in the samples, which is in good agreement with the XRD result shown in Figure 1. When the temperature rises above 370 K, three exothermic peaks appear on the DSC curves of the four samples, which are labeled as T_{p1}, T_{p2} and T_{p3}, as shown in Figure 2a, respectively. These exothermic events can be attributed to the complex crystallization transformation of the glass phase, which is also considered as a representative feature of the existence of the glassy phase [33,34].

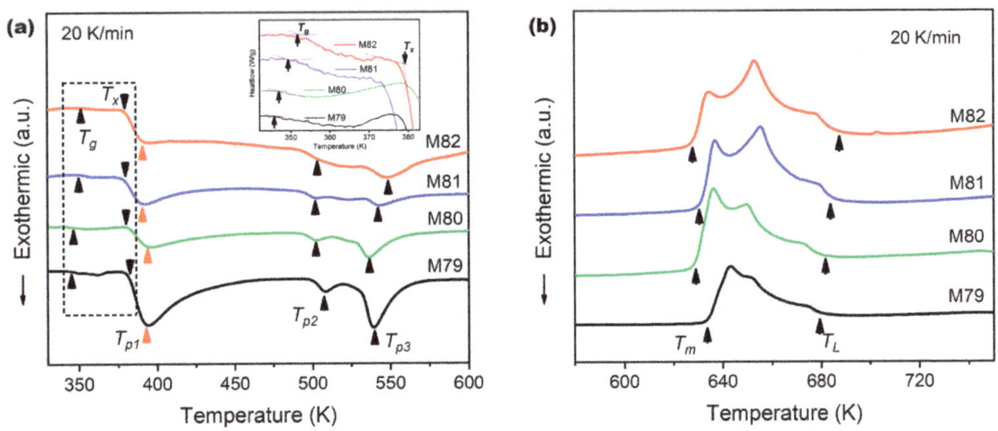

Figure 2. DSC curves of Mg-Zn-Ca-Sr MGMC samples with different Mg content showing: (**a**) low temperature range for glass transition and crystallization, with the inset showing a partially enlarged view of the temperature range marked by the dotted box; (**b**) high temperature range curves showing the melting of different samples.

The thermal parameters of as-prepared powder samples with different Mg content, including the glass transition temperature T_g, the onset crystallization temperature T_x, the three peak crystallization temperatures T_{p1}, T_{p2}, T_{p3}, and the supercooled liquid region

ΔT ($T_g - T_x$), are measured and calculated from the DSC curves and listed in Table 2, respectively. It can be seen that T_g is slightly improved with the increase in Mg content, which implies that the MG phase in different MGMC samples shows different thermal stability. However, T_x has a decreasing trend with the increased Mg content. It may be induced by the increase in α-Mg/MG interface area (α-Mg volume fraction increases), which significantly increases the number of heterogeneous nucleation sites and induces crystallization at lower temperatures [35,36]. For the supercooled liquid region ΔT, it decreases obviously with the increase in Mg content, indicating that the glass-forming ability (GFA) of the Mg-Zn-Ca-Sr alloy decreases. This is because the increase in Mg content makes the alloy composition leave the optimal GFA composition range, resulting in the precipitation of crystalline phases. In addition, with the increase in Mg content, the area of the three crystallization exothermic peaks (relates to crystallization exothermic enthalpy) has a trend of gradually decreasing, which indicates that the amount of glassy phase in the MGMC samples gradually decreases.

Table 2. Thermophysical parameters of as-prepared powder samples with different Mg content measured from DSC curves.

Sample	T_g (K)	T_x (K)	T_{P1} (K)	T_{p2} (K)	T_{p3} (K)	T_m (K)	T_L (K)	ΔT (K)
M79	346.6	382.6	394.1	507.1	539.4	632.1	679.2	36.0
M80	347.2	380.5	394.2	504.8	538.1	628.9	682.4	33.3
M81	349.9	379.6	389.9	503.7	543.8	629.8	686.1	29.7
M82	351.6	379.7	391.1	503.7	550.6	629.7	687.9	28.1

Figure 3a displays a typical SEM image of an as-spun $Mg_{82}Zn_{14}Ca_3Sr_1$ sample (Mg82), which shows that some snowflake-like patterns and white irregular particles are unevenly distributed on the gray matrix. It is very intuitive that the volume fraction of the snowflake phase is significantly higher than that of white particles. For further analysis, the EDS results of the gray matrix (point B), snowflake phase (point C), and white particles (point D) in Figure 3a are shown in Figure 3b–d. The Mg content in the snowflake area is significantly higher than that in the gray matrix and white particles, and it is also higher than the nominal composition of the M82 sample. It is well known that the composition of solvent elements in the solid solution phase can fluctuate in a large range, so the snowflake phase should be a solid solution phase with Mg as the solvent. In addition, the morphology of this snowflake structure is consistent with that of Mg dendrite in the literature [27]. Combined with the XRD results shown in Figure 1, it can be considered that the snowflake phase pattern comes from α-Mg dendrite crystals. For the white particles, the EDS results in Figure 3d show that the Mg:Zn atomic ratio is 53.8:40.1, which can only be achieved when the particles are composed of Mg-rich phase Mg_7Zn_3 and Zn-rich phase Mg_4Zn_7 or $MgZn_2$ in a certain proportion. Therefore, it is suggested that the white particle phase in Figure 3a is a particle cluster composed of Mg_7Zn_3, Mg_4Zn_7 and $MgZn_2$, which is consistent with the XRD test results shown in Figure 1. After excluding primary α-Mg and Mg-Zn composite particles, the gray matrix phase should be amorphous MG. To sum up, the M82 sample is a typical in situ formed Mg-based MGMC with the MG phase as the matrix and α-Mg dendrite and Mg-Zn compounds as the precipitation phases.

3.2. DB6 Degradation

The azo dye degradation efficiency of the MgZn-based MG alloy is significantly particle-size-dependent, according to the reports [14,17]. Therefore, particle size should be precisely measured before determining the azo dye degradation performance for any sample. Figure 4 shows the optical microscopic image and corresponding particle size statistical analysis of sample M82 after sieving between the 80-mesh sieve and the 100-mesh sieve. The particle size after screening fluctuates from ~200 to 850 μm, with an average value of 468.88 ± 16.17 μm, which is much larger than that of ball-milled [5] and gas-atomized [14]. In addition, the particles after grinding have sharp edges and corners, and their geometric shape shows the characteristics of irregular shape, which is consistent with the fracture of brittle materials. It is known that the properties of MG alloys and MGMC materials are obviously affected by various conditions such as thermal history [37,38], oxidation [39–41], and residual stress [42–44], etc. Relatively, large particles are less affected by the heating change, oxidation and other conditions due to low specific surface area. Therefore, the azo dye degradation performance tested with large particle samples is closer to its intrinsic degradation property than that of very small particle samples. The development of high-efficiency azo dye degradation materials with a large sample size is of great significance to the stability of the azo dye degradation performance and a long period of validity of sewage treatment agents.

Figure 5a shows the UV-visible absorption spectrum of the simulated dye wastewater solution, which is degraded with M82 powder at different times. The absorption spectrum of the DB6 solution usually has significant absorption between 500 and 700 nm, with a peak of about 570 nm [45,46]. After degradation for 0.5 min, the absorption peak height is significantly reduced by 84%, which shows the high efficiency of the M82 sample in the degradation of azo dyes. It can be seen from the locally enlarged image in Figure 5a that the difference between the absorption spectra of 0.5 and 1 min samples is not significant, indicating that DB6 dye molecules are mainly degraded or removed from the solution within 0.5 min. It can be verified from the macroscopic photo inserted in Figure 5b, which shows all degradation time solutions have a similar color of light purple.

Figure 5b shows the absorbance/degradation time curve (degradation curve) of the M82 sample and $Mg_{66}Zn_{30}Ca_4$ MG (MZC) tested under the same conditions, such as degradation temperature, pH value, dosage, and particle size. In order to improve the comparability of data, all measured absorbance values (termed as A) are divided by that of the original DB6 solution for normalization. It clearly shows that the degradation curve of M82 is below MG and has a higher slope before 100 s, indicating that the azo-dye degradation rate of the M82 sample is higher than MZC. The results also show that the DB6 solution is significantly decolorized within 120 s, and the removal rate ($R = (A_0 - A_t)/A_0$) is up to 85.2%, which is about 10% higher than that of the MZC sample (R = 77.5%) at the same time. Due to the fast degradation rate, the time window for manual operation is very small, and the sampling error is less than 60 s is large. Combined with the color change trend in the inset, the current results can prove that the azo dye degradation performance of M82 MGMC samples with high Mg content is better than that of the control sample MZC.

Figure 6 shows the effects of pH value, degradation temperature, powder dose, and recycling times of powder on the azo dye degradation efficiency of M82 samples, respectively. It is found that the pH value significantly affects the azo dye degradation rate of M82 samples, as shown in Figure 6a. When the pH is 4.0, the removal rate of dye molecules at 120 s reaches ~90%, which is significantly higher than that under neutral conditions (pH = 7.0) and alkaline conditions (pH = 10.0). We know that Mg can generate a relatively stable $Mg(OH)_2$ passivation film under a strongly alkaline environment. This film can block the contact between Mg^0 in MGMC and DB6 molecules in solution, which is not conducive to the continuous degradation of azo dyes. Figure 6b shows the influence of degradation temperature on the degradation curve of the M82 sample. When the temperature of the DB6 solution changes from 30 to 60 °C, all the degradation curves are mixed together, showing similar degradation rates without any regularity. This result indicates that the effect of temperature on the azo dye degradation is not significant. It is suggested that the reaction between Mg-based MGMC and azo dyes may not be a thermal activation process. Figure 6c shows the effect of powder dose on the degradation curve of the M82 sample. Obviously, a higher dose can provide a higher specific surface area and more Mg^0, which can improve the degradation rate of azo dyes. It is worth noting that when the dose exceeds 4 g/L, the difference between the degradation curves is small, indicating that high-speed degradation can achieve an economic balance between material consumption and degradation rate. Figure 6d shows the cyclic degradability of the recovered M82 powder. It can be clearly seen that after four recycles of M82 powder, the azo dye degradation efficiency is decreased in the early stage of the degradation experiment. The removal rate of DB6 molecule is only 25% at 120 s, but it can still reach 70% at 3600 s. It shows that the ability of M82 powder to rapidly degrade azo dyes is weakened after several recoveries. Although the recyclability and recycling rate of M82 cannot be comparable with that of a high-stability alloy (Cu/Fe/Co-based MG alloys [47–49]), it can still obtain potential applications in those situations without recycling, such as soil improvement, overall nitrogen removal of printing, and dyeing wastewater storage tank, etc. Benefiting from the degradability, lightness and non-toxicity of all component elements, Mg-Zn-Ca-Sr MGMC can degrade itself in nature without worrying about the secondary pollution of heavy metals contained in the sewage treatment agent.

Figure 7 shows the typical SEM images of the Mg82 MGMC sample after DB6 degradation for 1 h with different magnifications. After reacting with the DB6 solution, the shape and size of the M82 particle have no obvious change, as shown in Figure 7a. It is implied that the sample powder has no obvious consumption in the degradation reaction. However, through the highly magnified SEM images displayed in Figure 7b–d, it can be clearly seen that there are significant changes to the surface morphology of the sample: (1) some white sheet products stand upright on the substrate (Figure 7c); (2) the surface is covered by a light gray network (Figure 7d). The white sheet products have a size of ~700 nm and a thickness of ~5 nm, which are generally considered to be ZnO or $Zn(OH)_2$ deposits inferred from morphology. This is a common phenomenon in the azo dye degradation process of Mg-Zn-Ca MG alloy [5,20]. For the light gray network pattern, it can be found in different MG systems used for azo dye degradation [50–52], which is generally considered to be the degradation product of the reaction between dye molecules and the surface metal.

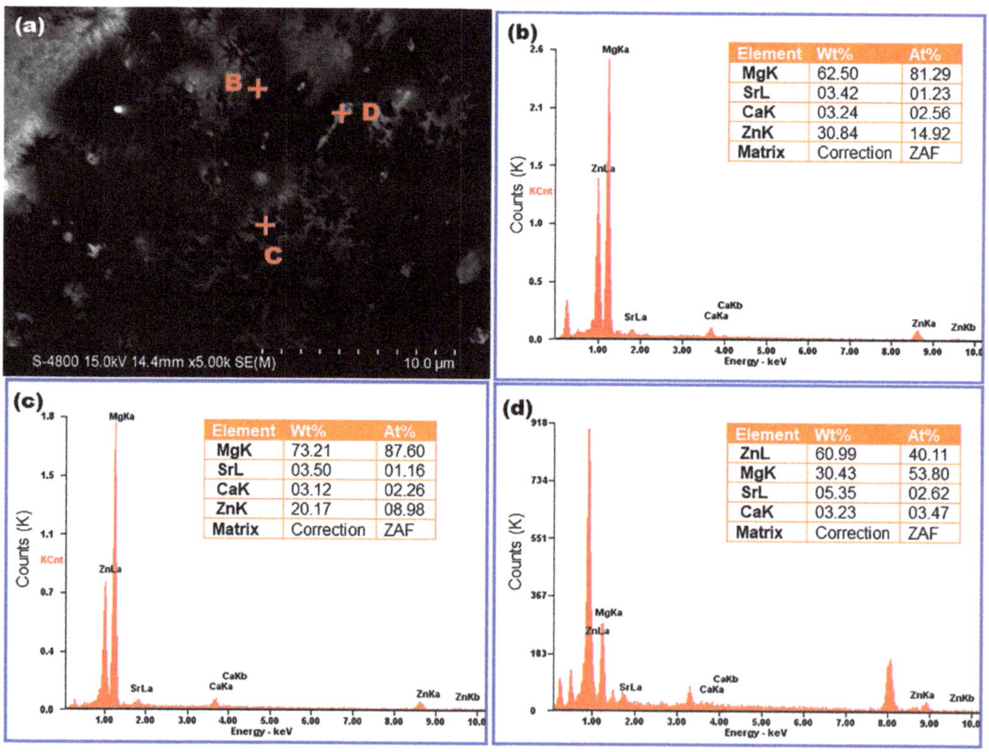

Figure 3. (**a**) Typical SEM image of an $Mg_{82}Zn_{14}Ca_3Sr_1$ MGMC sample; (**b–d**) EDS results of point B, C and D as marked in (**a**).

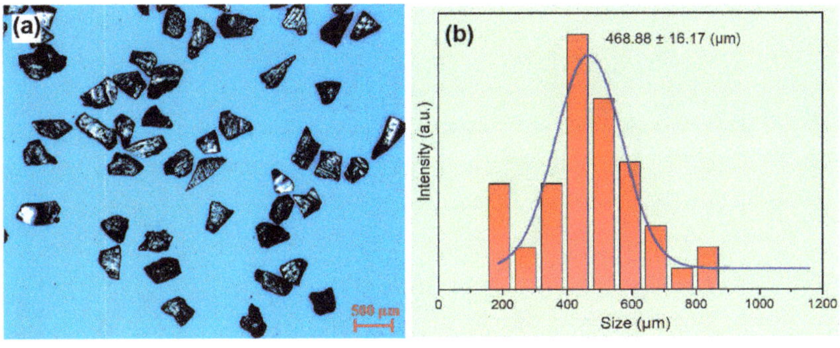

Figure 4. (**a**) Optical microscopic image of sample M82 power; (**b**) particle size statistical analysis of sample M82 using Image Pro plus 6.0 software.

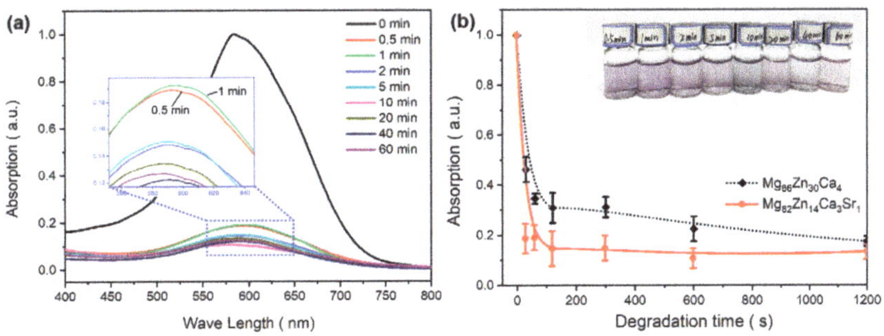

Figure 5. (**a**) Absorption spectra of the M82 sample after immersing in DB6 solution for different times; (**b**) comparison of the degradation curves of the Mg82 sample and $Mg_{66}Zn_{30}Ca_4$ MG alloy under similar particle size, with an inset showing the macroscopic optical photos of the DB6 solution degraded by using M82 at different times.

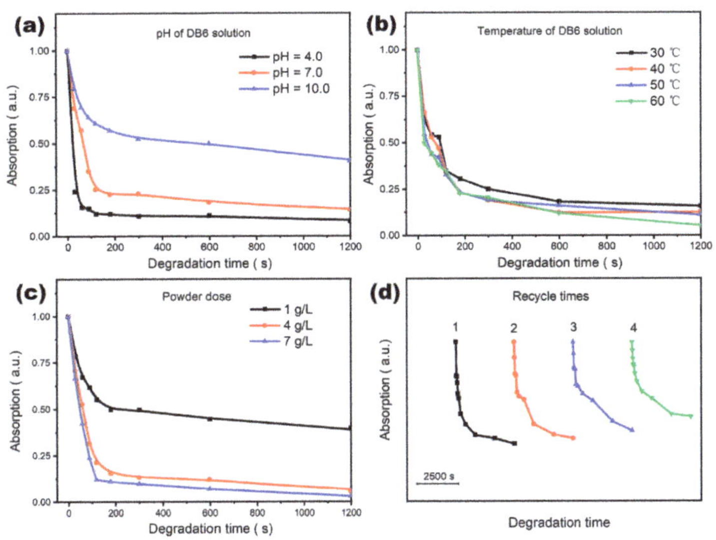

Figure 6. Degradation curves of Mg82 under different conditions of: (**a**) pH of DB6 solution; (**b**) treatment temperature; (**c**) dose; (**d**) recovery time.

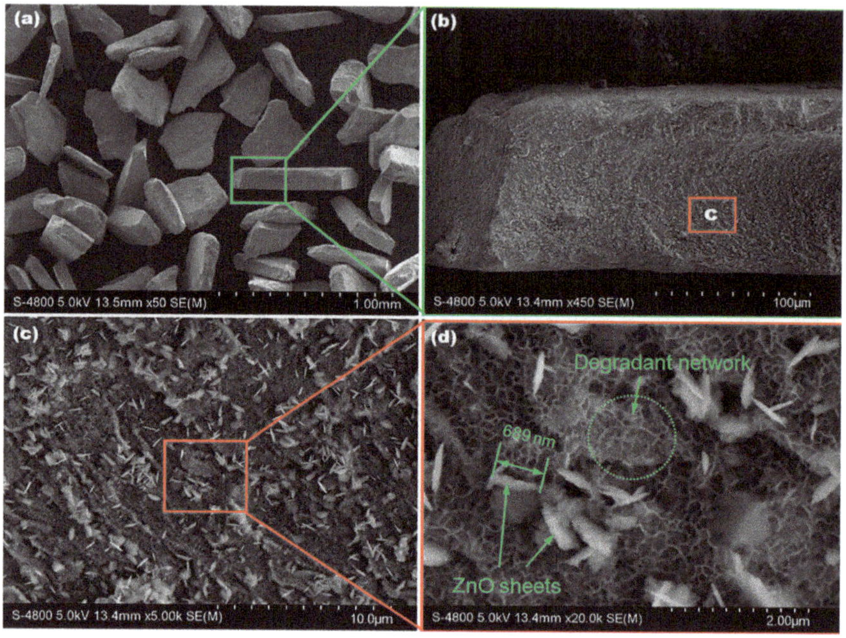

Figure 7. SEM images of the $Mg_{82}Zn_{14}Ca_3Sr_1$ MGMC sample after DB6 degradation for 60 min with different magnifications (**a**–**d**).

Figure 8 shows the XPS spectrum analysis of the recovered M82 powder after the azo dye degradation experiment. Figure 8a shows that the Mg 2p peak can be decomposed into two peaks with different bonding energy of 49.3 and 50.2 eV, corresponding to Mg^{2+} and Mg^0, respectively. The peak height of Mg^{2+} is about twice that of Mg^0, which indicates that after soaking in DB6 solution for 1 h, Mg atoms on the surface of M82 particles form bivalent Mg deposits, such as MgO, $Mg(OH)_2$ or $MgCO_3$. It should be noted that the existence of the Mg^0 peak implies that the surface metal of the particle still has the ability to degrade azo dyes, agreeing well with the results shown in Figure 6d. Figure 8b shows that the Ca 1s peak can be decomposed into three peaks with different bond energies, namely 347.0, 350.7 and 352.3 eV, corresponding to $CaCO_3$, CaO and Ca^0, respectively. It is very intuitive that the CaO peak is higher than $CaCO_3$ and Ca^0, which means that the sample surface contains a large amount of CaO. Considering that the XPS sample has undergone vacuum drying treatment, which may lead to the decomposition of hydroxide, these CaO may also come from the decomposition of $Ca(OH)_2$. In addition, carbonate ions $CO_3{}^{2-}$ may come from the decomposition of DB6 molecules or the dissolution of CO_2 in solution. Figure 8c shows that the Zn $2p^3$ peak can be decomposed into two peaks with similar bonding energies, 1021.2 and 1021.9 eV, corresponding to Zn^0 and Zn^{2+}, respectively. Compared with Mg and Ca, Zn is more stable in an aqueous solution, resulting in more Zn^0 remaining on the surface of the M82 sample after the 1 h azo dye degradation test. Figure 8d shows that the O 1s peak can be decomposed into four peaks with different bonding energies, 530.2, 530.9, 531.6 and 532.6 eV, corresponding to CaO, ZnO, MgO and $CO_3{}^{2-}$ respectively. Among these peaks, the peak heights of MgO and ZnO are significantly higher than other peaks, indicating that the degradation products on the surface of the M82 sample are mainly composed of MgO and ZnO.

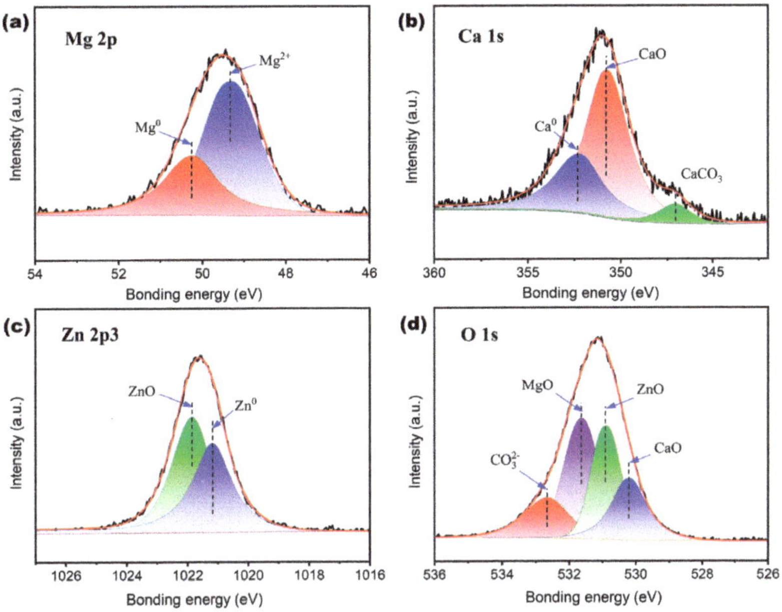

Figure 8. Partial peak analysis of local XPS spectrum of Mg82 sample after azo dye degradation: (**a**) Mg 2p; (**b**) Zn 2p; (**c**) Ca 1s and (**d**) O 1s.

4. Discussion

The rapid azo dye degradation behavior of Mg-Zn-Ca MG has been attributed to the disordered atomic arrangement of its amorphous structure, which improves the corrosion resistance and reduces the corrosion waste of zero-valent magnesium (Mg^0) [5]. Compared with Mg^0 and Zn^0 in crystalline alloys, zero-valent metals in MGs have a higher valence band and extended empty band structure, which can provide more active electrons for the degradation of azo dyes [53]. The electrons positioned at the conduction band can react with dissolved O_2 to generate $O_2^{\bullet -}$, accelerating the azo dye degradation [54–56]. However, the active zero-valent elements, such as Mg^0, Zn^0 and Ca^0 contained in MGs, can react with water in nature, so the amorphous structure can only delay the corrosion in aqueous solution to a certain extent but cannot completely prevent the corrosion loss of them. In other words, when Mg-Zn-Ca MG is used to treat azo dye wastewater, a certain amount of active metal will inevitably be wasted due to the intrinsic corrosion. Therefore, how to protect the Mg-Zn-Ca MG alloy from corrosion and reduce the loss of active metal atoms has become one of the key scientific issues to improve the azo dye degradation efficiency.

In this work, we produce more Mg^0 by increasing the Mg content in the alloy. Considering that Sr can improve the GFA of Mg-Zn-Ca MG alloys [26], we also introduce a small amount of Sr to reduce the adverse effect of Mg content change on the GFA of amorphous alloys. The results show that the Mg content of the MG matrix phase in Mg82 MGMC is effectively increased (Figure 3b), inducing improved azo dye degradation performance of the alloy. In addition, Mg-Zn-Ca-Sr MGMCs contain a small amount of Sr, which has a strong oxidation property. Therefore, it not only improves the GFA of the matrix phase but also helps to reduce the O content of the matrix alloy. Moreover, Sr and Ca have similar intrinsic properties and can react with water, resulting in an increase in [H] in the solution, which is also conducive to the degradation of azo dyes. However, the increase in Mg content eventually leads to the precipitation of crystalline α-Mg, which inevitably causes the reduction in glassy Mg atoms in the alloy. Interestingly, it is found that the azo dye degradation rate does not decrease due to α-Mg precipitation. The related mechanism can be understood by sacrificing anode protection.

It is well known that pure Mg and most of the Mg alloys can be used as anode materials to protect metal components from corrosion [57,58], according to the sacrificial anode theory. In our opinion, the α-Mg solid solution phase can play a key role in improving the azo dye degradation performance of the in situ MgZn-based MGMCs. Figure 9 shows the corresponding azo dye degradation mechanism based on the galvanic cell corrosion theory. The in situ MgZn-based MGMCs are composed of α-Mg, Zn-rich Mg-Zn intermetallic particles and MG matrix phase with high Mg content, in which the MG phase and α-Mg dendrite both have high volume fractions. When the composite sample is put into the DB6 solution, a small primary cell can be formed on the α-Mg/MG interface, and α-Mg dendrite acts as sacrificing anode with the reaction $Mg - 2e^- \rightarrow Mg^{2+}$. It forms a negative charge layer on the surface of the MG matrix, effectively preventing the discharge corrosion of MG as a cathode by sacrificing the α-Mg dendrite anode. In addition, the electrons e^- generated by the α-Mg dendrite combine with H^+ to generate [H] [21], which diffuses into the solution and reacts with the dye molecules. As a result, azo dyes with a macromolecular structure are reduced and decomposed into small molecules.

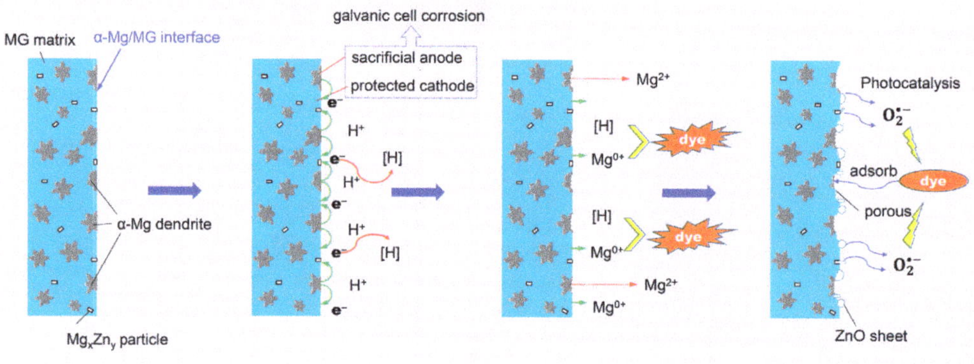

Figure 9. Schematic diagram showing the azo dye degradation mechanism by MgZn-based metallic glass composites with high Mg content.

In addition, MgZn-based MGMCs contain a certain amount of Zn-rich intermetallic particles, which may react with water as follows:

$$MgZn_2 + 4H_2O \rightarrow Mg^{2+} + 2OH^- + 2H^2\uparrow + Zn(OH)_2\downarrow \quad (1)$$

$$Mg_7Zn_3 + 20H_2O \rightarrow 7Mg^{2+} + 14OH^- + 10H^2\uparrow + 3Zn(OH)_2\downarrow \quad (2)$$

$$Mg_4Zn_7 + 22H_2O \rightarrow 4Mg^{2+} + 8OH^- + 11H^2\uparrow + 7Zn(OH)_2\downarrow \quad (3)$$

Through the chemical reaction revealed by Equations (1)–(3), $Zn(OH)_2$ or ZnO-like nanosheets can be formed, as shown in Figure 7d, which is consistent with the research results in the literature [20,54,56]. These hydroxide/oxide particles in aqueous solution usually attract hydroxyl groups with a negative charge, which is beneficial to the formation of the free radical $O_2^{•-}$ and the degradation of azo dyes. Of course, when the degradation time is extended to a certain time, the MG matrix region can also form porous and smaller ZnO/MgO-like nanosheets, which are usually assembled into porous networks on the surface (Figure 6d). This porous network structure can increase the physical adsorption effect of the MGMC surface on dye molecules, improve the probability of zero-valent metals, [H] and $O_2^{•-}$ contacting dye molecules, and thus promote the degradation of azo dyes. Moreover, under the irradiation of natural light or ultraviolet light, these oxides with different scales can also induce luminescence catalysis and further degrade azo dye molecules.

In short, MgZn-based MGMC with α-Mg dendrite structure can degrade azo dyes through various mechanisms such as zero-valent metal, photocatalysis, etc. Its degradation process is doped with additional processes, including corrosion, sacrificial anode dissolution, etc. The azo dye degradation reaction mainly occurs in the local area of the MG matrix, and its size and shape depend on the morphology of the α-Mg dendrite. Therefore, a suitable dendrite obtained by controlling the morphology, size and proportion of α-Mg can affect the geometric distribution of the negative charge region, which is conducive to improving the azo dye degradation performance. In addition, the strategy of adjusting Mg content mainly uses cheap metal elements as raw materials, and common instruments and equipment are used in alloy melting and powder preparation, so the preparation cost of MGMC material is relatively low. Meanwhile, the process of using MGMC to degrade dye wastewater is simple, without heating treatment and adding other chemicals, and without special production lines or expensive equipment, which is conducive to reducing costs and has great significance to the promotion of catalysts.

5. Conclusions

(1) In situ Mg-Zn-Ca-Sr metallic glass matrix composites with high magnesium content were successfully prepared by melt spinning. SEM, DSC and XRD analyses confirm that the composite is composed of primary α-Mg dendrites, a metallic glass matrix and a few Mg-Zn intermetallic particles.

(2) In-situ Mg-Zn-Ca-Sr MGMCs with different Mg content have similar thermal stability. With the increase in Mg content, the glass transition temperature T_g and the melting point are increased, but the onset crystallization temperature T_x is decreased, resulting in the reduced supercooled liquid region. It indicates that the thermal stability and glass-forming ability of the MG matrix decrease.

(3) The $Mg_{82}Zn_{14}Ca_3Sr_1$ MGMC sample with a particle size of up to 500 μm shows a high azo dye degradation rate, indicating that the intrinsic azo dye degradation performance is improved by turning the Mg content. It is sensitive to pH during the treatment of azo dye wastewater, and the acidic environment is conducive to a high degradation rate.

(4) The rapid degradation of azo dyes can be attributed to the sacrificial anode protection induced by a crystalline Mg dendrite, which improves the utilization rate of zero-valent Mg in the MG matrix phase. It is suggested that the azo dye degradation performance of MgZn-based MGMC can be improved by regulating the microstructure of MGMC, which provides a new basis for the design of new dye wastewater treatment materials.

Author Contributions: Investigation, writing—original draft preparation, R.J.; data curation, formal analysis, G.W.; supervision, writing—review and editing, project administration, X.W.; methodology, visualization, W.Y.; conceptualization, supervision, funding acquisition, Y.Q. All authors have read and agreed to the published version of the manuscript.

Funding: This research was funded by the "Natural Science Foundation of Hebei Province, China, grant number E2022202055" and "Scientific Research Projects of Colleges and Universities in Hebei Province, China, grant number ZD2022099".

Institutional Review Board Statement: Not applicable.

Informed Consent Statement: Not applicable.

Data Availability Statement: Any further detailed data may be obtained from the authors upon a reasonable request.

Conflicts of Interest: The authors declare no conflict of interest.

References

1. Wang, J. Chemical Functional Applications and Fundamental Researches of Metallic Glasses. *Mater. China* **2014**, 33, 270–281. [CrossRef]
2. Zhang, L.-C.; Jia, Z.; Lyu, F.; Liang, S.-X.; Lu, J. A review of catalytic performance of metallic glasses in wastewater treatment: Recent progress and prospects. *Prog. Mater. Sci.* **2019**, *105*, 100576. [CrossRef]

3. Selvaraj, V.; Karthika, T.S.; Mansiya, C.; Alagar, M. An over review on recently developed techniques, mechanisms and intermediate involved in the advanced azo dye degradation for industrial applications. *J. Mol. Struct.* **2021**, *1224*, 129195. [CrossRef]
4. Vandevivere, P.C.; Bianchi, R.; Verstraete, W. Review: Treatment and reuse of wastewater from the textile wet-processing industry: Review of emerging technologies. *J. Chem. Technol. Biot.* **1998**, *72*, 289–302. [CrossRef]
5. Wang, J.Q.; Liu, Y.H.; Chen, M.W.; Louzguine-Luzgin, D.V.; Inoue, A.; Perepezko, J.H. Excellent capability in degrading azo dyes by MgZn-based metallic glass powders. *Sci. Rep.* **2012**, *2*, 418. [CrossRef]
6. Zhao, Y.-Y.; Ma, E.; Xu, J. Reliability of compressive fracture strength of Mg–Zn–Ca bulk metallic glasses: Flaw sensitivity and Weibull statistics. *Scr. Mater.* **2008**, *58*, 496–499. [CrossRef]
7. Guo, F.; Joseph Poon, S.; Gu, X.; Shiflet, G.J. Low-density Mg-rich metallic glasses with bending ductility. *Scr. Mater.* **2007**, *56*, 689–692. [CrossRef]
8. Yin, H.L.; Yang, W.; Zhao, L.C.; Hu, X.M.; Liu, S.Q.; Cui, C.X.; Wang, X. Fabrication and mechanical property of three-dimensional carbon fiber reinforced Mg-based bulk metallic glass matrix composite. *Mater. Sci. Eng. A* **2022**, *839*, 142853. [CrossRef]
9. Yin, H.L.; Liu, S.Q.; Zhao, L.C.; Cui, C.X.; Wang, X. Vacuum infiltration molding and mechanical property of short carbon fiber reinforced Ti-based metallic glass matrix composite. *J. Mater. Process. Tech.* **2021**, *295*, 117151. [CrossRef]
10. Wang, X.; Zhao, L.; Hu, X.; Cheng, Y.; Liu, S.; Chen, P.; Cui, C. Fabrication and Mechanical Behavior of Ex Situ Mg-Based Bulk Metallic Glass Matrix Composite Reinforced with Electroless Cu-Coated SiC Particles. *Materials* **2017**, *10*, 1371. [CrossRef]
11. Shih, Y.-h.; Lin, C.-h. Effect of particle size of titanium dioxide nanoparticle aggregates on the degradation of one azo dye. *Environ. Sci. Pollut. Res.* **2012**, *19*, 1652–1658. [CrossRef]
12. Duarte, F.; Maldonado-Hodar, F.J.; Madeira, L.M. Influence of the Particle Size of Activated Carbons on Their Performance as Fe Supports for Developing Fenton-like Catalysts. *Ind. Eng. Chem. Res.* **2012**, *51*, 9218–9226. [CrossRef]
13. Becker, J.; Raghupathi, K.R.; St Pierre, J.; Zhao, D.; Koodali, R.T. Tuning of the Crystallite and Particle Sizes of ZnO Nanocrystalline Materials in Solvothermal Synthesis and Their Photocatalytic Activity for Dye Degradation. *J. Phys. Chem. C* **2011**, *115*, 13844–13850. [CrossRef]
14. Zhao, Y.F.; Si, J.J.; Song, J.G.; Yang, Q.; Hui, X.D. Synthesis of Mg-Zn-Ca metallic glasses by gas-atomization and their excellent capability in degrading azo dyes. *Mater. Sci. Eng. B-Adv.* **2014**, *181*, 46–55. [CrossRef]
15. Iqbal, M.; Wang, W.H. Synthesis and characterization of Mg-based amorphous alloys and their use for decolorization of Azo dyes. In Proceedings of the 13th International Symposium on Advanced Materials (ISAM), Islamabad, Pakistan, 23–27 September 2014.
16. Wang, X. Surface Crystallization in Mg-Based Bulk Metallic Glass during Copper Mold Casting. *Adv. Mater. Sci. Eng.* **2014**, *2014*, 798479. [CrossRef]
17. Ramya, M.; Karthika, M.; Selvakumar, R.; Raj, B.; Ravi, K.R. A facile and efficient single step ball milling process for synthesis of partially amorphous Mg-Zn-Ca alloy powders for dye degradation. *J. Alloys Compd.* **2017**, *696*, 185–192. [CrossRef]
18. Chen, Q.; Pang, J.; Yan, Z.C.; Hu, Y.H.; Guo, L.Y.; Zhang, H.; Zhang, L.C.; Wang, W.M. MgZn-based amorphous ribbon as a benign decolorizer in methyl blue solution. *J. Non-Cryst. Solids* **2020**, *529*, 119802. [CrossRef]
19. Luo, X.; Li, R.; Zong, J.; Zhang, Y.; Li, H.; Zhang, T. Enhanced degradation of azo dye by nanoporous-copper-decorated Mg-Cu-Y metallic glass powder through dealloying pretreatment. *Appl. Surf. Sci.* **2014**, *305*, 314–320. [CrossRef]
20. Chen, P.; Hu, X.; Qi, Y.; Wang, X.; Li, Z.; Zhao, L.; Liu, S.; Cui, C. Rapid Degradation of Azo Dyes by Melt-Spun Mg-Zn-Ca Metallic Glass in Artificial Seawater. *Metals* **2017**, *7*, 485. [CrossRef]
21. Chen, Q.; Yan, Z.; Guo, L.; Zhang, H.; Zhang, L.; Kim, K.; Li, X.; Wang, W. Enhancing the acid orange dye degradation efficiency of Mg-based glassy alloys with introducing porous structure and zinc oxide. *J. Alloys Compd.* **2020**, *831*, 154817. [CrossRef]
22. Xia, Q.; He, S.-Y.; Zhang, W.; Xiang, Q.-C.; Qu, Y.-D.; Ren, Y.-L.; Qiu, K.-Q. Degradation efficiency of Mg65Cu25-xAgxY10 nanoporous dealloyed ribbons on pesticide wastewater. *Trans. Nonferrous Met. Soc.* **2022**, *32*, 1472–1484. [CrossRef]
23. Zhang, C.; Zhu, Z.; Zhang, H. Mg-based amorphous alloys for decolorization of azo dyes. *Results Phys.* **2017**, *7*, 2054–2056. [CrossRef]
24. Gu, X.; Shiflet, G.J.; Guo, F.Q.; Poon, S.J. Mg–Ca–Zn Bulk Metallic Glasses with High Strength and Significant Ductility. *J. Mater. Res.* **2005**, *20*, 1935–1938. [CrossRef]
25. Ren, Y.; Yue, C.; Wang, T.; Qiu, K.; Li, R. Degrading ability and recycle use characteristics of Mg-Zn-Ca amorphous ribbons in azo dye solution treatment. *Trans. Nonferrous Met. Soc.* **2016**, *26*, 2152–2159.
26. Li, H.; Pang, S.; Liu, Y.; Sun, L.; Liaw, P.K.; Zhang, T. Biodegradable Mg–Zn–Ca–Sr bulk metallic glasses with enhanced corrosion performance for biomedical applications. *Mater. Des.* **2015**, *67*, 9–19. [CrossRef]
27. Gao, J.; Sharp, J.; Guan, D.; Rainforth, W.M.; Todd, I. New compositional design for creating tough metallic glass composites with excellent work hardening. *Acta Mater.* **2015**, *86*, 208–215. [CrossRef]
28. Lee, J.I.; Ryu, W.H.; Yoon, K.N.; Park, E.S. In-situ synthesis of Mg-based bulk metallic glass matrix composites with primary α-Mg phases. *J. Alloys Compd.* **2021**, *879*, 160417. [CrossRef]
29. Yan, K.; Bai, J.; Liu, H.; Jin, Z.-Y. The precipitation behavior of MgZn2 and Mg4Zn7 phase in Mg-6Zn (wt.%) alloy during equal-channel angular pressing. *J. Magnes. Alloy.* **2017**, *5*, 336–339. [CrossRef]
30. Xie, Y.-P.; Wang, Z.-Y.; Hou, Z.F. The phase stability and elastic properties of MgZn2 and Mg4Zn7 in Mg–Zn alloys. *Scr. Mater.* **2013**, *68*, 495–498. [CrossRef]
31. Lin, X.-p.; Kuo, Y.; Wang, L.; Ye, J.; Zhang, C.; Wang, L.; Guo, K.-y. Refinement and strengthening mechanism of Mg−Zn−Cu−Zr−Ca alloy solidified under extremely high pressure. *Trans. Nonferrous Met. Soc.* **2021**, *31*, 1587–1598. [CrossRef]
32. Luo, Z.P.; Zhang, S.Q.; Tang, Y.L.; Zhao, D.S. On the stable quasicrystals in slowly cooled Mg-Zn-Y alloys. *Scr. Metall. Mater.* **1995**, *32*, 1411–1416. [CrossRef]

33. Ma, Y.; Tang, X.; Wang, X.; Zhang, M.; Hu, H.; Gong, P.; Wang, X. Preparation and mechanical properties of tungsten-particle-reinforced Zr-based bulk-metallic-glass composites. *Mater. Sci. Eng. A* **2021**, *815*, 141312. [CrossRef]
34. Wang, X.; Gong, P.; Yao, K.-F. Mechanical behavior of bulk metallic glass prepared by copper mold casting with reversed pressure. *J. Mater. Process. Tech.* **2016**, *237*, 270–276. [CrossRef]
35. Wang, Y.S.; Tan, M.J.; Chua, B.W.; Bayraktar, E. YSZ-Reinforced Mg-Based Amorphous Composites: Processing, Characterisation & Corrosion. *Adv. Mater. Res.* **2014**, *939*, 122–129. [CrossRef]
36. Wat, A.; Lee, J.I.; Ryu, C.W.; Gludovatz, B.; Kim, J.; Tomsia, A.P.; Ishikawa, T.; Schmitz, J.; Meyer, A.; Alfreider, M.; et al. Bioinspired nacre-like alumina with a bulk-metallic glass-forming alloy as a compliant phase. *Nat. Commun.* **2019**, *10*, 961. [CrossRef] [PubMed]
37. Wang, X.; Shao, Y.; Gong, P.; Yao, K.F. The effect of simulated thermal cycling on thermal and mechanical stability of a Ti-based bulk metallic glass. *J. Alloys Compd.* **2013**, *575*, 449–454. [CrossRef]
38. Wang, X.; Shao, Y.; Gong, P.; Yao, K. Effect of thermal cycling on the mechanical properties of $Zr_{41}Ti_{14}Cu_{12.5}Ni_{10}Be_{22.5}$ alloy. *Sci. China Phys. Mech.* **2012**, *55*, 2357–2361. [CrossRef]
39. Wang, X.; Shao, Y.; Yao, K.-F. Chemical composition dependence of atomic oxygen erosion resistance in Cu-rich bulk metallic glasses. *Chin. Sci. Bull.* **2012**, *57*, 4801–4804. [CrossRef]
40. Lin, B.; Yang, K.; Bao, X.; Liu, J.; Guo, Q.; Zhang, L.; Wang, Q.; Hua, N. Enhanced wear, corrosion, and corrosive-wear resistance of the biocompatible Ti-based bulk metallic glass by oxidation treatment. *J. Non-Cryst. Solids* **2022**, *576*, 121231. [CrossRef]
41. Jia, Q.; He, W.; Hua, D.; Zhou, Q.; Du, Y.; Ren, Y.; Lu, Z.; Wang, H.; Zhou, F.; Wang, J. Effects of structure relaxation and surface oxidation on nanoscopic wear behaviors of metallic glass. *Acta Mater.* **2022**, *232*, 117934. [CrossRef]
42. Xu, H.; Shao, Y.; Chen, S.; Yao, K. Stress-induced activation of the commercial Fe-based metallic glass ribbons for azo dye degradation. *J. Non-Cryst. Solids* **2021**, *572*, 121117. [CrossRef]
43. Zine, P.U.; Joshi, P.K. Inducing compressive residual stress to minimize effect of crack generation by using controlled shot peening process. *Mater. Today Proc.* **2023**, *72*, 870–877. [CrossRef]
44. Chen, S.-Q.; Shao, Y.; Cheng, M.-T.; Yao, K.-F. Effect of residual stress on azo dye degradation capability of Fe-based metallic glass. *J. Non-Cryst. Solids* **2017**, *473*, 74–78. [CrossRef]
45. Wang, F.; Wang, H.; Zhang, H.; Dan, Z.; Weng, N.; Tang, W.; Qin, F. Superior azo-dye degradation of Fe-Si-B-P amorphous powders with graphene oxide addition. *J. Non-Cryst. Solids* **2018**, *491*, 34–42. [CrossRef]
46. Wu, S.; Pan, Y.; Lu, J.; Wang, N.; Dai, W.; Lu, T. Effect of the addition of Mg, Ti, Ni on the decoloration performance of AlCrFeMn high entropy alloy. *J. Mater. Sci. Technol.* **2019**, *35*, 1629–1635. [CrossRef]
47. Tang, M.; Lai, L.; Ding, D.; Liu, T.; Kang, W.; Guo, N.; Song, B.; Guo, S. Rapid degradation of Direct Blue dye by Co-based amorphous alloy wire. *J. Non-Cryst. Solids* **2022**, *576*, 121282. [CrossRef]
48. Si, J.; Gu, J.; Luan, H.; Yang, X.; Shi, L.; Shao, Y.; Yao, K. Porous composite architecture bestows Fe-based glassy alloy with high and ultra-durable degradation activity in decomposing azo dye. *J. Hazard. Mater.* **2020**, *388*, 122043. [CrossRef]
49. Zhao, B.; Zhu, Z.; Qin, X.D.; Li, Z.; Zhang, H. Highly efficient and stable CuZr-based metallic glassy catalysts for azo dye degradation. *J. Mater. Sci. Technol.* **2020**, *46*, 88–97. [CrossRef]
50. Peng, X.; Han, J.; Wang, Y.; Bo, Z.; Nie, A.; Li, P.; Li, Y.; Wu, H.; Liu, P.; Lu, Z.; et al. Unexpected enhanced catalytic performance via highly dense interfaces in ultra-fine amorphous-nanocrystalline biphasic structure. *Appl. Mater. Today* **2022**, *29*, 101689. [CrossRef]
51. Sun, H.; Zheng, H.; Yang, X. Efficient degradation of orange II dye using Fe-based metallic glass powders prepared by commercial raw materials. *Intermetallics* **2021**, *129*, 107030. [CrossRef]
52. Miao, F.; Wang, Q.; Di, S.; Yun, L.; Zhou, J.; Shen, B. Enhanced dye degradation capability and reusability of Fe-based amorphous ribbons by surface activation. *J. Mater. Sci. Technol.* **2020**, *53*, 163–173. [CrossRef]
53. Zhang, C.; Zhang, H.; Lv, M.; Hu, Z. Decolorization of azo dye solution by Fe-Mo-Si-B amorphous alloy. *J. Non-Cryst. Solids* **2010**, *356*, 1703–1706. [CrossRef]
54. Bharti, D.B.; Bharati, A.V. Photocatalytic degradation of Alizarin Red dye under visible light using ZnO & CdO nanomaterial. *Optik* **2018**, *160*, 371–379. [CrossRef]
55. Micheal, K.; Ayeshamariam, A.; Devanesan, S.; Bhuvaneswari, K.; Pazhanivel, T.; AlSalhi, M.S.; Aljaafreh, M.J. Environmental friendly synthesis of carbon nanoplates supported ZnO nanorods for enhanced degradation of dyes and organic pollutants with visible light driven photocatalytic performance. *J. King Saud Univ. Sci.* **2020**, *32*, 1081–1087. [CrossRef]
56. Reddy, I.N.; Reddy, C.V.; Shim, J.; Akkinepally, B.; Cho, M.; Yoo, K.; Kim, D. Excellent visible-light driven photocatalyst of (Al, Ni) co-doped ZnO structures for organic dye degradation. *Catal. Today* **2020**, *340*, 277–285. [CrossRef]
57. Kekedy-Nagy, L.; Teymouri, A.; Herring, A.M.; Greenlee, L.F. Electrochemical removal and recovery of phosphorus as struvite in an acidic environment using pure magnesium vs. the AZ31 magnesium alloy as the anode. *Chem. Eng. J.* **2020**, *380*, 122480. [CrossRef]
58. Yu, B.L.; Uan, J.Y. Sacrificial Mg film anode for cathodic protection of die cast Mg-9 wt.%Al-1 wt.%Zn alloy in NaCl aqueous solution. *Scr. Mater.* **2006**, *54*, 1253–1257. [CrossRef]

materials MDPI

Article

Effect of Multi-Element Microalloying on the Structure and Properties of High Chromium Cast Iron

Tao Liu [1], Jibing Sun [1,*], Zhixia Xiao [1], Jun He [1], Weidong Shi [2] and Chunxiang Cui [1]

1 Key Laboratory for New Type of Functional Materials in Hebei Province, School of Materials Science and Engineering, Hebei University of Technology, No. 5340 Xiping Road 1, Beichen District, Tianjin 300401, China; 18406580738@163.com (T.L.); xiaozhixia@hebut.edu.cn (Z.X.); 13512452879@163.com (J.H.)
2 Tianjin Lixinsheng New Material Technology Co., Ltd., Tianjin 301602, China
* Correspondence: hbgdsjb@hebut.edu.cn; Tel.: +86-22-60202006

Abstract: High chromium cast iron (HCCI) has been widely used as wear-resistant material in the industry. Alloying is an effective way to improve the microstructure and mechanical properties of HCCI. This paper added multi-component V-Fe-Ti-Nb-C-Zr-B alloy (VFC) to HCCI, showing a significant synergistic solution-strengthening effect. The results show that the added V-Ti-Nb-B are dissolved in M_7C_3 carbide to form the (Cr, Fe, V, Ti, Nb)$_7$(C, B)$_3$ alloy carbide, and a small amount of V and all Zr are dissolved in austenite and martensite. Adding VFC into HCCI improved the hardenability of HCCI, decreased the residual austenite content from 6.0 wt% to 0.9 wt%, increased the martensite content from 70.7 wt% to 82.5 wt%, and changed the structure and content of M_7C_3 carbide. These changes increased the hardness of as-cast and heat-tread HCCI by 1.4% and 4.1%, increased the hardness of austenite and martensite by 7.9% and 7.0%, increased the impact toughness by 16.9%, and decreased the friction coefficient and wear loss by 2.3 % and 7.0 %, respectively. Thus, the hardness, toughness, wear resistance, and friction resistance of HCCI alloy are improved simultaneously.

Keywords: iron alloys; high chrome cast iron; alloying; mechanical property

1. Introduction

High chromium cast iron (HCCI) has much higher wear resistance and is widely used in manufacturing wear-resistant parts [1,2]. HCCI consists of the essential components of Fe, Cr, and C, and its microstructure is composed of martensite or austenite and M_7C_3 carbide. M_7C_3 carbide possesses high hardness (1300–1800 HV), and its chemical composition, quantity, and morphology are mainly determined by the alloying elements in HCCI [3]. Researchers [4–8] have studied the role of alloying elements in HCCI. After adding 1.19 wt% V to HCCI containing 19 wt% Cr, M_7C_3 content increased from 30.97 vol% to 31.96 vol%, the carbide size decreased from 7.48 μm to 6.74 μm, and M_6C_5 carbide is formed [4]. The MC carbide formed when V content was higher than 2 wt% [5]. VC has a higher hardness (2600–3000 HV) and certain plasticity than the 1300–1800 HV of M_7C_3 carbide in HCCI. Kopycinski et al. [9] added 0–0.2 wt% Ti to HCCI (Cr 21.2 wt%, C 1.7 wt%). The bending strength of as-cast HCCI increased from 821MPa to 995MPa with the addition of 0.1 wt% Ti and decreased to 987 MPa with the addition of 0.2 wt% Ti. When 0.97 wt% Ti was added to HCCI containing 28 wt% Cr and 3 wt% C, the primary carbide size decreased from 1 mm to 50–350 μm [6,7]. The formed TiC can be used as the heterogeneous nucleation center of Cr_7C_3, refining M_7C_3 carbide significantly. NbC formed when Nb was added to hyper-eutectic HCCI containing 19–20 wt% Cr and 2.8 wt% C. NbC distributed at the grain boundaries can hinder the growth of primary M_7C_3 carbide and refine the M_7C_3 carbide size. Maja et al. [10] added 0.1–0.4 wt% Nb to HCCI (Cr 10–30 wt%, C 2–3.5 wt%) and observed that the hardness of as-cast HCCI increased from 432 HB to 485 HB. After adding 0.4–0.6 wt% Nb and heat treatment, the hardness of cast iron slightly decreases, but the bending strength increases by 40% [8]. With increasing the Nb content from 0 to 1.79 wt%, the formed NbC perform a second phase strengthening role in HCCI containing 15 wt% Cr and 3 wt% C and increases the hardness of as-cast HCCI from 48 HRC to 55 HRC [11]. Zr is a strong carbide-forming element in steel. Additionally, Zr has a high binding affinity with C, O, and N and can perform a role in the deoxidation and purification of molten steel [12]. According to the Fe-Zr phase diagram [13], the solubility of Zr in α-Fe and γ-Fe is 0.1 at% and 0.2 at%, respectively. When Zr forms the ZrO_2 in steel [14] or precipitates as (Fe, Zr) at the ferrite/martensite boundaries [15], these precipitates can hinder the growth of austenite grains. Adding 0.11 wt% Zr to ferrite/martensitic two-phase steel containing 19 wt% Cr and 0.2 wt% C [15] can reduce the austenite grain size from 15.8 μm to 12.8 μm. B can be solidly dissolved in the M_7C_3 carbide of HCCI to increase the hardness of $M_7(C, B)_3$ [16]. Adding 1.5 wt% B can improve the wear resistance of HCCI while adding 3 wt% B harms the wear resistance of HCCI [17].

Compared with adding one alloying element to HCCI, adding multiple alloying elements to HCCI can better refine the microstructure and improve the mechanical properties [5,18]. Bedolla et al. [5] added 2.02 V-1.84 Ti-1.95 Nb (wt%) to HCCI, decreasing the wear loss by 21.5%. After adding V-Ti-Nb-Mo alloy to HCCI, You et al. [18] found that V can dissolve in M_7C_3 carbide to improve its hardness by solution strengthening and can form VCr_2C_2 compound, TiC and NbC particles were precipitated in austenite, and Mo was soluted in austenite to improve the hardenability of the alloy. Finally, adding 0.60V-0.60Ti-0.60Nb-0.35Mo (wt%) alloying elements increased the toughness and hardness of cast iron by 45.4% and 13.8%, respectively, and the synergistic effect of these elements led to the fine eutectic structure.

According to the heat treatment processes in references [19–22], for HCCI containing 2.76% C and 17.91% Cr, the commonly used quenching process is 960 °C × 3 h, tempering process is 450 °C × 2h.

Therefore, adding a strong carbide-forming element can improve the properties of HCCI alloys through a solution or second-phase strengthening. Multi-element microalloying has better synergistic strengthening effects. However, few people have studied the following two scientific issues so far. One is that Zr has a more vital ability to form carbides than Ti, Nb, V, Cr, and Fe, which is rarely considered when adding multiple alloying elements. The second is that the added multiple alloying elements mainly form different carbides. Few people have discovered and discussed the impact of these alloying elements on the crystal structure of M_7C_3. V is the most effective alloying element. According to reference [5], we added a high content of V and a small amount of C to the multi-element alloy, hoping to compare the effect of MC and M_2C phases formed by adding V alone in the ingot. Simultaneously, according to the specific content of microalloying elements added in references [4–10,13,16,18], multi-element 75.9V-12.6Fe-5Ti-3.1Nb-2.2Zr-0.6C-0.6B alloy is added to HCCI to explore multiple synergistic strengthening effects.

2. Materials and Methods

2.1. Alloy Preparation

A 75.9V-12.6Fe-5Ti-3.1Nb-2.2Zr-0.6C-0.6B (for short, VFC) ingot was vacuum arc-melted using pure Fe, V, Ti, Nb, Zr, pig iron, and Fe-17 wt% B alloy. The HCCI ingot with a composition of 2.76C-0.27Si-0.52Mn-17.91Cr-0.12Mo-0.22Ni (wt%) was prepared using the medium-frequency induction melting furnace. New alloyed HCCI (for short, HCCI-VFC) was prepared by melting the HCCI ingot with 3.1 wt% VFC ingot (=2.29V-0.46Fe-0.15Ti-0.10Nb-0.07Zr-0.02C-0.02B) (wt%) in the medium-frequency induction melting furnace. HCCI and HCCI-VFC alloys are vacuum-heated at 960 °C for 3 h, cooled to room temperature, tempered at 450 °C for 2 h, then cooled to room temperature in the furnace.

2.2. Structural and Mechanical Property Test

Rigaku Dmax 2500Pc X-ray diffractometer was used for phase characterization, and the XRD results were refined using FullProf software (version: January-2023) to calculate the cell parameters and phase content. Then, a Hitachi S-4800 scanning electron microscope

(SEM) was used to observe the microstructure, the KathMatic KC series laser confocal microscope and SEM were used to analyze the fracture morphology, and the Nano Measurer software (version: 1.2.5) was used to calculate the grain size. The samples were etched with 5 % nitrate alcohol for 40 s before SEM observation.

We used the HR-150A Rockwell hardness tester to test the hardness of samples, take the average value of five tests on each sample, and used the HWV-2T Vickers hardness tester to test the microhardness. According to GB/T229-2007 of Chinese national standard, the impact sample size was processed to 10 mm × 10 mm × 55 mm. We used the JB-200 pendulum impact testing machine to test the impact toughness at 150 J impact energy and take the average value of three samples. According to GB/T3960-2016 standard, the wear samples were processed to "30 mm × 7 mm × 6 mm"—in engineering drawings of all sizes, and the M-200 model friction and wear testing machine was used to conduct the wear resistance test. The counterpart's materials in the sliding friction and wear test are GCr15 bearing steel. The loading test force was 350 N, the rotation speed of the friction pair was 200 r/min, and the test time was 1 h. The weight loss was measured using an analytical balance with an accuracy of 0.00001 g, and the wear loss averaged the three tests.

3. Results and Discussion

3.1. Microstructure

Figure 1a–d shows OM metallographic photos of as-cast and heat-treated HCCI and HCCI-VFC. The microstructure of as-cast HCCI is composed of elliptical austenite and M_7C_3 carbide distributed at grain boundaries. After heat treatment, fine particles or fine lath eutectic carbide at the grain boundaries and strip- or block-like primary carbide form a network structure. Figure 1e is an SEM image of the as-cast HCCI alloy composed of M_7C_3 carbide, austenite (for short, A), and a small amount of martensite (for short, M). According to the EDS results in Table 1, region A has a composition of $(Cr_{0.51}Fe_{0.46}M_{0.03})_7C_3$ ($M_{0.03} = Si_{0.004}Mn_{0.022}Ni_{0.004}$) without V, Ti, and Nb, and corresponds to the eutectic M_7C_3 carbide with a petal microstructure composed of granular (~1.0 μm) and rod-shaped (4.6 ± 1.8 μm long and 1.1 ± 0.2 μm wide) grains. Region B is a blocky primary M_7C_3 carbide with a composition of $(Cr_{0.49}Fe_{0.45}M_{0.07})_7C_3$ ($M_{0.07} = Si_{0.021}Mn_{0.021}Ni_{0.013}Mo_{0.004}$), rich in Si and Mo. Its average size is 6.1 ± 2.6 μm, and some are more than 10 μm long. Region C presents a nearly hexagonal shape and has a composition of $(Fe_{0.83}Cr_{0.13}M_{0.04})C_x$ (M = Si, Mn, Mo, Ni) and a Vickers hardness of 430.7 HV of austenite in HCCI (400–500 HV). The average grain size of austenite in HCCI ingot is 18.6 ± 13.3 μm, including large grains of 28.0 ± 9.8 μm, as shown in a red circle, and small grains of 7.5 ± 2.5 μm, as shown in purple circles.

Table 1. EDS results of each region in Figure 1 (at%).

Element	Fe	Cr	Si	Mn	Mo	Ni	V	Ti	Nb	Zr	B	C	Possible Phases
A	24.9	27.6	0.2	1.2	0	0.2	0	0	0	0	0	45.9	M_7C_3
B	24	26	1.1	1.1	0.2	0.7	0	0	0	0	46.9	0	M_7C_3
C	58.1	9.1	0.8	0.8	0	1.1	0	0	0	0	0	30.1	A
D	24.2	28.6	0	1.4	0.3	0.8	0	0	0	0	0	44.7	M_7C_3
E	50.9	8.5	0.9	1.3	0	0.9	0	0	0	0	0	37.5	M
F	52.3	8.3	0.7	1.4	0	1	0	0	0	0	0	36.3	M
G	17.41	4.56	0.22	0.45	0.12	0.26	0.28	0.16	0	0	50.75	25.72	M_7C_3
H	0.7	0	0	0	0	0	68.8	3.9	1.5	0.1	0	25.0	$(V, M)_2C$
I	36.2	0	0	0	0	0	54.0	0.5	0.9	0.1	8.3	0	V(Fe, M)
J	0.6	0	0	0	0	0	3.4	11.1	1.5	29.9	35.3	18.2	(Zr, M)(C, B)
K	9.4	11.9	0.3	0.5	0	0.2	0.5	0.1	0.1	0	44.9	32.1	M_7C_3
L	24.5	3.8	0.3	0.4	0	0.1	0.1	0	0	0.1	16.2	54.5	A
M	23.5	3.7	0.2	0.4	0	0.2	0.2	0.1	0	0.1	19.8	51.8	M
N	10.1	11.9	0.2	0.6	0.1	0.2	0.6	0.1	0.1	0	45.9	30.2	M_7C_3

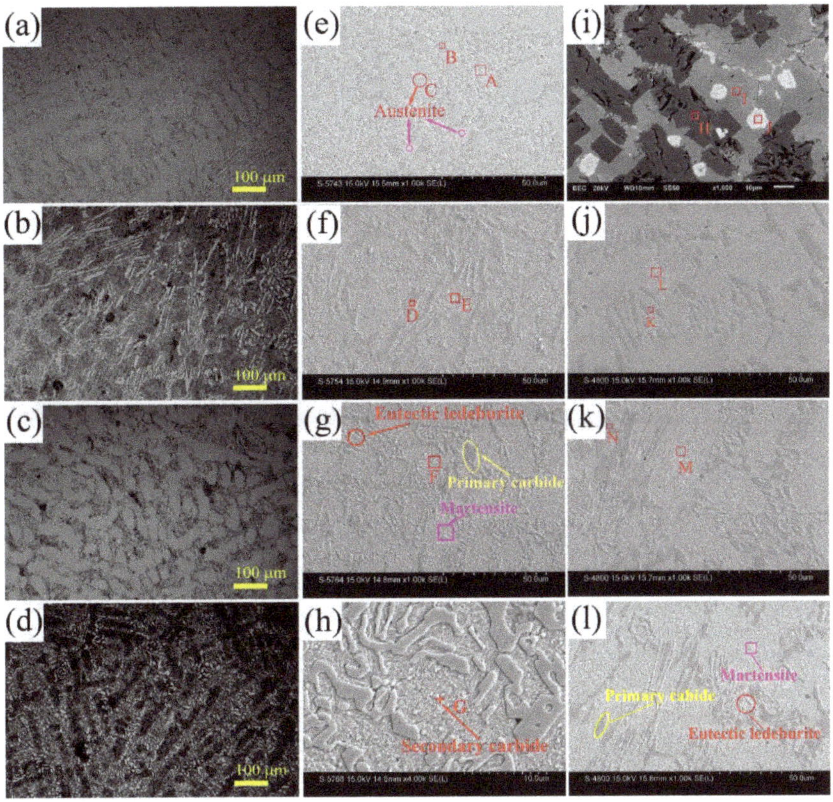

Figure 1. OM and SEM images of HCCI, VFC, and HCCI-VFC alloys. (**a–d**): OM images: (**a**) as-cast HCCI; (**b**) heat-treated HCCI; (**c**) as-cast HCCI-VFC; (**d**) heat-treated HCCI-VFC; (**e–l**): SEM images: (**e**) as-cast HCCI; (**f**) quenched HCCI; (**g**) tempered HCCI; (**h**) enlarged view of tempered HCCI; (**i**) VFC alloy; (**j**) as-cast HCCI-VFC; (**k**) quenched HCCI-VFC; (**l**) tempered HCCI-VFC.

Figure 1f–h shows the SEM images of HCCI after heat treatment. Figure 1f shows that as-quenched HCCI is composed of M_7C_3 carbide, M, and a small amount of residual austenite (for short, A′). Region D is M_7C_3 carbide with a composition close to $(Cr_{0.52}Fe_{0.44}M_{0.04})_7C_3$ ($M_{0.04}$ = $Si_{0.022}Mn_{0.005}Ni_{0.013}$). Region E is lath M grain with a composition close to $(Fe_{0.81}Cr_{0.14}M_{0.05})C_x$ (M = Si, Mn, Mo, and Ni). Vickers hardness of region E is 690.3 HV, proving that the matrix has changed from A to M. Figure 1g shows that the fine eutectic carbide and the long-strip or blocky primary carbide are distributed on the grain boundaries, forming a network structure in the tempered HCCI. The granular eutectic carbide is ~1 μm, and the stipe-shaped eutectic carbide is 5.5 ± 2.7 μm long and 1.0 ± 0.2 μm wide. EDS results show that the matrix composition in region F of tempered HCCI alloy is similar to that in region C but slightly rich in Mn. The averaged Vickers hardness of region F is 713.4 HV, improved by 23.1 HV, meaning that a part of A′ has transformed into M after tempering, and the increased M content further increases the hardness of the HCCI matrix. In addition, the average grain size of the tempered alloy is 18.1 ± 8.5 μm, equivalent to as-cast alloy. Additionally, Figure 1h shows that white granular secondary carbides are dispersed on the matrix. Region G is rich in Fe, Cr, B, and C (Table 1). Theoretically, it is possible to form M_3C, M_7C_3, and $M_{23}C_6$ carbide. According to EDS results, the white granular secondary carbide should be M_7C_3 carbide.

Figure 1i shows the SEM image of the VFC alloy. The ingot shows three contrasts. Black region H is rich in V, Ti, and C, and its composition is $(V_{68.8}Zr_{0.1}Ti_{3.9}Fe_{0.7}Nb_{1.5})C_{25.0}$ = $(V, M)_3C$. V-C compounds have no stable V_3C structure, while V_2C, Fe_2C, and Nb_2C have similar stable M_2C structures. Therefore, the added trace elements can dissolve in the V_2C structure to form a $(V, M)_2C$ alloy solid solution. Gray region I is rich in V and Fe, corresponding to the V (Fe, M) matrix. White region J is $(V_{3.4}Zr_{29.9}Ti_{11.1}Fe_{0.6}Nb_{1.5})(B_{18.2}C_{35.3})$ = $(Zr, M)_{0.9}(B, C)$, rich in Zr, B, and C. ZrC, VC, TiC, and NbC can form MC structures. Therefore, V, Ti, and Nb replace Zr, and B replaces C to form the (Zr, M)(B, C) phase.

Figure 1j is the SEM image of the as-cast HCCI-VFC alloy. It is also composed of M_7C_3 carbide, A, and a small amount of M. Region K has the composition of $(Cr_{0.54}Fe_{0.43}M_{0.03})_7C_3$ ($M_{0.03}$ = $Si_{0.009}Mn_{0.015}Ni_{0.006}$), containing 0.5 at% V, 0.1 at% Ti, and 0.1 at% Nb, indicating that V-Ti-Nb has dissolved into the M_7C_3 carbide with a length of 5.4 ± 2.1 μm and width of 1.3 ± 0.3 μm. The composition of Region L is similar to that of region C but is slightly rich in Si and poor in Ni, containing 0.1 at% V and 0.1 at% Zr. Its Vickers hardness is 464.8 HV, 7.3% higher than that of region L (430.7 HV), indicating that a small amount of V and Zr has solid-solution strengthened the A matrix with a size of 18.5 ± 13.3 μm. Strong carbide-forming elements (Ti, Zr, V, Nb) can shift the C curve to the right, improving the hardenability of the HCCI alloy. The V and Zr dissolved into austenite improve the hardenability of the HCCI-VFC alloy. In short, compared to Figure 1e, the dissolved alloying elements did not refine the size of M_7C_3 carbide and austenite in as-cast HCCI alloy but improved the hardness of the matrix.

Figure 1j–l shows the SEM images of HCCI-VFC after heat treatment. Region M is the M matrix with a composition of $(Fe_{0.82}Cr_{0.13}M_{0.04}(VFC)_{0.01})C_x$ (M = Si, Mn, Mo, Ni, $(VFC)_{0.01}$ = $V_{0.007}Ti_{0.003}$). V-Ti-Nb and B elements are mainly dissolved in M_7C_3 carbides, while Zr is dissolved in austenite and martensite. The Vickers hardness of the martensite in Figure 1l is 763.2 HV, which is 7.0% higher than that of the matrix in region F. This further proves that the matrix hardness is improved by solution strengthening. Additionally, the eutectic carbide has transformed into clump- and worm-shape. Region N is M_7C_3 carbide with a composition of $(Cr_{0.50}Fe_{0.42}M_{0.05}(VFC)_{0.03})_7C_3$ ($M_{0.04}$ = $Si_{0.007}Mn_{0.022}Mo_{0.004}Ni_{0.007}$), $(VFC)_{0.063}$ = $V_{0.047}Ti_{0.010}Nb_{0.006}$). Figure 1l shows that white secondary carbides are dispersed on the HCCI-VFC matrix. After HCCI are tempered, carbon and chromium in residual austenite will precipitate in secondary carbides [23].

Thus, the alloying process of adding VFC ingot to HCCI alloy can be described as follows: first, we melt alloy elements into the VFC alloy (Figure 2a). When the HCCI alloy was completely melted in the smelting furnace, VFC ingot was added into the HCCI melt (Figure 2b). After that, V-Ti-Nb alloying elements are rapidly dispersed in the melt through electromagnetic stirring (Figure 2c). During pouring, V-Ti-Nb-Fe-Zr-C-B alloying elements were dissolved into the austenite and M_7C_3 carbide (Figure 2d).

3.2. Phase Transformation

Figure 3 and Table 2 show the XRD patterns and refinement results of the tempered HCCI and HCCI-VFC alloys. The XRD results are refined by the Rietveld method using the FullProf software. The phase content of HCCI and HCCI-VFC is calculated directly using FullProf software. They are both composed of martensite, residual austenite, and M_7C_3 carbide. Additionally, HCCI alloy comprises 70.7 wt% M, 6.0 wt% A′, and 23.3 wt% M_7C_3 carbide with a hexagonal structure. In HCCI-VFC alloy, the A′ content decreased to 0.9 wt%, and the M content increased to 82.5 wt%, indicating that VFC addition improves the hardenability of austenite, increases the martensite content, and reduces the A′ content. Notably, the HCCI-VFC alloy contains 7.9 wt% M_7C_3 with an orthogonal structure and 8.7 wt% M_7C_3 with a hexagonal structure. In other words, the VFC addition changed the crystallization process of the HCCI alloy. At the same time, the dissolved alloying elements expand the cell volume of M, A′, and M_7C_3 carbide.

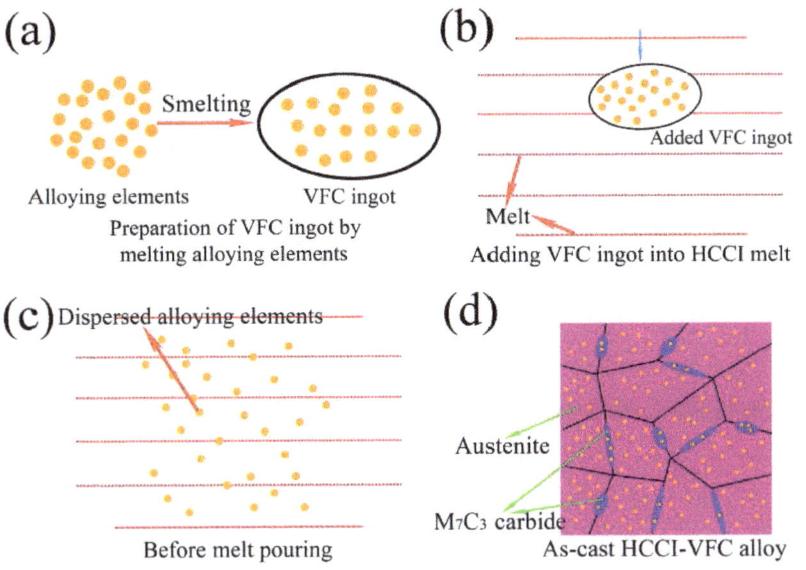

Figure 2. Melting micro-process models (**a**) melting alloy elements into VFC ingot; (**b**) adding VFC into HCCI melt; (**c**) VFC dispersing in the melt; (**d**) as-cast HCCI-VFC alloy.

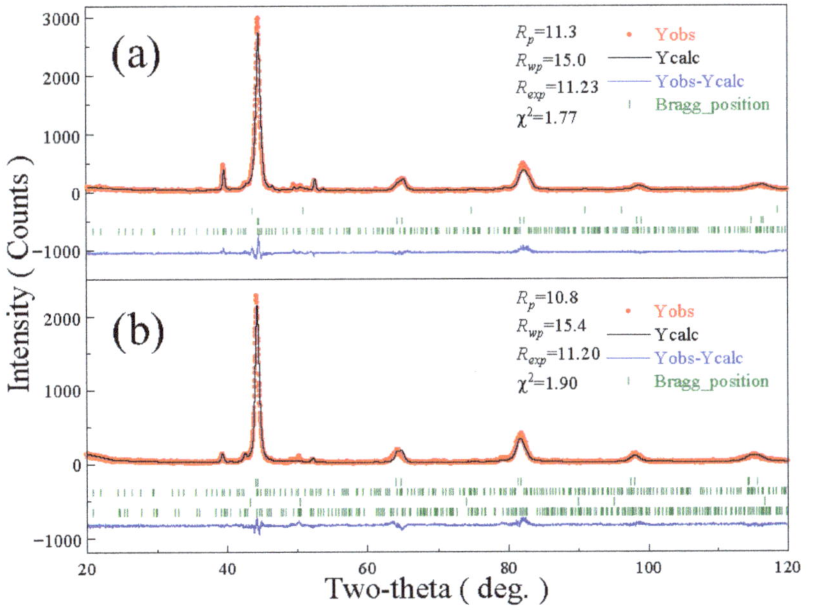

Figure 3. XRD patterns and refinement results of the heat-treated alloys. (**a**) HCCI; (**b**) HCCI-VFC.

Table 2. Refinement results corresponding to Figure 2.

Phase/System/SG	Parameters	HCCI	HCCI-VFC
Martensite /Tetragonal /I4/mmm (139)	a (nm)	0.286250(13)	0.289869(17)
	c (nm)	0.289249(20)	0.287309(18)
	V ($\times 10^{-3}$ nm^3)	23.7	24.1
	Content (wt%)	70.7	82.5
Austenite /Cubic /Fm$\bar{3}$m (225)	a (nm)	0.35829(7)	0.3617(2)
	V ($\times 10^{-3}$ nm^3)	45.9	47.3
	Content (wt%)	6.0	0.9
M_7C_3 /Orthogonal /Pnma (62)	a (nm)	—	0.45332(9)
	b (nm)	—	0.7014(3)
	c (nm)	—	1.189(2)
	V ($\times 10^{-3}$ nm^3)	—	378.2
	Content (wt%)	—	7.9
M_7C_3 /Hexagonal /P6_3mc (186)	a (nm)	1.39084(7)	1.4008(2)
	c (nm)	0.45028(7)	0.454385(0)
	V ($\times 10^{-3}$ nm^3)	754.3	772.1
	Content (wt%)	23.3	8.7

Cr_7C_3 has two crystal structures, which are a hexagonal structure with a space group (SG) of P6_3mc (186) at lower temperatures and an orthogonal structure with the Pnma (62) SG at high temperatures [24]. The formation enthalpy for Ortho-Cr_7C_3 and Hexa-Cr_7C_3 are −0.115 eV/atom and −0.087 eV/atom, respectively, so they are thermodynamically stable. Moreover, Ortho-Cr_7C_3 is more stable than Hexa-Cr_7C_3 [25]. Xing et al. [16] reported the chemical stability of (Cr, Fe)$_7C_3$ carbide at high temperatures could be improved by adding the alloying elements of Mo, W, and B into Cr-Ni cast irons to form (Cr, Fe, Mo, W)$_7$(C, B)$_3$ compound. Moreover, the added alloying elements of Mo and W partially replaced the Cr and Fe atoms to form stronger metal bonds, reduced the Gibbs free energy of the M_7C_3 carbide, improved the atomic binding energy in the carbide, and enhanced the chemical stability. In addition, the chemical bonds formed by B and (Cr, Mo, W) is more potent than those by C and (Cr, Mo, W), and the chemical stability of the formed M_7(C, B)$_3$ carbide is also greatly improved. Table 1 shows that V-Ti-Nb and B elements are mainly dissolved into M_7C_3 carbides to form the more stable (Cr, Fe, V, Ti, Nb)$_7$(C, B)$_3$ compound. In other words, they are first dissolved in Ortho-M_7C_3, which is stable at high temperatures, then part of Ortho-M_7C_3 is transformed into Hexa-M_7C_3, which is stable at low temperatures, thereby forming two M_7C_3 structures with similar content in HCCI-VFC alloy. Compared to HCCI alloy, it can be concluded that V-Ti-Nb-B addition can stabilize Ortho-M_7C_3 carbide. The solubility of Zr in α-Fe and γ-Fe is 0.1 at% and 0.2 at%, respectively.

Moreover, we only added 3.1 wt% VFC alloy, equivalent to only adding 0.07% Zr. Therefore, Zr can indeed be dissolved in γ-Fe and α-Fe and can also be dissolved in A and M. However, because Zr has the largest atomic radius in the alloy (0.145 nm), Zr is not conducive to stabilizing M_7C_3 compounds and tends to dissolve entirely in A and M, causing unit cell volume expansion. The atomic radii of V (0.135 nm) and Nb (0.148 nm) are also larger than those of Fe (0.127 nm) and Cr (0.127 nm), resulting in the volume expansion of the M_7C_3 cell in HCCI-VFC alloy.

3.3. Alloying Effect on Mechanical Properties

3.3.1. Hardness

The hardness of the matrix and carbide in the alloy determines the hardness of high chromium cast iron. Table 3 shows HCCI-VFC alloy has a higher hardness than HCCI alloy. Moreover, the effect of alloying on improving the hardness of heat-treated alloys is more pronounced. EDS results in Table 1 show that V-Ti-Nb and B elements are mainly dissolved in M_7C_3 carbides, while a small amount of V and all Zr are dissolved in austenite and martensite, resulting in a solid solution strengthening effect. After quenching, most of

A is transformed into M, increasing the hardness of HCCI and HCCI-VFC by 20.0% and 20.9%, respectively. After tempering, the A′ content decreases, and the granular secondary carbides precipitate on the matrix, increasing the hardness of HCCI and HCCI-VFC alloy by 4.5% and 6.4%, and matrix hardness by 3.3% and 3.6%, respectively. Hexa-Cr_7C_3 has a higher shear modulus (G = 117.2 GPa) and Young's modulus (E = 311.2 GPa) than Ortho-Cr_7C_3 with G = 109.4 GPa and E = 293.6 GPa [25]. Therefore, changes in the M_7C_3 structure should have little impact on strength. The total M_7C_3 content in both structures is 16.6 wt%, less than 23.3 wt% in HCCI alloy. However, due to an increase of 11.5 wt% in the M content, a decrease of 5.1% in the A′ content, and an increase of more than 7% in the matrix hardness, the hardness of the HCCI-VFC alloy increased to 63.4 HRC, which is 4.1% higher than that of HCCI alloy.

Table 3. Rockwell hardness (HRC) of high chromium cast iron alloy and Vickers hardness (HV) of the matrix in HCCI and HCCI-VFC alloys.

Sample	Alloy Hardness (HRC)			Matrix Hardness (HV)		
	As-Cast	As-Quenched	Tempered	As-Cast	As-Quenched	Tempered
HCCI	48.6 ± 0.2	58.3 ± 0.4	60.9 ± 0.3	430.7 ± 4.1	690.3 ± 7.3	713.4 ± 5.4
HCCI-VFC	49.3 ± 0.6	59.6 ± 0.3	63.4 ± 0.2	464.8 ± 4.7	736.7 ± 7.4	763.2 ± 5.6
Increased by (%)	1.4	2.2	4.1	7.9	6.7	7.0

3.3.2. Toughness

The impact toughness of the heat-treated HCCI-VFC alloy increased from 7.1 J/cm^2 of HCCI alloy to 8.3 J/cm^2, increasing by 16.9%. High chromium cast iron is composed of M_7C_3 carbide and the M matrix and can be regarded as an in situ composite material in which M_7C_3 carbide is the reinforcing phase, and M is the matrix. Table 2 shows that the M content increased by 11.5 wt% and the M_7C_3 content decreased to 16.6 wt% in HCCI-VFC alloy, meaning adding VFC alloy weakens the continuity of carbide at grain boundaries and improves the impact toughness of HCCI-VFC.

To further analyze the reason for the improvement of impact toughness, Figure 4 shows the two-dimensional and three-dimensional impact fracture morphology of the heat-treated HCCI and HCCI-VFC alloys. The black region in Figure 4a,c represents the crack source, and the blue, green-yellow, and red-orange areas correspond to the fracture's fiber area, radiation area, and shear lip area, respectively. The area ratios of the three areas were measured by Image-Pro-Plus software (version: November-2022), as shown in Table 4. The cracks in both alloys originate from the fiber zone. The fiber area ratio of HCCI-VFC alloy is 39.0% less than that of HCCI alloy, while the radiation area ratio is 217.9% higher than that of HCCI alloy. The propagation from the fiber area to the radiation area indicates a transition from slow crack growth to rapid unstable crack growth. The larger the radiation area ratio, the higher the toughness of the alloys. The final fracture area forms a shear lip. The shear lip area ratio of HCCI-VFC samples increased by 5.8%. When the material's toughness is better and the crack propagation speed is lower, the shear lip area ratio in the fracture surface is more prominent. For example, the shear lip may disappear on the brittle material's fracture. From the three fracture area ratios, the HCCI-VFC has more ductile fracture characteristics, improving impact toughness.

Figure 4b,d show the fracture height of the heat-treated HCCI-VFC alloy is higher than the HCCI alloy, meaning the HCCI-VFC alloy has a higher elongation along the longitudinal axis perpendicular to the fracture surface than that of HCCI.

Figure 4e,f shows the SEM images of the impact fracture surface corresponding to Figure 4a,d. The fracture morphology of HCCI alloy (Figure 4e) is mainly composed of river pattern and cleavage plane. The tearing fiber of martensite can be observed in region A, indicating that the martensite matrix is the main phase that bears impact force. Figure 4f shows tear edges in HCCI-VFC alloy, and the fracture morphology is significantly different from Figure 4e. For example, the carbides in region B are orderly and linearly distributed, surrounded by martensite. The level of martensite tearing is slightly lower than in Figure 4e. This result proves that the strengthened M_7C_3 carbide and the high content of the M matrix help the alloy absorb impact energy and improve the interfacial strength between the carbide and martensite, which is beneficial to improving the impact toughness of HCCI-VFC.

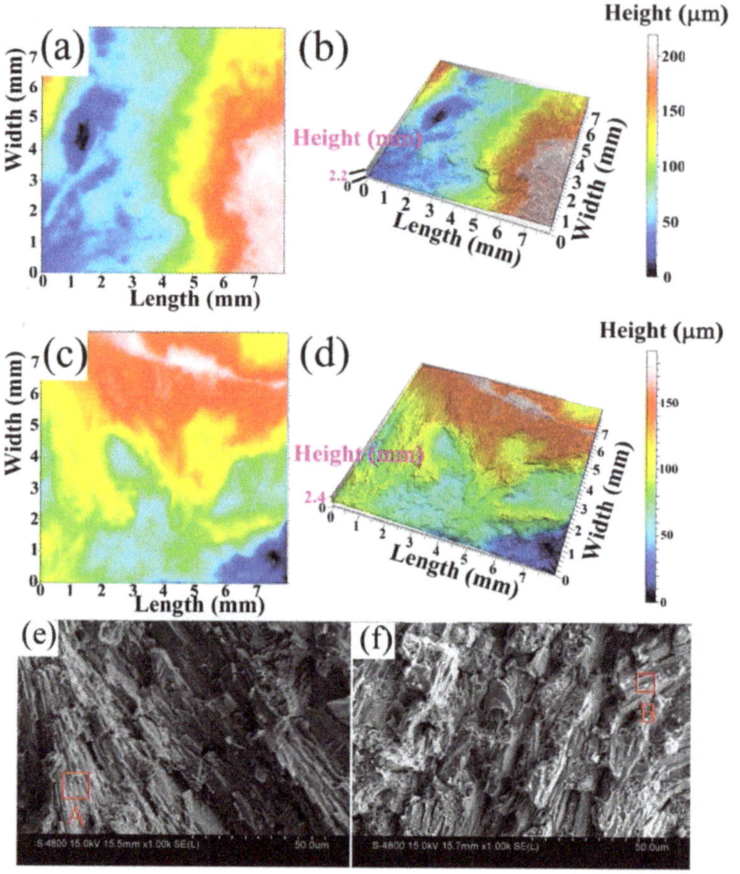

Figure 4. Macro morphology of impact fracture of the heat-treated HCCI and HCCI-VFC alloy. (**a**) two-dimensional fracture morphology of HCCI; (**b**) three-dimensional fracture morphology of HCCI; (**c**) two-dimensional fracture morphology of HCCI-VFC; (**d**) three-dimensional fracture morphology of HCCI-VFC; (**e**) microstructure of impact fracture of HCCI; (**f**) microstructure of impact fracture of HCCI-VFC.

Table 4. Impact toughness (J/cm^2) and composition of impact fracture of heat-treated alloys.

Sample	Impact Toughness (J/cm^2)	Composition of Impact Fracture			
		Fiber Zone Ratio (%)	Radiation Zone Ratio (%)	Shear Lip Zone Ratio (%)	Fracture Height (mm)
HCCI	7.1	55.1	8.9	36.0	2.2
HCCI-VFC	8.3	33.6	28.3	38.1	2.4
Increased by (%)	16.9	-39.0	218.0	5.8	9.1

3.3.3. Wear

The friction coefficient and wear loss of HCCI-VFC are 0.433 and 0.0185 g/h, smaller than the corresponding 0.443 and 0.0199g/h of HCCI and decreased by 2.3% and 7.0%, respectively. Figure 5a,b shows the SEM images of the wear surface. Many deep ditches, deep exfoliation (region A), deep scratches (region B), and rough scratches are found in HCCI alloy (Figure 5a). In Figure 5b, HCCI-VFC alloy has a fractured M_7C_3 carbide and deformed matrix (region C), small ditches, superficial peeling, and deep scratches (region D). The wear mechanism mainly behaves as abrasive wear and adhesive wear. The hardness of the carbide and matrix of HCCI-VFC is higher than that of HCCI, and HCCI-VFC has better toughness, which endows the HCCI-VFC alloy with better wear resistance and friction reduction.

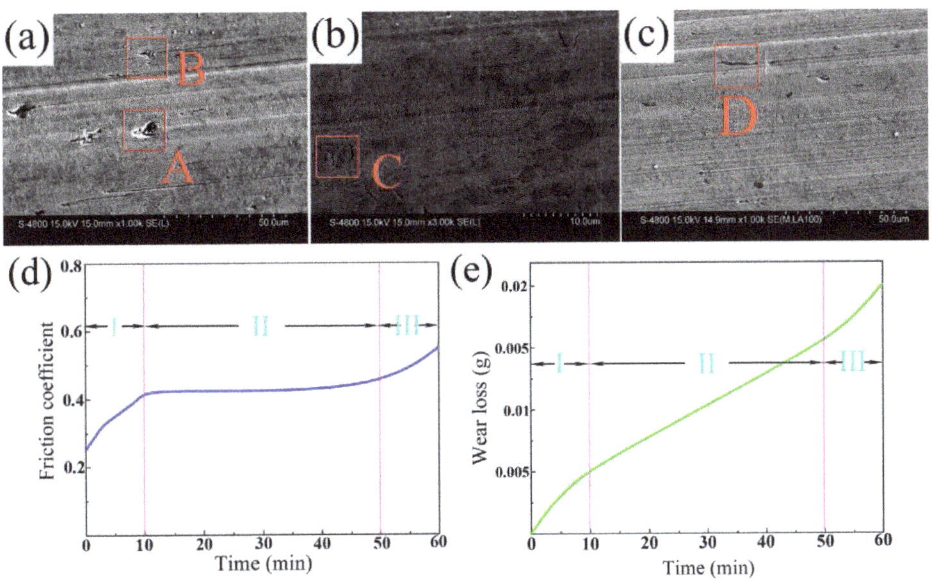

Figure 5. SEM images of the wear surface of heat-treated alloys and the friction coefficient and wear loss curves versus time. (**a**) wear surface of HCCI; (**b**) amplified view of (**a**); (**c**) wear surface of HCCI-VFC; (**d**) the friction coefficient curves versus time; (**e**) wear loss curves versus time.

Figure 5d,e shows the friction coefficient and wear loss curves versus time. The first stage is the run-in stage. As the contact area between the sample and the friction counterpart increases gradually, the friction coefficient increases gradually, and the wear loss of the sample also increases gradually. The second stage is the stable wear stage. The friction coefficient remains unchanged, and the wear loss increases almost linearly. The third stage is the severe wear stage. With an increase in wear time, the friction coefficient and the wear loss increase rapidly.

4. Conclusions

In this paper, the multi-element microalloying effect of adding 75.9V-12.6Fe-5Ti-3.1Nb-2.2Zr-0.6B-0.6C alloy ingot (VFC) into 2.99C-0.57Si-0.55Mn-17.76Cr-0.11Mo-0.12Ni (HCCI) was studied. The main conclusions are as follows:

(1) The added V-Ti-Nb-B are dissolved in M_7C_3 carbide to form the (Cr, Fe, V, Ti, $Nb)_7(C, B)_3$ alloy carbide, and a small amount of V and all Zr are dissolved in austenite and martensite. Adding VFC has little effect on M_7C_3 carbide and austenite or martensite size. Heat-treated HCCI and HCCI-VFC alloys are composed of martensite, residual austenite, and M_7C_3 carbide. After tempering, secondary carbides are dispersed and

precipitated on the martensite matrix. Adding VFC into HCCI improved the hardenability of HCCI, decreased the residual austenite content from 6.0 wt% to 0.9 wt%, increased the martensite content from 70.7 wt% to 82.5 wt%, and changed the structure and content of M_7C_3 carbide from 23.3 wt% hexagonal M_7C_3 in HCCI to 7.9 wt% orthogonal M_7C_3 and 8.7 wt% orthogonal M_7C_3 carbide in HCCI-VFC.

(2) The solid-soluted V-Ti-Nb-B strengthened the M_7C_3 carbide, and the solid-soluted V-Zr strengthened austenite and martensite, increasing the hardness of as-cast and heat-tread HCCI alloy to 49.3 HRC and 63.4 HRC, increased the hardness of austenite and martensite to 464.8 HV and 763.2 HV, increased the impact toughness from 7.1 J/cm^2 to 8.3 J/cm^2, and reduced the friction coefficient and wear loss by 2.3% and 7.0%, respectively. The wear mechanism behaves as abrasive wear and adhesive wear. Thus, the hardness, toughness, wear resistance, and friction resistance of HCCI alloy have been improved simultaneously by adding a multi-element VFC alloy, showing a significant synergistic effect.

Author Contributions: Conceptualization, J.S.; data curation, Z.X.; J.H. and W.S.; investigation, C.C.; methodology, J.S.; supervision, J.S.; writing—original draft, T.L.; writing—review and editing, J.S. All authors have read and agreed to the published version of the manuscript.

Funding: This research was funded by the Central Government Guides Local Funds for Science and Technology Development (No. 216Z1008G), and the Natural Science Foundation of Hebei province, China (No. E2022202017).

Institutional Review Board Statement: Not applicable.

Informed Consent Statement: Not applicable.

Data Availability Statement: All data included in this study are available upon request by contact with the corresponding author.

Conflicts of Interest: The authors declare that they have no known competing financial interest or personal relationships that could have appeared to influence the work reported in this paper.

References

1. Jain, A.S.; Chang, H.W.; Tang, X.H.; Hinckley, B.; Zhang, M.X. Refinement of primary carbides in hypereutectic high-chromium cast irons: A review. *J. Mater. Sci.* **2020**, *56*, 999–1038. [CrossRef]
2. Siekaniec, D.; Kopycinski, D.; Tyrala, E.; Guzik, E.; Szczesny, A. Optimisation of Solidification Structure and Properties of Hypoeutectic Chromium Cast Iron. *Materials* **2022**, *15*, 6243. [CrossRef] [PubMed]
3. Hadji, A.; Bouhamla, K.; Maouche, H. Improving wear properties of high-chromium cast Iron by manganese alloying. *Int. J. Metalcast.* **2016**, *10*, 43–55. [CrossRef]
4. Filipovic, M.; Kamberovic, Z.; Korac, M. Solidification of high chromium white cast iron alloyed with vanadium. *Mater. Trans.* **2011**, *52*, 386–390. [CrossRef]
5. Bedolla-Jacuinde, A.; Guerra, F.V.; Mejía, I.; Zuno-Silva, J.; Rainforth, M. Abrasive wear of V–Nb–Ti alloyed high-chromium white irons. *Wear* **2015**, *332–333*, 1006–1011. [CrossRef]
6. Kana, A.Z.V.; Krutis, V. Effect of alloying elements on properties and structure of high chromium cast irons. *Arch. Metall. Mater.* **2018**, *63*, 609–614.
7. Zhou, Y.; Yang, Y.; Yang, J.; Hao, F.; Li, D.; Ren, X.; Yang, Q. Effect of Ti additive on (Cr, Fe)$_7$C$_3$ carbide in arc surfacing layer and its refined mechanism. *Appl. Surf.* **2012**, *258*, 6653–6659. [CrossRef]
8. Penagos, J.J.; Pereira, J.I.; Machado, P.C.; Albertin, E.; Sinatora, A. Synergetic effect of niobium and molybdenum on abrasion resistance of high chromium cast irons. *Wear* **2017**, *376–377*, 983–992. [CrossRef]
9. Kopyciński, D.; Siekaniec, D.; Szczęsny, A.; Guzik, E.; Nowak, A. The effect of Fe-Ti inoculation on solidification, structure and mechanical properties of high chromium cast iron. *Arch. Metall. Mater.* **2017**, *62*, 2183–2187. [CrossRef]
10. Maja, M.E.; Maruma, M.G.; Mampuru, L.A. Effect of niobium on the solidification structure and properties of hypoeutectic high-chromium white cast irons. *J. S. Afr. Inst. Min. Metall.* **2016**, *116*, 981–986. [CrossRef]
11. Arnoldo, B.J.; Guerra, F.; Mejia, I.; Vera, U. Niobium additions to a 15%Cr–3%C white iron and its effects on the microstructure and on abrasive wear behavior. *Metals* **2019**, *9*, 1321.
12. Zhou, X.; Li, G.; Shen, X.; Liu, Y. Tensile strength improvement of martensitic ODS steels with Zr and Hf additions. *Mat. Sci. Engi. A* **2022**, *829*, 142071. [CrossRef]
13. Yang, Y.; Tan, L.; Bei, H.; Busby, J.T. Thermodynamic modeling and experimental study of the Fe–Cr–Zr system. *J. Nucl. Mater.* **2013**, *441*, 190–202. [CrossRef]

14. Lei, J.W.; Wu, K.M.; Li, Y.; Hou, T.P.; Xie, X.; Misra, R.D.K. Effects of Zr addition on microstructure and toughness of simulated CGHAZ in high-strength low-alloy steels. *J. Iron Steel Res. Int.* **2019**, *26*, 1117–1125. [CrossRef]
15. Zeng, W.; Zhou, M.; Yang, M.; Qiu, R. Effect of Zr on the microstructure and mechanical properties of 12Cr ferritic/martensitic steels. *Fusion Eng. Des.* **2022**, *177*, 113084. [CrossRef]
16. Xing, J.D.; Gao, Y.M.; Wang, E.Z. Effect of phase stability on the wear resistance of white cast iron at 800 °C. *Wear* **2002**, *252*, 755–760. [CrossRef]
17. Purba, R.H.; Shimizu, K.; Kusumoto, K.; Gaqi, Y.; Todaka, T. Effect of boron addition on three-body abrasive wear characteristics of high chromium based multi-component white cast iron. *Mater. Chem. Phys.* **2022**, *275*, 125–136. [CrossRef]
18. You, M.A.; Xiulan, L.I.; Liu, Y.G.; Zhou, S.Y. Microstructure and properties of Ti-Nb-V-Mo alloyed high chromium cast iron. *Natl. Acad. Sci. Lett.* **2013**, *36*, 839–844.
19. Gong, L.; Fu, H.; Zhi, X. Corrosion wear of hypereutectic high chromium cast iron: A review. *Metals* **2023**, *13*, 308. [CrossRef]
20. Zhi, X.H.; Xing, J.D.; Gao, Y.M.; Fu, H.G.; Peng, J.Y.; Xiao, B. Effect of heat treatment on microstructure and mechanical properties of a Ti-bearing hypereutectic high chromium white cast iron. *Mater. Sci. Eng. A* **2007**, *487*, 171–179. [CrossRef]
21. Liu, Q.; Zhang, H.; Wang, Q.; Nakajima, K. Effect of heat treatment on microstructure and mechanical properties of Ti-alloyed hypereutectic high chromium cast iron. *ISIJ Int.* **2012**, *52*, 2288–2294. [CrossRef]
22. Chen, Z.; Guo, Q.; Fu, H.; Zhi, X. Effect of heat treatment on microstructure and properties of modified hypereutectic high chromium cast iron. *Mater. Test.* **2022**, *64*, 33–54. [CrossRef]
23. Wiengmoon, A.; Chairuangsri, T.; Chomsang, N.; Poolthong, N.; Pearce, J.T.H. Effects of heat treatment on hardness and dry wear properties of a semi-solid processed Fe-27wt%Cr-2.9wt%C cast Iron. *J. Mater. Sci. Technol.* **2008**, *24*, 330–334.
24. Zhang, D.; Hou, T.P.; Liang, X.; Zheng, P.; Zheng, Y.H.; Lin, H.F.; Wu, K.M. The structural, magnetic, electronic, and mechanical properties of orthogonal/hexagonal MC (M = Fe and Cr) carbides from first-principles calculations. *Vacuum* **2022**, *203*, 111–125. [CrossRef]
25. Yangzhen, L.; Yehua, J.; Rong, Z. First-principles study on stability and mechanical properties of Cr_7C_3. *Rare Metal Mat. Eng.* **2014**, *43*, 2903–2907. [CrossRef]

materials

MDPI

Article

Migration Behavior of Inclusions at the Solidification Front in Oxide Metallurgy

Chunliang Yan [1,2], Fengming Wang [3], Wenling Mo [4,*], Pengcheng Xiao [1] and Qingjun Zhang [1,2,5,*]

1 College of Metallurgy and Energy, North China University of Science and Technology, Tangshan 063210, China; yanchunlinag@ncst.edu.cn (C.Y.); xiaopc@ncst.edu.cn (P.X.)
2 Comprehensive Testing and Analyzing Center, North China University of Science and Technology, Tangshan 063210, China
3 College of Science, North China University of Science and Technology, Tangshan 063210, China; wfm77@ncst.edu.cn
4 College of Yisheng, North China University of Science and Technology, Tangshan 063210, China
5 Hebei Province High-Quality Steel Continuous Casting Engineering Technology Research Center, Tangshan 063210, China
* Correspondence: Correspondence: mwl@ncst.edu.cn (W.M.); zhangqingjun@ncst.edu.cn (Q.Z.); Tel.: +86-315-8816-213 (Q.Z.)

Abstract: Distribution of inclusions plays an essential role in inducing intracrystalline ferrite, and the migration behavior of inclusions during solidification has a significant influence on their distribution. The solidification process of DH36 (ASTMA36) steel and the migration behavior of inclusions at the solidification front were observed in situ using high-temperature laser confocal microscopy. The annexation, rejection, and drift behavior of inclusions in the solid–liquid two-phase region were analyzed, providing a theoretical basis for regulating the distribution of inclusions. Analysis of inclusion trajectories showed that the velocity of inclusions decreases significantly as they near the solidification front. Further study of the force on inclusions at the solidification frontier shows three situations: attraction, repulsion, and no influence. Additionally, a pulsed magnetic field was applied during the solidification process. The original dendritic growth mode changed to that of equiaxed crystals. The compelling attraction distance for inclusion particles with a diameter of 6 μm at the solidification interface front increased from 46 μm to 89 μm, i.e., the effective length for the solidification front engulfing inclusions can be increased by controlling the flow of molten steel.

Keywords: in situ observation; solidification front; inclusion migration behavior; microstructure

1. Introduction

Oxide metallurgy involves the use of micro inclusions in steel to induce intragranular acicular ferrite in crystals, forming microstructures with different orientations and cross-interlocking to improve the strength and toughness of the steel [1–4]. Presently, the excellence of acicular ferrite is unanimously recognized. Some progress has been made in the composition design of steel and the mechanism of inclusion-induced intragranular ferrite [5–8]. Although oxide metallurgy has been applied to large-wire welding technology [9], the nucleation and growth of nano- and micron-scale inclusions are not truly coupled to the macroscopic flow field.

The objective of oxide metallurgical steel is to form a uniformly dispersed system of inclusions distributed in the steel grains. The flow of molten steel and the migration behavior of inclusions during the solidification process directly determine their distribution in the solidified structure and ultimately affect the properties of the bulk material [10]. The engulfment or pushing of inclusion particles by the advancing melt–solid interface in steel is an important issue. Therefore, the migration behavior of leading-edge inclusions at the solidification interface has received widespread attention.

To enhance the inclusion of particles captured by the solidification interface into the crystal, rather than aggregating to the end of solidification and causing inclusion aggregations [11–14], many researchers have devoted themselves to studying particle behavior at the solidification frontier. Moreover, a variety of models have been developed to explain the transition between particle absorption and frontier pushing, and the dominant factors affecting particle behavior have been identified, thereby aiming to control particle behavior and distribution.

However, there are many restrictions on the application of these models owing to the complexity of the influencing system and its parameters. Research on how inclusions migrate to the solidification interface and their influencing factors during the solidification stage has been limited. Thus far, artificially controlling the behavior of particles at the forefront of the solidification interface has not been possible, significantly hindering the application of oxide metallurgy in industrial production.

Compared to traditional methods for studying the solidification process, high-temperature confocal laser scanning microscopy (HTCLSM) has a higher resolution and allows real-time in situ observation of transient phenomena at high temperatures. HTCLSM plays a vital role in research on welding, materials, and steel [15–21]. Dippenaar used HTCLSM to study the phase transformation processes of slag, flux, and steel during heating and cooling, correlating them with theoretical predictions [22–24]. He developed a concentric solidification technology based on high-temperature confocal experiments, which can be used to study the surface tension, interface morphology, and high-temperature physical properties of metallic materials [25,26]. Hiroyuki et al. discussed the mechanism of trapping and repelling inclusions using a moving solid–liquid interface during the solidification of Al-Sikilled steels [27]. However, the use of HTCLSM to study inclusion movement, behavior, and influencing factors during the solidification process of oxide metallurgical steels has not yet been reported.

Therefore, this study used HTCLSM to observe the migration behavior of inclusions in situ during the solidification process of molten steel, especially the migration at the solidification interface. From theoretical and experimental points of view, the feasibility of changing the particle distribution and the behavior of particles being engulfed or repelled at the solidification interface by changing the flow rate is discussed. It is important for improving the quality of non-modulated steel, pipeline steel, welding steel, and especially large-line energy welding steel.

2. Materials and Methods

2.1. Materials and Sample Preparation

The raw material selected was DH36 (ASTM A36) steel smelted using an oxide metallurgy process, whose chemical composition (mass fraction, %) is measured by X-ray fluorescence spectrometer (XRF) ZSX Primus II from Nippon Institute of Sciences; its composition is listed in Table 1. To obtain more explicit images, the surface of a specimen was polished to a mirror finish.

Table 1. Chemical composition of the experimental steel (wt%).

Element	C	Si	Mn	P	S	Al	Ti	Cr	Mg	Nb	Mo
Content (%)	0.06	0.34	1.46	0.0067	0.001	0.03	0.017	0.014	0.003	0.04	0.07

2.2. Experimental Apparatus and Procedure

In this study, the solidification behavior was observed in situ using a confocal laser scanning microscope equipped with an infrared furnace (VL2000DX, Lasertec, Tokyo, Japan). A halogen lamp (typically 1.5 kW) irradiated the gold-coated surface of the reaction chamber with light. The gold coating reflected energy towards the focal plane of the crucible to rapidly heat the sample, as depicted in Figure 1.

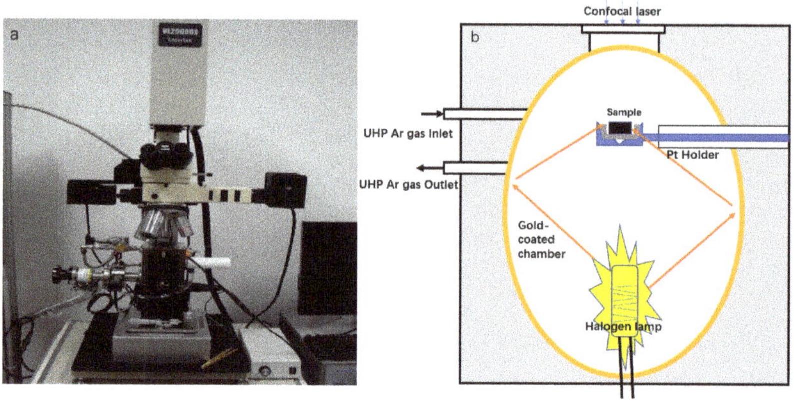

Figure 1. (**a**) Confocal scanning laser microscope (CLSM); (**b**) schematic of experimental apparatus.

Thermocouples were installed within the Pt sample holder, which provided feedback to the halogen lamp power-supply controller. Inert crucibles are known to act partially as thermal barriers, resulting in thermocouple readings that do not reflect the actual temperature of the specimen. Thus, the experimental target temperature was calibrated using a material with a known melting point in the temperature range of interest.

Though the position temperature measured by the thermocouple of the high-temperature laser confocal microscope had a specific error with respect to the actual temperature on the surface of the steel sample, the final temperature of the experiment was determined to be 1873 K. To ensure a uniform internal temperature for the modelling, the sample was slowly heated to 1873 K at a rate of 30 K/min and held for 600 s. The cooling rate was too high, and the solidification time was too short to effectively observe the migration behavior of inclusion particles at the front of the solidification interface. Conversely, the cooling rate was too low for the induction of needle-like ferrite by oxide particles; the final cooling rate was determined to be 50 K/min. The temperature control curve of the experiment is shown in Figure 2.

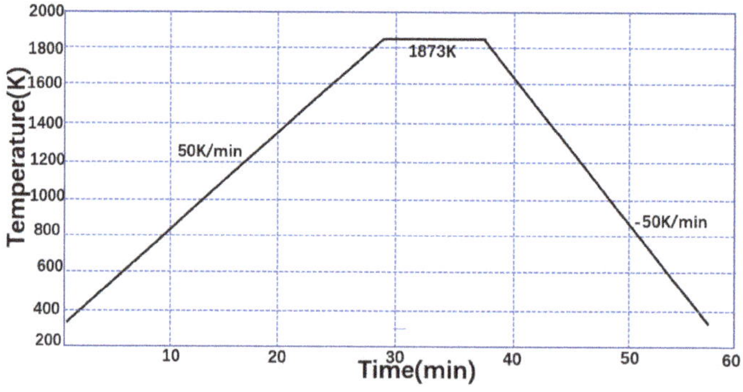

Figure 2. Temperature profile of the CLSM experiment.

The cleaned sample was placed in an alumina crucible (inner diameter of 8 mm and height of 3 mm) on a Pt holder in an infrared furnace. Before heating, the infrared furnace was alternately evacuated and backfilled with argon (Ar > 99.999%) three times to avoid the oxidation of the sample during the experiment.

A confocal laser scanning microscope was installed above the infrared image oven to observe the solidification process and inclusion behavior in the DH36 steel at high temperatures.

3. Results

3.1. Solidification Process of Molten Steel

Crystallization occurred within a specific temperature range applicable for ship plate steel with a large crystallization temperature interval. When the temperature was above the liquidus, the molten steel flowed continuously in the crucible owing to the influence of the surface tension drive; the field of view was all liquid, as shown in Figure 3a. At 1805 K, the molten steel at the observation position began to solidify. A steel cell crystal grew in a preferentially convex manner at 1805 K, forming a distinct bump, as shown in Figure 3b. New quasiparticles appeared, promoting crystal growth. The crystals that precipitated from the liquid state continued to increase in size, became thicker, and further extended in specific directions to form various dendritic morphologies, as shown in Figure 3c–e. In contrast, the amount of liquid phase in the system continued to decrease. Owing to dendrite expansion, a staggered network structure was formed, which hindered the flow of the liquid; molten steel flowed to the area with the least crystal nucleation. As the precipitated crystals in each physical region grew to form a network, the flow of liquid metal was obstructed, and the space occupied by the liquid metal continued to shrink. As the crystal grew, the liquid state completely disappeared, and the entire area eventually became solid, as shown in Figure 3f. At this time, the temperature decreased to 1773 K; the temperature difference during the solidification process was 32 K. According to the theory of component supercooling in the solidification process, the interface of the solidification front was unstable at a cooling rate of 50 K/min, perturbing the unit cell section to form a secondary dendrite arm, and finally growing as dendrites. Perturbations at the solid–liquid interface can be clearly distinguished, and coarsening of the dendrite arm can be observed. The dendrite trend is chaotic, and the spacing size is large.

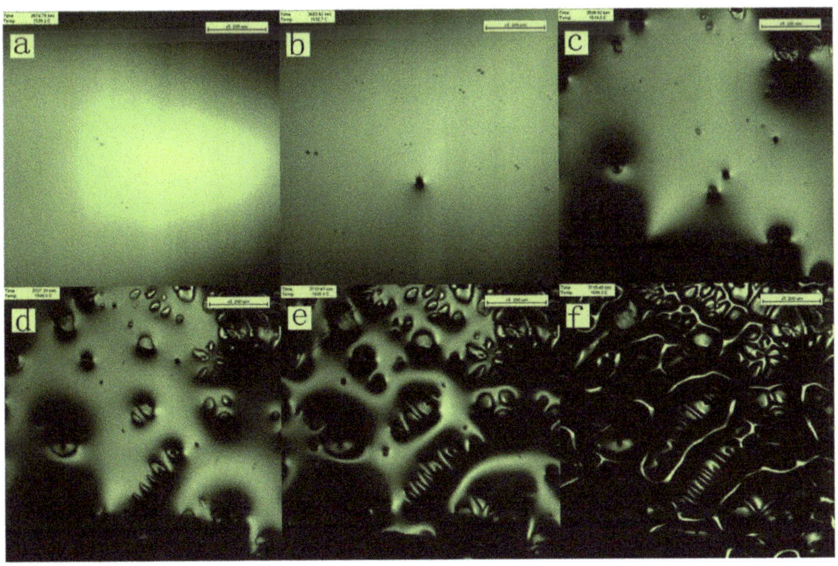

Figure 3. In situ observation of the solidification process.

During the solidification process, the faster the liquid steel flows, the easier it is for inclusions to accumulate at the end of solidification along with the remaining liquid surface flow. Inclusions do not quickly enter the austenite grains during solidification but are located at the grain boundaries. A magnetic field effectively inhibits the flow of the melt and reduces the diffusion coefficient [28–32]. To study the effect of the melt flow rate on the solidification process and the behavior of the inclusions, a pulsed magnetic field was applied during the solidification process [33,34]. As in the previous experiment, we observed a planar-to-dendritic transition during the solidification of the DH36 alloy melt, as shown in Figure 4a–d. However, the dendrites in the field of view precipitated almost simultaneously, the dendrite spacing increased simultaneously, and the dendrite grains became refined and orderly as shown in Figure 4e,f.

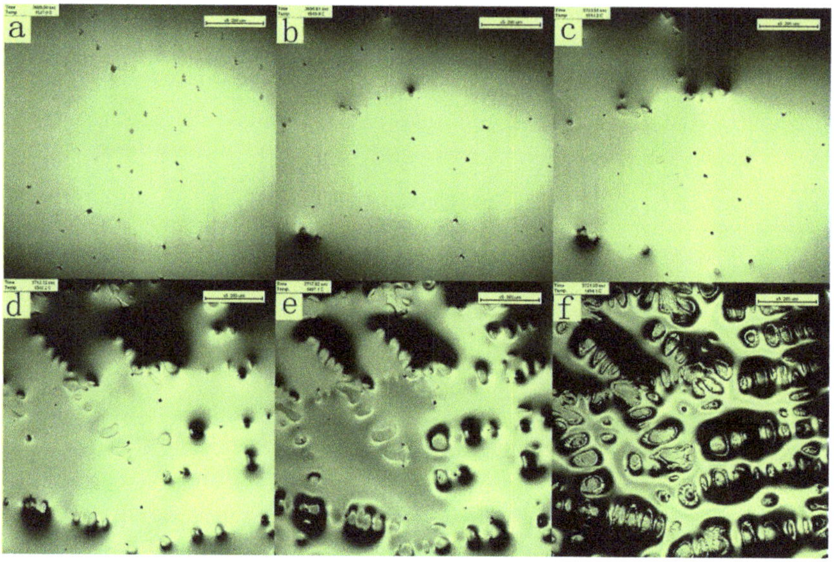

Figure 4. In situ observation of the solidification process when applying a magnetic field.

The temperature at which solidification begins was different between the two observations, though a small region was observed during the CLSM experiment. The temperature at which the crystal first appears in the field of view may not be the actual crystallization temperature. The primary phase may have been generated below the liquid surface or outside the field of view. Quasi-particles arranged on the surface likely originated from the tip of a branch of the same dendrite arm. The crystals observed in the image were most likely cross sections of the dendrites.

Comparing the in situ observations of solidified dendrites (Figures 3f and 4f), the solidified microstructure of the two comparative samples is shown in the Figure 5. The untreated process produced relatively coarse dendrites. After the pulsed magnetic field treatment, the morphology changed significantly; the grains were refined, and the dendrites were neatly arranged and uniform in size.

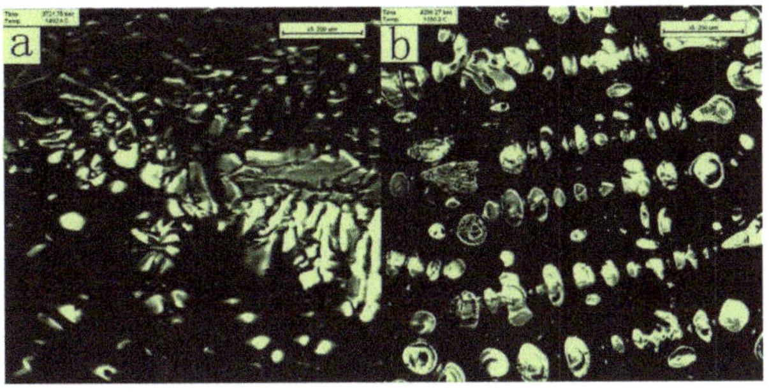

Figure 5. The solidified microstructure of sample (**a**) without magnetic and (**b**) with magnetic field.

3.2. Migration Behaviour of Inclusions at the Solidification Front

Using a confocal scanning laser microscope, we observed the movement of inclusions in the solid–liquid two-phase zone during an initial stage of solidification. The protrusion is a solidified phase that forms a liquid channel, in which inclusions drift along the fluid medium.

Five inclusion particles were selected for observation, inclusions a, b, c, d, and f, respectively. The positions were recorded every 0.4 s until particle motion ceased. ImageJ 1.51j8 image processing software was used to establish a coordinate axis based on the image scale and record the coordinates of the particles in each image to obtain the behavior of the particles during solidification. The trajectories of the inclusions at the solidification front are shown in Figure 6. In Figure 6d1,d2,e1,e2 refer to the two segment behavior trajectories of inclusion d and e. The inclusion particles were approximately spherical in shape; the diameters of inclusions a, b, and c are 6 μm, 5 μm, and 6 μm, respectively. The three inclusions were not immediately engulfed upon approaching the solidification interface but moved around the solidification interface. They were then engulfed either by the solidification interface or at a position away from the interface movement. Inclusion d twice approached the solidification interface frontier, but there was no solidification interface front capture. The inclusion exhibited a circular motion around the solidification interface front, and when the solid-phase region continued to expand as solidification progressed, the inclusion was finally engulfed by the solidification interface frontier. Inclusion f did not move in a circular manner around the solidification interface, maintaining a relatively long distance from the solidification interface. Instead, it moved randomly in the molten steel. As solidification progressed, new solidification interfaces continued to form and grow from the liquid surface, and inclusion e was eventually engulfed by the front of the solidification interface.

A pulsed magnetic field was applied to the molten steel to change tehe flow intensity of the liquid phase during the solidification process. The migration behavior of inclusions at the front edge of the solidification interface was observed. The migration behavior trajectories of inclusions v, w, x, and y were continuously recorded at the front edge of the solidification interface, as shown in Figure 7. Inclusions v and w exhibited irregular motion and were observed colliding and aggregating in the molten steel, migrating to form a trajectory until they were captured by the solidification interface.

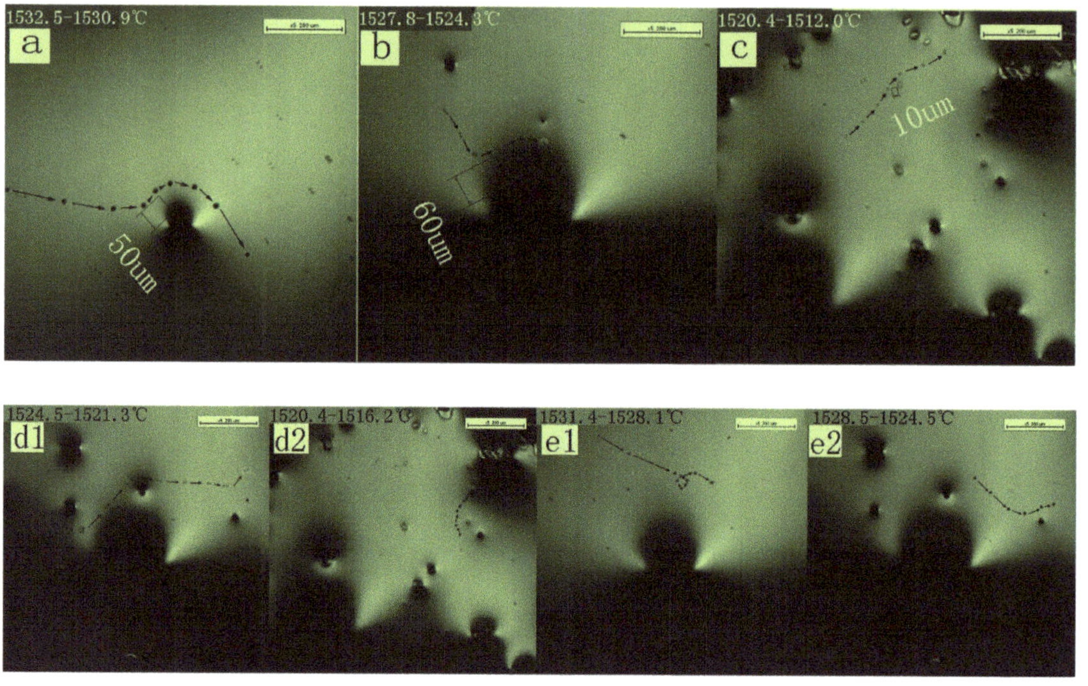

Figure 6. Behavioral trajectories of inclusions (**a–c**,**d1**,**d2**), and (**e1**,**e2**) migrating at the solidification front.

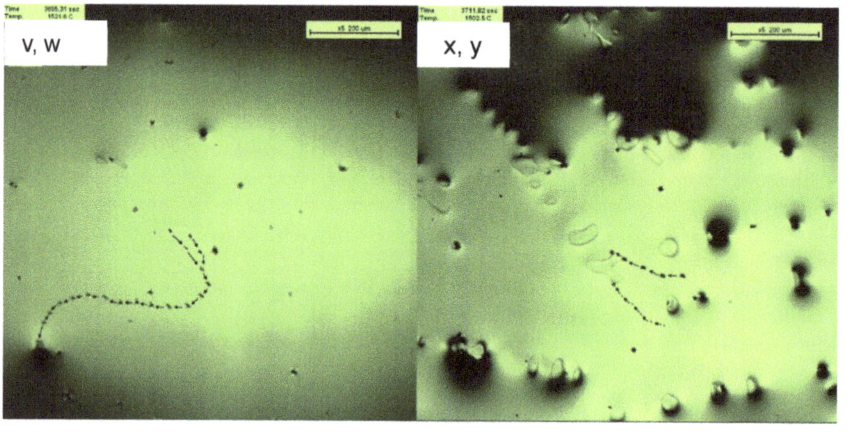

Figure 7. Behavioral trajectories of inclusions v, w, x, and y at the front edge of the solidification interface under pulsed magnetic field conditions.

The motion trajectories of the inclusions after applying a pulsed magnetic field were significantly different from those without the applied magnetic field.

4. Discussions

4.1. Analysis of the Inclusion Movement at the Solidification Front

The particles were assumed to move at a uniform speed within a time of 0.4 s; the velocity and acceleration of the inclusions were calculated. The migration velocity and acceleration of the inclusions in the liquid phase with no applied magnetic field are shown in Figure 8.

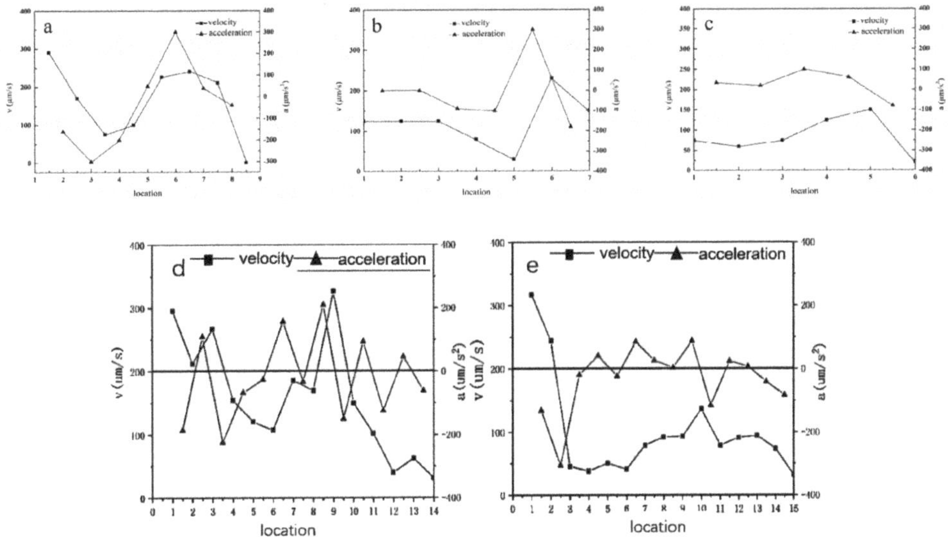

Figure 8. Velocity and acceleration of inclusions (**a**–**d**), and (**e**) at the solidification front.

During the solidification process, the volume fraction of the solid phase increased, the solid-phase surface gradually became uneven, and the drift rate of the inclusions decreased. The average drift rate of inclusions far from the solid–liquid interface was 200 μm/s. In contrast, the drift rate near the solid–liquid interface was relatively low, approximately 100 μm/s. The velocities of inclusions A and B decreased as the inclusions approached the solidification interface, yielding a negative acceleration. The maximum absolute value of the acceleration occurred at 50 μm from the solidification interface, where the velocity decreased to its minimum. The velocity around the solidification interface gradually increased with positive acceleration. Then, the velocity approaching the solidification interface was zero. The velocity of inclusion C increased through the solidification interface, where the acceleration was positive and maximum at 5 μm from the solidification interface. The acceleration decreased but maintained inclusion motion. Subsequently, inclusion C was attracted to another solidification interface.

Inclusion D passed through the front edge of the solidification interface many times; each time it passed through the solidification interface, the particle underwent deceleration; when approaching the solidification interface, the acceleration was negative, and the particle decelerated. When moving around the solidification interface, the acceleration was positive and increased inclusion motion. In the molten steel, inclusion F was relatively distant from the solidification interface and could not be attracted by the interface. As the solid-phase fraction continued to increase, the particles around the interface decelerated.

The small particles moved towards the solid at a velocity higher than the speed caused by the fluid flow. The effect of attraction is very clear, and the distance from which the force was effective was approximately 40 μm to 50 μm. After the inclusion approached the solid–liquid interface, it moved circularly around the precipitated solid phase and gradually moved closer to the solid phase zone. It was quickly captured, indicating that the attraction effect of the solid–liquid interface to the nearby drifting inclusions was sufficient to resist the impulse caused by inclusion drift.

The velocity and acceleration of the inclusion particles at the solid–liquid interface under the action of a pulsed magnetic field are shown in Figure 9.

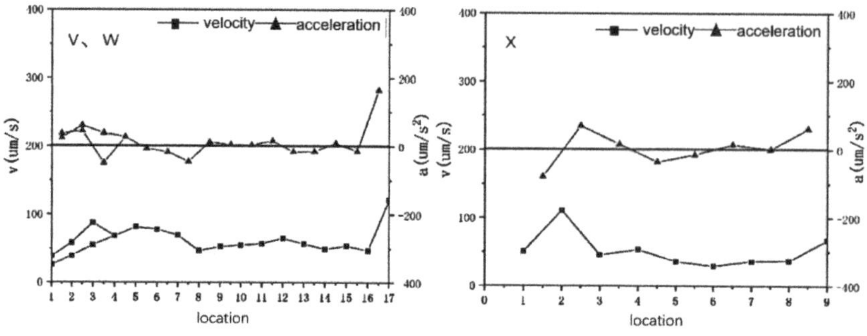

Figure 9. Velocity and acceleration of inclusions v, w, and x at the solidification front under pulsed magnetic field conditions.

The velocity and acceleration of inclusions v and w were stable during inclusion movement; the velocity suddenly increased when approaching the solidification interface until the inclusion was captured by the solidification interface. Inclusion x was eventually captured by the solidification interface when the molten steel flowed. Its velocity was nearly identical to that of the merged particles and suddenly increased when it approached the solidification interface.

This indicates that inclusion motion was affected by the force at the solidification front. When no magnetic field was applied during the solidification process, the velocity of the inclusion particles at the solidification front was approximately 200 μm/s, and the amplitude of change was large. After suppressing the melt flow by applying a magnetic field, the particle velocity was below 100 μm/s, exhibiting constant velocity. The inclusions moved with the molten steel in the initial stage of solidification. As the temperature decreased, the solidification process continued, and the inclusion particles accelerated when passing through the solidification interface. By applying a magnetic field, the flow of the melt was inhibited, and the motion, speed, and acceleration of the inclusions decreased, indicating that the flow of molten steel plays a very important role in the migration behavior of the inclusions.

4.2. *Nature of the Long-Range Attraction of Inclusions at the Solidification Front*

The force analysis on the inclusion particles is based on the laws of motion behavior. The forces on the inclusions in the fluid are those: (1) not related to the relative motion of the fluid particles, including gravity, buoyancy, etc.; (2) correlated with fluid particle relative motion or direction along relative direction of motion, including viscous resistance Stokes force and Basset force; and (3) correlated to the relative motion of the fluid particle and the direction perpendicular to the direction of motion, including the (i) Magnus force generated by the particle rotation due to the different flow velocities at different positions during the solidification of the molten steel, (ii) Magnus force from low speed to high speed when the particle is in a flow field graded by the velocity, and (iii) lateral lift Saffman force, even if there is no particle rotation. At high temperatures, the influence of the Brownian force should also be considered for inclusion particles of micron size and smaller. The forces acting on the inclusions at the solidification interface are shown in Figure 10 [35].

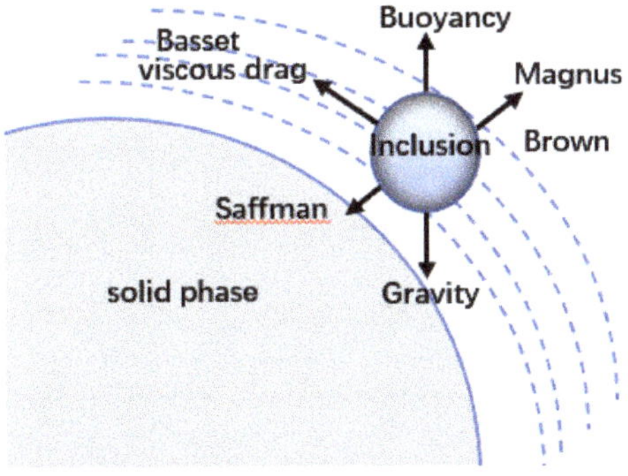

Figure 10. Schematic of particle–interface interaction.

In this high-temperature experiment, the inclusions floated and moved on the surface regardless of gravity and buoyancy. As the inclusion size decreased, the specific surface energy increased, and the action of gravity, buoyancy, and turbulent entrainment decreased. The interaction of the particles with the solid and fluid phases plays an important role. When the inclusion particles are smaller than 10 μm, the Brownian motion is more intense, and the particles are subjected to greater Basset and Brownian forces. The velocity of the inclusions at the front of the solidification interface changed significantly, and the Saffman lift should be considered.

The Stokes viscous resistance between the inclusion particles in the molten steel and the fluid is one of the leading forces during movement. It is given by [35]:

$$F_D = C_D \times \frac{3\rho_m}{4\rho_p d_p}\left|V_m - V_p\right|\left(V_m - V_p\right), \tag{1}$$

where C_D is the drag coefficient (a function of Reynolds number), ρ_m and ρ_p represent the density of molten steel and inclusions, respectively (kg/m^3), d_p represents the diameter of the inclusion particles (μm), and V_m and V_p represent the instantaneous velocity vectors of the molten steel fluid and inclusions, respectively.

When the inclusions move in a viscous fluid at variable speeds, the Basset force on the particle surface from the surrounding unstable fluid is given by [36]:

$$F_B = C_B \cdot \frac{9}{\rho_p d_p}\sqrt{\frac{\rho_m \cdot \mu_{eff}}{\pi}}\int_0^t \frac{\frac{d(V_m - V_p)}{d\tau}}{\sqrt{t-\tau}}\,d\tau, \tag{2}$$

where C_B is the Basset coefficient, μ_{eff} is the effective dynamic viscosity of molten steel (Pa/s), and τ is the calculation time step (s).

The velocity of the fluid in the molten steel varied, and a pressure gradient formed between the fluids. The Saffman force applied to an inclusion under a pressure gradient is given by [37]:

$$F_{LS} = C_{LS} \cdot \frac{6K_S\mu_{eff}}{\rho_p \pi d_p}\sqrt{\frac{\rho_m \xi}{\mu_{eff}}} \cdot \left(V_m - V_p\right), \tag{3}$$

where ξ denotes the pressure gradient, K_s is the Saffman force coefficient (1.615) [38], and C_{LS} is the Saffman force correction coefficient.

Under high-temperature conditions, the Brownian motion of micron-sized inclusions in molten steel cannot be ignored, and the Brownian force is expressed as [39]:

$$F_R = \frac{12\delta}{\rho_p}\sqrt{\frac{3\mu_{eff}k_B T}{\pi d_p^5 \Delta\tau}}, \tag{4}$$

where K_B is the Boltzmann constant (1.38×10^{-23} J/K), T is the temperature of the molten steel (K), $\Delta\tau$ is the time step (s), and δ is a random variable that follows a standard normal distribution.

From the above analysis, the drifting, engulfing, and pushing of inclusions during the solidification process of molten steel is shown to be a very complex process. During the solidification process, the volume fraction of the solid–liquid dual-phase changes, meeting the force balance of the sample where the shape of the specimen is constantly adjusted. Therefore, the molten steel in the paste area flows, promoting inclusion drift. In addition, the absorption and repulsion of relatively solid inclusions affect their movement. The accelerated movement of the inclusions indicates a strong attraction between the inclusions and the solid phase, almost perpendicular to the melt flow direction; this long-distance attraction is unrelated to surface flow [40,41].

The investigation revealed that the origin of the long-range attraction of inclusions to the solid phase could not be attributed to van der Waals forces, Rayleigh Benard flow of the liquid steel bath, or macroscopic local surface flow of liquid steel, but rather to the capillary pressure effect around the particles on the surface of the molten steel [42–44].

Using a schematic of the meniscus around a spherical particle at the steel–gas interface, the force between them is shown to be affected by the size of the particle and the contact angle between the particle and the liquid surface. Attraction occurs if the contact angle is greater than 90°. A single particle partially immersed in the steel melt lowers the liquid level around it. If the particle and the solid come very close together, the liquid surface between them will be further depressed or drawn downward (Figure 11).

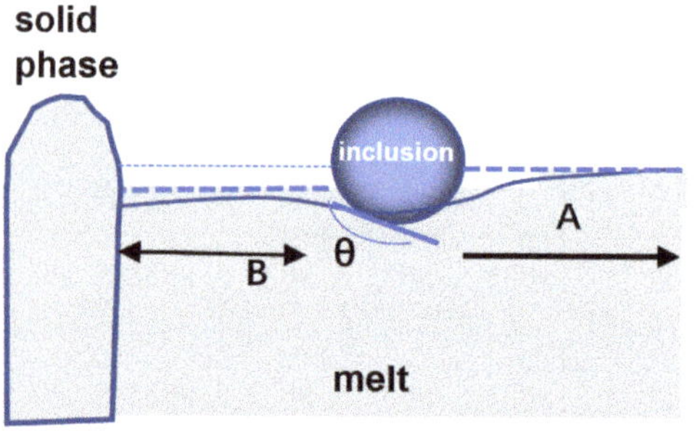

Figure 11. Schematic of capillary attraction between inclusions and the solid at the solid–liquid fronts.

This leads to a change in the capillary pressure between the outer region (A in Figure 11) and the inner region of the inclusion (B in Figure 11). The stress in region B is lower, and the particles are attracted to the solid–liquid interface, a mechanism known as capillary attraction.

The capillary effect is shown to be responsible for the long-range attraction of inclusions on the surface of molten steel. The contact angle between the solid particle; the steel melt, particle size, and density; and surface tension of the steel melt influences the capillary attraction force.

Until now, the capillary force between two small particles partly immersed in a liquid and its acting length have not been identified.

The total attraction force, F, can be given as [45]:

$$F = 0.5 \cdot g\ (\rho_L - \rho_P)\omega \Delta h^2. \tag{5}$$

Force is proportional to Δh^2, where Δh is of the equilibrium difference in liquid surface height between the inside (region B) and outside (region A) of the particle. Δh is given by [45]:

$$\Delta h = 2\gamma cos\theta / g(\rho_L - \rho_P)d, \tag{6}$$

where ρ are the densities of the liquid and inclusion; γ is the surface tension of liquid; d is the separation between the solid phase and inclusion; and θ is the contact angle between liquid and solid inclusion. Evidently, Δh is inversely proportional to d, and Δh determines the strength of attraction.

As d decreases, Δh increases, i.e., when the inclusions are close to the solid–liquid interface, the liquid level between them is dragged downward, and the attraction gradually increases.

Near the solid–liquid interface, the viscous resistance F_d experienced by the inclusion can be expressed as [27]:

$$F_d = 6\pi\eta v \frac{R^2}{d} \tag{7}$$

The viscous force also increases with a decrease in d. However, when the increase in attraction greatly exceeds the viscosity force, the inclusions are engulfed by the solid–liquid interface. There is a critical distance d_0 between the inclusions and the solid–liquid interface. When the inclusions drift to a position less than d_0, they are absorbed by the solid–liquid interface. In contrast, the absorption effect is not apparent.

Solid spherical inclusions were observed on the surface of the molten steel. They were driven by capillary pressure and eventually engulfed and aggregated at the solid–liquid interface. However, from the analysis of the velocity and acceleration of the inclusions, it can be seen that the forces and motions of different particles are different. Concentration gradients, temperature gradients, and surface tension at the front of the solidification interface can lead to the flow of liquids, which causes inclusions to be pushed away from the solidification interface.

The inclusions are attracted to the solidification interface when they are sufficiently close, which manifests as inclusion motion in the direction of the solid–liquid interface or flowing around the solid stage and then arranging on the solid phase. The direction of attraction to the solidification interface is determined by inclusions, and the closer to the solidification interface, the greater the attraction. The solidification interface attraction is limited, and within the effective range d_0, we call it region I. In region II, the molten steel passes through the solidification interface owing to the difference in speed and the Bernoulli effect: the larger the flow velocity of the fluid, the smaller the pressure. The inclusion particles are subjected to stress along the solidification interface, resulting in a circular movement. They then leave the solidification interface front or are attracted to the solidification interface. Region III is the area sufficiently distant from the solidification interface. The inclusions are less affected by the solidification interface, instead acted upon with the same force as that for the molten steel. The force analysis of the inclusions at the front of the solidification interface and its trajectory at the solidification interface front can be inferred from the movement of the leading-edge inclusions, as shown in Figure 12a. The force analysis and actual migration behavior trajectories of the inclusions at the leading edge of the solidification interface were consistent (Figure 12b).

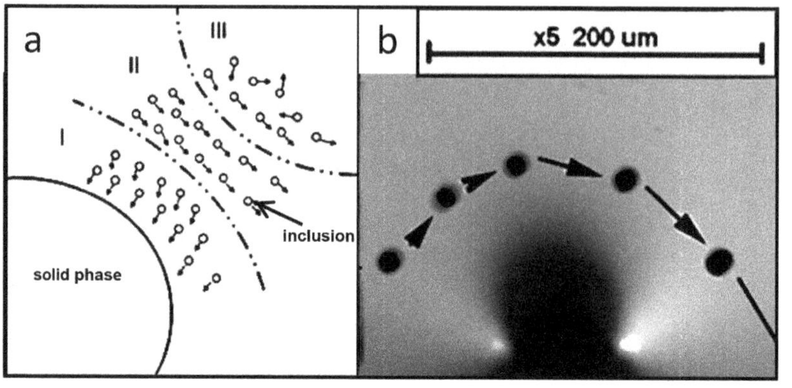

Figure 12. (**a**) Zoning of the frontier area of the solid–liquid interface; (**b**) movement behavior of inclusions at the front of the solidification interface.

The effective attraction distance from the solidification interface can be measured based on the movement trajectory and motion of the inclusions at the front of the solidification interface, as listed in Table 2.

Table 2. Effective distance from inclusion particles to solidification interface.

Particle	Diameter (μm)	Time (s)	Distance (μm)
a	6	3691.0	62
b	5	3692.5	60
c	6	3700.9	44
d	5.6	3700.4	22
e	5.8	3705.4	40
v	5.8	3696.8	93
x	6.8	3701.3	89
y	6.2	3701.8	86

In the absence of a pulsed magnetic field, the effective attraction distance for inclusion particles with a diameter of approximately 6 μm at the front of the solidification interface was 20–60 μm. The flow velocity slowed after the application of a pulsed magnetic field. Then, the effective attraction distance for inclusions with a diameter of approximately 6 μm was 85–95 μm, nearly twice that when no magnetic field was applied. Thus, the application of a pulsed magnetic field to suppress the flow of molten steel can increase the effective attraction distance.

Enhancing the driving force and increasing the effective attraction distance can engulf inclusions at the solid–liquid interface, ensuring their uniform distribution within the solidified grain, rather than preferential accumulation at the grain boundaries. The improved mechanical properties of these materials can be fully exploited with a homogeneous distribution of the reinforcement phase. Consequently, it is necessary to control the flow and cooling rate conditions to control entrapment. Thus, the inclusion particles were engulfed by growing dendrites rather than being repelled by the interface and squeezed to the final solidification site.

5. Conclusions

In this study, the behavior of inclusions during the solidification of DH36 steel was observed in situ using a high-temperature laser confocal microscope. The most important conclusions of this study are as follows.

1. During the solidification of molten steel, heterogeneous nucleation particles were formed. As solidification progressed, the solidification interface became interconnected and staggered to create a network, and the liquid phase gradually disappeared until solidification was complete.
2. The migration behavior of inclusions at the solidification interface can be divided into three situations: attraction, repulsion, and no effect, occupying three regions. These are region I near the interface, where the inclusions are attracted and engulfed; region II, where the inclusions are pushed by the solidification interface; and region III, where the inclusions are sufficiently far from the solidification interface, and there is no effect on the migration of inclusions. The velocity of the inclusions decreased significantly when they approached the solidification front, easing entry of the inclusions into the front.
3. A pulsed magnetic field was applied during the solidification process to slow the melt flow speed. The solidification front interface tended to be stable, changing from the original growth mode of the dendrite to an equiaxed crystal. The compelling attraction distance of inclusions with a diameter of 6 μm at the solidification interface front increased from 46 μm to 89 μm. Therefore, the effective distance for engulfing inclusions in the solidification front can be increased by controlling the flow of the molten steel.

Author Contributions: Research and analysis, C.Y.; design of experiments and morphology analysis, Q.Z.; image processing, F.W. and P.X.; data calculation, W.M. All authors have read and agreed to the published version of the manuscript.

Funding: This research was funded by the Key Project of the National Natural Science Foundation of China—Joint Fund for Regional Innovation and Development, grant number: U21A20114. Key Project of Science and Technology Research of Hebei Provincial Department of Education, ZD2014012.

Institutional Review Board Statement: Not applicable.

Informed Consent Statement: Not applicable.

Data Availability Statement: Not applicable.

Acknowledgments: The authors would like to thank the Technology of Hebei Engineering Research Center of High-Quality Steel Continuous Casting.

Conflicts of Interest: The authors declare no conflict of interest.

References

1. Gregg, J.M.; Bhadeshiah, K.D.H. Solid-state nucleation of acicular ferrite on minerals added to molten steel. *Acta Mater.* **1997**, *45*, 739–748. [CrossRef]
2. Ma, Z.T.; Janke, D. Characteristics of oxide precipitation solidification of deoxidized steel. *ISIJ Int.* **1998**, *38*, 46–52. [CrossRef]
3. Shim, J.H.; Byun, J.S.; Cho, Y.W.; Oh, Y.J.; Shim, J.D.; Lee, D.N. Hot deformation and acicular ferrite microstructure in C-Mn steel containing Ti_2O_3 inclusions. *ISIJ Int.* **2000**, *40*, 819–823. [CrossRef]
4. Kim, H.S.; Lee, H.G.; Oh, K.S. Evolution of size, composition, and morphology of primary and secondary inclusions in Si/Mn and Si/Mn/Ti deoxidized steels. *ISIJ Int.* **2002**, *42*, 1404–1411. [CrossRef]
5. Yokomizo, T.C.; Enomoto, M.; Umezawa, O.; Spanos, G.; Rosenberg, R.O. Three-dimensional distribution, morphology, and nucleation site of intragranular ferrite formed in association with inclusions. *Mater. Sci. Eng. A* **2003**, *344*, 261–267. [CrossRef]
6. Sarma, D.S.V.; Karasev, A.V.; Jönsson, P.G. On the role of non-metallic inclusions in the nucleation of acicular ferrite in steels. *ISIJ Int.* **2009**, *49*, 1063–1074. [CrossRef]
7. Fattahi, M.; Nabhani, N.; Hosseini, M.; Arabian, N.; Rahimi, E. Effect of Ti-containing inclusions on the nucleation of acicular ferrite and mechanical properties of multiphases weld metals. *Micron* **2013**, *45*, 107–114. [CrossRef] [PubMed]
8. Hu, J.; Du, L.X.; Wang, J.J. Effect of V on intragranular ferrite nucleation of high Ti bearing steel. *Scr. Mater.* **2013**, *68*, 953–956. [CrossRef]
9. Li, X.B.C.; Min, Y.; Yu, Z.; Liu, C.J.; Jiang, M.F. Effect of Mg addition on nucleation of intra-granular acicular ferrite in Al-killed low carbon steel. *J. Iron Steel Res. Int.* **2016**, *23*, 415–421. [CrossRef]
10. Potschke, J.; Rogge, V. On the behaviour of foreign particles at an advancing solid-liquid interface. *J. Cryst. Growth* **1989**, *94*, 726–738. [CrossRef]

11. Casses, P.; Azouni, M.A. Critical velocity of a freezing front interacting with spherical-particles. *Int. Commun. Heat Mass Transfer.* **1995**, *22*, 605–615. [CrossRef]
12. Layachi, H. Effect of shrinkage flow on particle engulfment. *Scr. Mater.* **2007**, *56*, 57–60.
13. Tao, Y.T.; Yeckel, A.; Derby, J.J. Analysis of particle engulfment during the growth of crystalline silicon. *J. Cryst. Growth* **2016**, *452*, 1–5. [CrossRef]
14. Friedrich, J.; Reimann, C.; Jauss, T.; Cröll, A.; Sorgenfrei, T.; Tao, Y.; Derby, J.J.; Friedrich, J.; Reimann, C.; Jauss, T.; et al. Engulfment and pushing of Si3N4 and SiC particles during directional solidification of silicon under microgravity conditions. *J. Cryst. Growth* **2017**, *475*, 33–38. [CrossRef]
15. Lee, J.-M.; Shibata, K.; Asakura, K.; Masumoto, Y. Observation of γ-α transformation in ultralow-carbon steel under a high temperature optical microscope. *ISIJ Int.* **2002**, *42*, 1135–1143. [CrossRef]
16. Yoho, Y.; Takeuchi, H.; Ishikawa, T.; Iwata, N.; Nakanishi, K. Strain-induced boundary migration of carbon steel at high temperatures. *Scr. Mater.* **2009**, *61*, 1001–1003.
17. Shao, X.; Wang, X.; Jiang, M.; Wang, W.; Huang, F. Effect of heat treatment conditions on shape control of large-sized elongated MnS inclusions in desulfurized free-cutting steels. *ISIJ Int.* **2011**, *51*, 1995–2001. [CrossRef]
18. Miao, Z.-J.; Shan, A.-D.; Wand, W.; Lu, J.; Xu, W.-L.; Song, H.-W. Solidification process of conventional superalloy by confocal scanning laser microscope. *Trans. Nonferrous Met. Soc. China* **2011**, *21*, 236–242. [CrossRef]
19. Park, J.-Y.; Jae Wook, R.; Sohn, I.L. In-situ crystallization of highly volatile commercial mold flux using an isolated observation system in the confocal laser scanning microscope. *Metall. Mater. Trans. B* **2014**, *45*, 1186–1191. [CrossRef]
20. Yago, Y.; Tanaka, K.; Nakanishi, K. In-situ observation of grain growth and recrystallization of steel at high temperature. *Mater. Sci. Forum* **2011**, *638–642*, 1077–1082. [CrossRef]
21. Chen, R.C.; Hong, C.; Li, J.J.; Zheng, Z.Z.; Li, P.C. Austenite grain growth and grain size distribution in isothermal heat-treatment of 300 M steel. *Procedia Eng.* **2017**, *207*, 663–668. [CrossRef]
22. Sohn, I.; Dippenaar, R. In-Situ Observation of Crystallization and Growth in High-Temperature Melts Using the Confocal Laser Microscope. *Metall. Mater. Trans.* **2016**, *47*, 2083–2094. [CrossRef]
23. Dippenaar, R.; Bernhard, C.; Schider, S.; Wieser, G. Austenite Grain Growth and the Surface Quality of Continuously Cast Steel. *Miner. Met. Mater. Soc. ASM Int.* **2013**, *45*, 409–418. [CrossRef]
24. Phelan, D.; Reid, M.H.; Dippenaar, R. High Temperature Laser Scanning Confocal Microscopy for real-time studies of phase transformations. *Microsc. Microanal.* **2005**, *11* (Suppl. S2), 670–672. [CrossRef]
25. Griesser, S.; Dippenaar, R. Enhanced Concentric Solidification Technique for High-Temperature Laser-Scanning Confocal Microscopy. *ISIJ Int.* **2014**, *54*, 533–535. [CrossRef]
26. Reid, M. Dominic Phelan and Rian Dippenaar, Concentric Solidification for High Temperature Laser Scanning Confocal Microscopy. *ISIJ Int.* **2004**, *44*, 565–572. [CrossRef]
27. Shibata, H.; Yin, H.; Yoshinaga, S.; Emi, T.; Suzuki, M. In-situ Observation of Engulfment and Pushing of Nonmetallic Inclusions in Steel Melt by Advancing Melt/Solid Interface. *ISIJ Int.* **1998**, *38*, 149–156. [CrossRef]
28. Zhang, L.; Li, W.; Yao, J.P.; Qiu, H. Effects of pulsed magnetic field on microstructures and morphology of the primary phase in semisolid A356 Al slurry. *Mater. Lett.* **2012**, *66*, 190–192. [CrossRef]
29. Li, Y.J.; Tao, W.Z.; Yang, Y.S. Grain refinement of Al–Cu alloy in low voltage pulsed magnetic field. *Mater. Process. Technol.* **2012**, *212*, 903–909. [CrossRef]
30. Chen, H.; Jie, J.C.; Fu, Y.; Ma, H.J.; Li, T.J. Grain refinement of pure aluminium by direct current pulsed magnetic field and inoculation. *Trans. Nonferrous Met. Soc. China* **2014**, *24*, 1295–1300. [CrossRef]
31. Zhang, L.; Hu, P.H.; Zhou, Q.; Zhan, W.; Jin, F. Effects of pulsed magnetic field on microstructure, mechanical properties and bio-corrosion behavior of Mg-7Zn alloy. *Mater. Lett.* **2017**, *193*, 224–227. [CrossRef]
32. Shao, Q.; Wang, G.; Wang, H.D.; Xing, Z.G.; Fang, C.Y.; Cao, Q.L. Improvement in uniformity of alloy steel by pulsed magnetic field treatment. *Mater. Sci. Eng. A* **2021**, *799*, 140–143. [CrossRef]
33. Zhang, M.N.; Zhang, Q.J. Mini high temperature device with micro area strong magnetic field for microscopic observation. CN203437628U, 19 February 2014.
34. Cui, Z.M. The Oxide Metallurgy Behaviour in Medium Carbon Steel and the Effect of Pulsed Magnetic Field. Ph.D. Thesis, University of Science and Technology Beijing, Bejing, China, 2017.
35. Jiang, Y.X. *Research and Analysis of a Spherical Particle Rising in the Liquid*; Dalian University of Technology: Dalian, China, 2009.
36. Hatami, M.; Ganji, D.D. Motion of a spherical particle on a rotating parabola using Lagrangian and high accuracy Multi-step Differential Transformation Method. *Powder Technol.* **2014**, *258*, 94–98. [CrossRef]
37. Saffman, P.G. The lift on a small sphere in a slow shear flow. *J. Fluid Mech.* **1965**, *22*, 385. [CrossRef]
38. Wang, Y.; Li, H.; Guo, L. Numerical simulation of the force condition of spherical inclusion particles in liquid steel. *J. Univ. Sci. Technol. Beijing* **2013**, *35*, 1437–1442.
39. Kim, M.M.; Zydney, A.L. Effect of electrostatic, hydrodynamic and Brownian forces on particle trajectories and sieving in normal flow filtration. *J. Colloid Interface Sci.* **2004**, *269*, 425. [CrossRef]
40. Uhlmann, D.R.; Chalmers, B.; Jackson, K.A. Interaction between particles and a solid-liquid interface. *Appl. Phys.* **1964**, *35*, 2986–2993. [CrossRef]

41. Rempel, A.W.; Worster, M.G. The interaction between a particle and an advancing solidification front. *J. Cryst. Growth* **1999**, *205*, 427–440. [CrossRef]
42. Garvin, J.W.; Udaykumar, H.S. Effect of a premelted film on the dynamics of particle–solidification front interactions. *J. Cryst. Growth* **2006**, *290*, 602–614. [CrossRef]
43. Chen, X.H.; Yan, H. Solid–liquid interface dynamics during solidification of Al 7075–Al_2O_3np based metal matrix composites. *Mater. Des.* **2016**, *94*, 148–158. [CrossRef]
44. Liu, Y.; Han, Q. Interaction between nuclear particles and a solid-liquid interface in Al-4.5Cu alloy. *Acta Mater.* **2021**, *213*, 116955–116963. [CrossRef]
45. Yin, H.; Shibata, H.; Emi, T.; Suzuk, M. ln-Situ Observation of Collision Agglomeration and Cluster Formation of Alumina Inclusion Particles on Steel Melts. *ISIJ Int.* **1997**, *37*, 936–945. [CrossRef]

materials

MDPI

Article

Microstructure and Mechanical Properties of Dual Scaled NbC/Ti_2AlC Reinforced Titanium–Aluminum Composite

Sen Cui [1,2,*], Chunxiang Cui [2] and Xin Wang [2]

1 School of Materials Science and Engineering, Shijiazhuang Tiedao University, Shijiazhuang 050043, China
2 School of Materials Science and Engineering, Hebei University of Technology, Tianjin 300401, China; hutcui@hebut.edu.cn (C.C.); ahaxin@hebut.edu.cn (X.W.)
* Correspondence: cuisen0624@126.com

Abstract: A TiAl composite containing hybrid particles and whisker reinforcements is fabricated by vacuum melting. The results of this study show that the comprehensive mechanical properties and refining effect of the material are best when the content of reinforcement is 1 wt.%, and then the mechanical properties begin to deteriorate as the content increases further. Finely dispersed NbC particles and uniformly dispersed Ti_2AlC whiskers are the ideal second phases. The synergistic strengthening effect of NbC particles and in situ Ti_2AlC whiskers are key to the improvement of mechanical properties. Compared with the TiAlNb matrix, the fracture stress/strain of the composite at 1073 K is improved from 612 MPa/19.4% to 836 MPa/26.6%; the fracture toughness at room temperature is improved from 18.8 MPa/m^2 to 27.4 MPa/m^2.

Keywords: strengthening; TiAl; metal matrix composite

1. Introduction

Titanium–Aluminum (TiAl) has attracted the attention of scholars due to its high specific strength and good high-temperature performance [1–4]. The urgent problem to be solved for the TiAl alloy is to improve its brittleness at room temperature and enhance the high-temperature strength. By adding β stabilizing elements such as Nb, Mo and Cr, the phase composition can be adjusted and the degree of segregation of TiAl alloy can be reduced, which has been a widely used method [5–7]. On this basis, a composite method is an efficient way to further improve the mechanical performance of cast TiAl alloys. By introducing ceramic phases into the alloy matrix, the comprehensive mechanical properties of metal matrix composites (MMCs) could be significantly improved when comparing with the matrix [8,9]. This is due to the intrinsic mechanical properties of these ceramic reinforcements and their refining effects on the microstructure of the alloy matrix. In the past few decades, different scaled ceramic phases, such as particles [10] and whiskers [11], have both been extensively studied as reinforcements in MMCs. In terms of the strengthening mechanisms, the emphases of particles and whiskers are different. Particle reinforcements have a relatively small size. Therefore, particles with the same volume fraction can provide more heterogeneous nucleation cores than whiskers, and thus have a better refining effect. Moreover, small particles have a greater tendency to distribute inside the grains, which can play a better role in Orowan strengthening in the process of deformation. Whiskers are relatively large in size and can bear the load effectively. Hence, whiskers can play a larger role in the load transferring mechanism. Meanwhile, the bridging effect of whiskers between multiple grains can also delay the fracture of the alloy and improve the comprehensive mechanical properties.

Based on the above background, both particle reinforcements and whisker reinforcements have specific advantages. It is of great research value to study TiAl matrix composites containing the hybrid particle and whisker reinforcements.

NbC possesses high strength and a high melting point, and is widely used in alloys to enhance their high temperature strength and abrasion performance [12,13]. Hence, NbC can be the particle reinforcement in TiAl alloys. In the casting process, NbC particles can act as the heterogeneous nucleation cores to refine and strengthen TiAl matrix. MAX phase is a kind of reinforcement with excellent high temperature mechanical properties [14–16]. Meanwhile, the generation and growth of Ti_2AlC/Ti_2AlN in TiAl alloys have already been proven [17–20]. MAX phases such as Ti_2AlC and Ti_2AlN have good mechanical properties at the service temperature of TiAl alloys, including their high temperature strength and high temperature ductility. Hence, they are ideal whisker reinforcements in TiAl alloys.

This work prepares a TiAl composite containing hybrid particle&whisker reinforcements by casting strategy. NbC is the ex situ particle reinforcement and graphene is used as the reaction source of the Ti_2AlC phase, which can form in the TiAl melt and grow into whiskers. By controlling the content of NbC and graphene, the second phases of ideal morphology can generate in the TiAl composite, and the synergistic effect of multiplying strengthening mechanisms brought by the hybrid reinforcements helps to improve the mechanical properties of TiAl matrix.

2. Materials and Methods

2.1. Preparation of TiAl Composite

Ti-48Al-6Nb (TiAlNb alloy) is used as the matrix of the TiAl composite, which is fabricated by a vacuum melting furnace. Titanium (99.9%) bulks, Aluminum (99.9%) bulks and commercial Aluminum–Niobium (Al-70 wt.%Nb) intermediate alloy are used as the raw materials, which are prepared by vacuum melting. NbC particles and graphene (mass ratio 6:1) mixed powders are used as the raw materials of reinforcements. NbC particles are about 2 μm in size; flake-like graphene is about 10 μm in width. The mixed powders are ball-milled for 20 h by a GN-2 high energy ball-milling machine; the milled powders are less than 1 μm in size. Then, the mixed powders and TiAlNb alloy are used to prepare the TiAl composite by the vacuum melting furnace. TiAlNb alloy (cuboid, about 100 g in weight) is placed in the crucible and the mixed powders are hung above the crucible. The mixed powders are placed into the TiAl melt after the TiAlNb is fully melted. A high purity Mo stirrer is used to stir the melt containing the mixed powder. The melt is then poured into a copper mold to obtain the final NbC/Ti_2AlC reinforced TiAl composite (NTTC, φ20 in size, about 100 g in weight, as shown in Supplementary Materials Figure S1).

2.2. Structural Characterization

A Zeiss metalloscope (AxioCam ICc5, Jena, Germany) is used for the observation of the metallographic structure of the TiAlNb matrix and TiAl composite. The microstructure of the different specimens is observed by a scanning electron microscope (SEM, S-4800, Hitachi Ltd., Tokyo, Japan) with EDAX energy-dispersive spectrometry (EDS). Ion beam thinning is used for the preparation of the electron transparent foils of the TiAl composite. High-resolution transmission electron microscopy (HRTEM) images and the selected area electron diffraction (SAED) patterns of the specimens are obtained by JEM-2100 transmission electron microscopy (TEM).

2.3. Mechanical Properties Test

Room-temperature (RT) compression tests are carried out by an SHT5305 universal tester (MTS Ltd, Shanghai, China) with electronic extensometer. The strain rate of compression is $\varepsilon' = 2.5 \times 10^{-4}\ s^{-1}$. High-temperature (HT) compression tests are carried out by a Gleeble 3180 (DSi Ltd., Minneapolis, MN, USA) thermal mechanical simulator at 1073 K (a typical service temperature of TiAl alloy) with a strain rate of 0.01 s^{-1}. The specimens of RT compression and HT compression tests are $\varphi 2 \times 4$ mm in size and $\varphi 8 \times 12$ mm in size, respectively. A microhardness test was carried out by an HMV-2T microhardness tester (SHIMADZU Ltd., Kyoto, Japan) with a diamond indenter under 980 mN load for 10 s. Each specimen was tested 10 times and the average values were regarded as the results. All the specimens are obtained by wire-electrode cutting followed by mechanical polish. Three-point bending tests are carried out by the SHT5305 universal tester at room temperature. The specimens are 2 × 4 × 20 mm in size with a span length of 16 mm and are all cut by electrical discharge machining followed by mechanical polishing. The bending samples are processed with a slit notch (2 mm in depth and 0.3 mm in width) to calculate the fracture toughness. The fracture toughness is calculated by the following equation [21]:

$$K_{IC} = 3PL(10a)^{1/2}/(200bh^2)[1.93 - 3.07(a/h) + 14.53(a/h)^2 - 25.07(a/h)^3 + 25.80(a/h)^4] \quad (1)$$

where K_{IC} is the fracture toughness and a is the depth of the slit notch (mm). P is the maximum load (N), L is the span length (mm) and b and h refer to the width (mm) and thickness of the specimens (mm), respectively. The compression speed of cross head is 0.1 mm/min.

3. Results

The morphology and size of NbC/graphene mixed powders are determined by SEM. As shown in Figure 1, it is easy to distinguish the graphene flake and the NbC particles in the mixed powders in Figure 1a. Figure 1b,c show the EDS results of NbC and graphene, respectively. The flakes of graphene are about 10 μm in width. The spherical particles of NbC are about 2 μm in size. The flake-like graphene has a large size and floats on top of the NbC particles. Hence, this mixed powder needs to be processed by ball-milling to refine the size and reduce the segregation of the graphene.

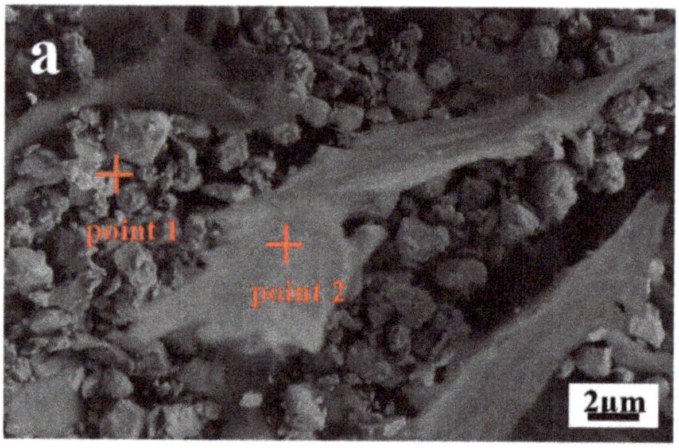

b

Element	Wt%	At%
CK	5.23	29.95
NbL	94.77	70.05
Matrix	Correction	ZAF

c

Element	Wt%	At%
CK	99.76	99.97
NbL	00.24	00.03
Matrix	Correction	ZAF

Figure 1. SEM image and EDS patterns of mixed powders. (**a**) SEM image showing the morphology of NbC and graphene. (**b**) and (**c**) are the EDS results of point 1 and point 2 in Figure 1a, respectively.

Figure 2 shows the morphology of the mixed powders after ball-milling. Figure 2a,c,e show the 500× micrographs of powders after 6 h, 12 h and 20 h of ball-milling. Figure 2b,d,e show the corresponding images of the microstructure at 5000×, respectively. The evolution of powder morphology at 500× shows that the size of the graphene flakes decreases gradually. Meanwhile, the 5000× micrographs show that the dispersion degree increases as the ball-milling time increases. The morphology evolution from Figure 2b,d,f shows that the uniformity of the mixed powders is improved significantly; meanwhile, the size of the powders is decreased significantly.

The metallographic microstructures of the TiAlNb alloy and NTTC are exhibited in Figure 3. The grain size of the TiAlNb alloy is about 400 μm, as shown in Figure 3a. Moreover, micrometer-scale lamellas can be observed inside the grains. NTTC samples with different mass fraction (0.5%, 1%, 2%, 3%) are named as T0.5, T1, T2 and T3, respectively. The grain size of the TiAlNb matrix is about 400μm. With the addition of NbC and graphene, T0.5 in Figure 3b shows a much-refined grain size of nearly 100 μm. As the powder amount increases to 1%, the grain size of the T1 sample (Figure 3c) is slightly decreased compared with T0.5. However, as the mass fraction of NbC–graphene powders increases further, the grain size of the T2 sample in Figure 3d shows no significant difference with T1. Figure 3e shows the microstructure of the T3 sample, which differs greatly from those of the other samples. The content of the second phase increases significantly, and its grain size cannot be determined. In the high magnification image in Figure 3e, the segregation of the second phase could be observed.

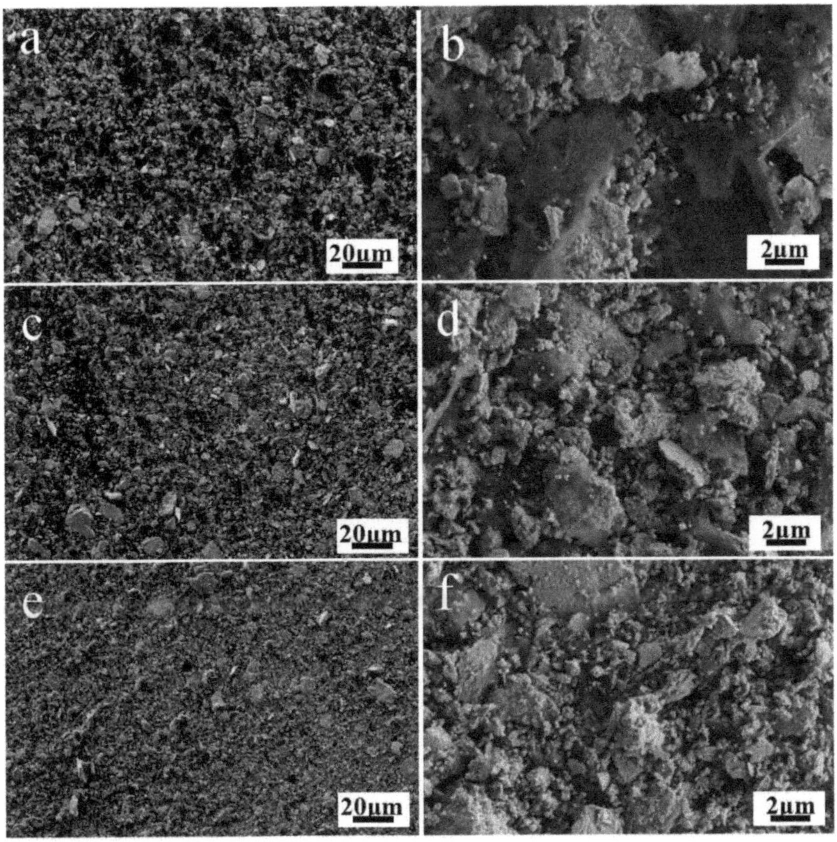

Figure 2. SEM images showing the NbC–graphene mixed powders after different ball-milling times. (**a**,**c**,**e**) are the microstructure of NbC–graphene after 6 h, 12 h and 20 h ball-milling at 500×, respectively. (**b**,**d**,**f**) are the micrographs at 5000× of powders in Figure 2a,c,e, respectively.

The SEM images of NTTC are shown in Figure 4. The microstructure of T0.5, T1, T2 and T3 is displayed in Figure 4a–d, respectively. The rod-like phases in T0.5 and T1 show uniform distribution. As the mass fraction reaches 2%, the content of the second phases in T2 increases obviously; meanwhile, some particles tend to gather, which denotes that the segregation in the T2 sample has already begun. After this, the segregation of the second phase becomes more obvious in the T3 samples, as shown in Figure 3f; the second phase loses its rod-like appearance and irregular bulk appears.

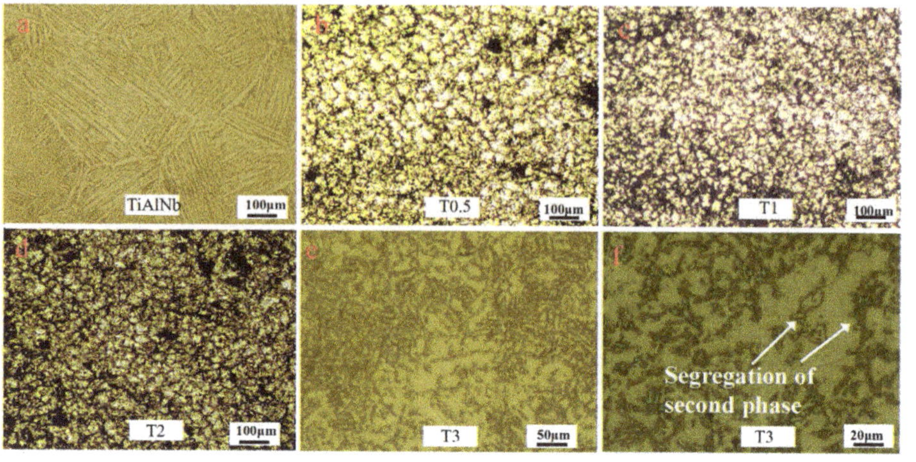

Figure 3. Microstructure of TiAlNb and NTTC. (**a**) Coarse TiAlNb grains. (**b**,**c**,**d**,**e**) are the microstructure of NTTC with 0.5 wt.%, 1 wt.%, 2 wt.% and 3 wt.% NbC–graphene powders (named as T0.5, T1, T2 and T3, respectively). (**f**) refers to the high magnification images of the T3 sample.

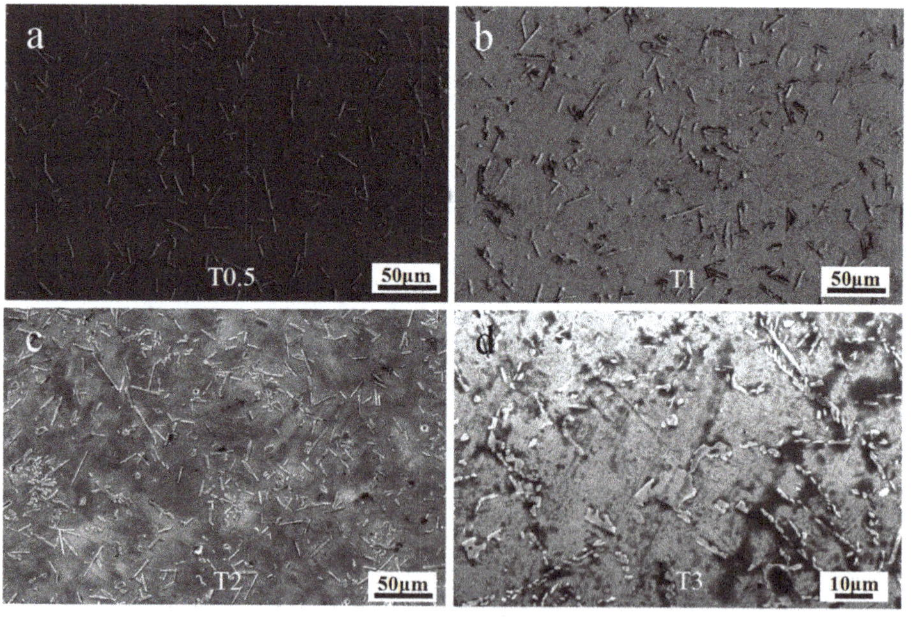

Figure 4. SEM images showing the microstructure of NTTC. (**a**) T0.5. (**b**) T1. (**c**) T2. (**d**) T3.

For qualitative analysis of the rod-like second phase, the T1 samples were selected for further SEM and EBSD analysis, as shown in Figure 5. Figure 5a shows the SEM images showing the distribution of the rod-like second phases. The relative EBSD image of Figure 5a is shown in Figure 5b. The EBSD result shows that each rod has a single crystallographic orientation, which denotes that the rod-like second phases are in situ whiskers.

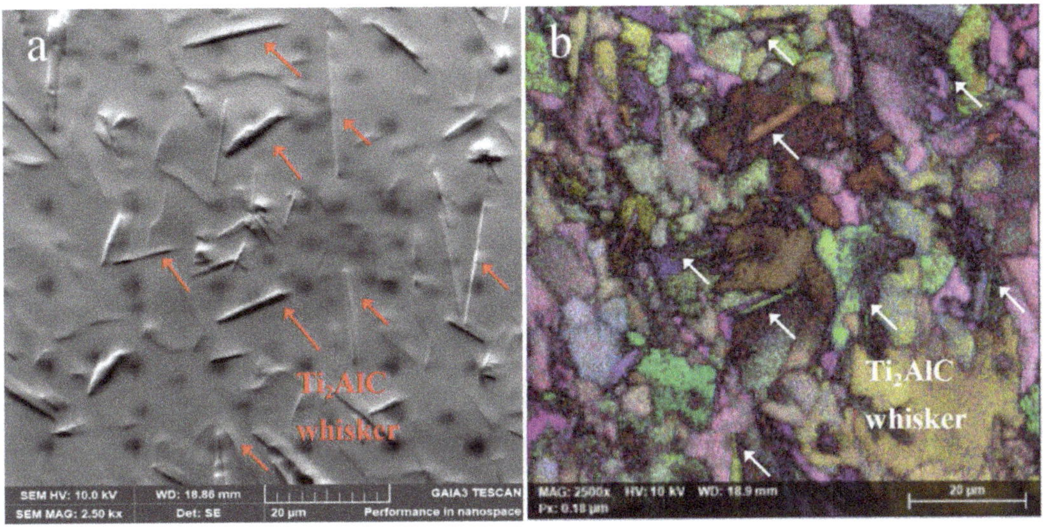

Figure 5. SEM image and the relative EBSD results of T1 sample. (**a**) Microstructure with abundant rod-like second phase. (**b**) EBSD image of (**a**).

To further determine the phase formation of the whiskers and observe the distribution of ex situ NbC particles, the microstructure of the T1 sample inside the grains is analyzed, as shown in Figure 6. Figure 6a exhibits that the whiskers can be located both inside the grain and in the grain boundary. The insert is the EDS result of Point 1 in Figure 6a. According to this result, the atomic ratio of element Ti, Al and N is nearly 2:1:1; hence, the whiskers can be identified as Ti_2AlC phase. Figure 6b shows the NbC particles which distribute in the lamellas inside the grain. Meanwhile, the refined lamellas can be observed in Figure 6, which shows roughly dozens of nanometers.

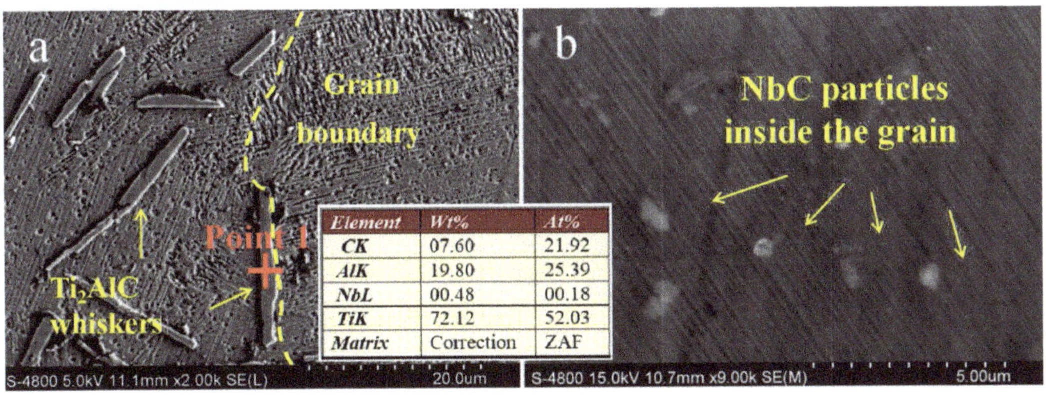

Element	*Wt%*	*At%*
CK	07.60	21.92
AlK	19.80	25.39
NbL	00.48	00.18
TiK	72.12	52.03
Matrix	Correction	ZAF

Figure 6. Microstructure of T1 samples. (**a**) Ti_2AlC whiskers inside the grain and in the grain boundary. (**b**) NbC particles in the lamellas.

Figure 7 shows the TEM images of the T1 sample. Figure 7a shows the α_2/γ lamellas in T1, which are less than 100 nm. Compared with the micrometer scale lamellas in the TiAlNb matrix, the lamellas in T1 have been significantly refined. Figure 7b shows the SAED patterns of the diaphasic lamellas in Figure 7a; the calibrate results show the typical Blackburn orientation relationship, which is the parallel relation between $(111)_\gamma$ and

$(0001)_{\alpha2}$. Figure 7c shows a Ti_2AlC whisker located in the lamellas. The HRTEM image of the interface between this whisker and the TiAl matrix is shown in Figure 7d, in which the parallel relationship between the two phases can be observed. The FFT patterns of Figure 7d are exhibited in Figure 7e. According to the calibration results, the specific orientation relationship between the two phases can be determined, which is $(111)_{\gamma}//(0006)_{Ti2AlC}$.

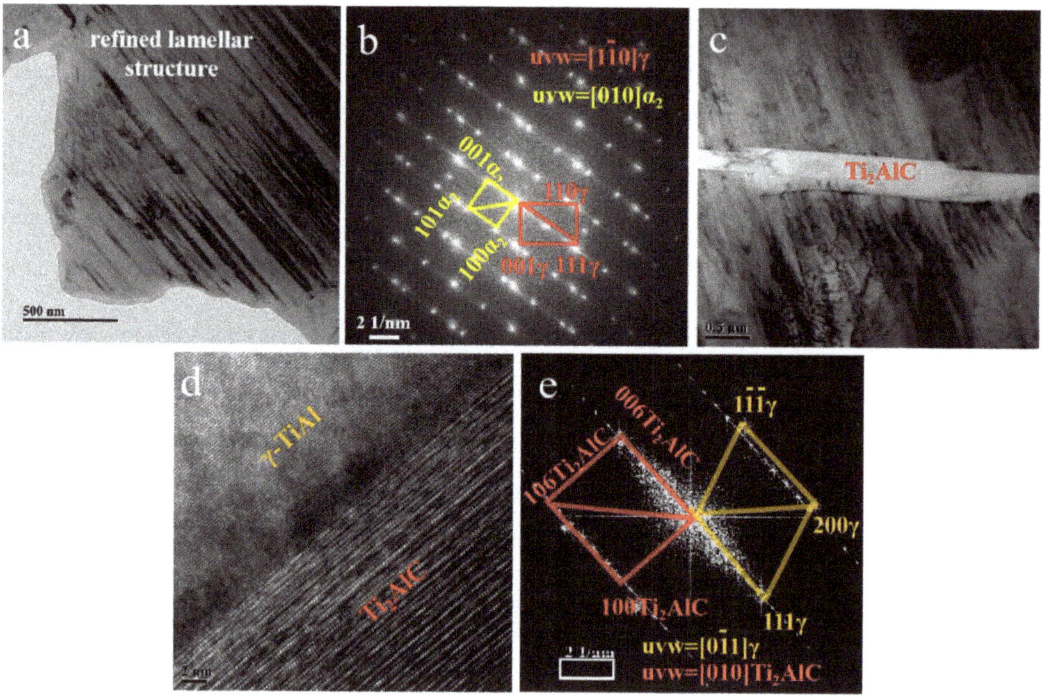

Figure 7. TEM image of T1 sample. (**a**) Refined α_2/γ lamellar structure in T1 sample. (**b**) SAED patterns of α_2/γ in (**a**). (**c**) TEM image showing the location of Ti_2AlC whisker in the lamellas. (**d**) HRTEM image showing the interface between Ti_2AlC and γ-TiAl. (**e**) FFT patterns of (**d**).

In order to analyze the mechanical properties of the NTTC and TiAlNb matrix, different tests are carried out, as shown in Figure 8. Figure 8a shows the room-temperature compression curves of the different samples. The fracture stress and strain of TiAlNb, T0.5, T1, T2 and T3 are 1453 MPa/17.8%, 1887 MPa/22.8%, 2247 MPa/24.4%, 2183 MPa/22.0% and 2063 MPa/18.6%, respectively. Figure 8b shows the compression curves of the different samples at 1073 K, the fracture stress and strain of TiAlNb, T0.5, T1, T2 and T3 are 612 MPa/19.4%, 769 MPa/24.9%, 836 MPa/26.6%, 798 MPa/20.2% and 753 MPa/16.7%, respectively. The fracture toughness of TiAlNb, T0.5, T1, T2 and T3 are 18.8 MPa/m^2, 25.0 MPa/m^2, 27.4 MPa/m^2, 25.6 MPa/m^2 and 21.1 MPa/m^2, respectively, which are exhibited in Figure 8c. All the values are also presented in Table 1. (The Parallel test results of room temperature compression tests are shown in Supplementary Materials Tables S1 and S2) The mechanical properties of samples show that, with the addition of reinforcements, the mechanical properties of the TiAl alloy firstly improve and then deteriorate. T1 shows the best comprehensive mechanical properties. The microhardness of the different samples is shown in Table 2. The microhardness increases as the amount of reinforcements increases.

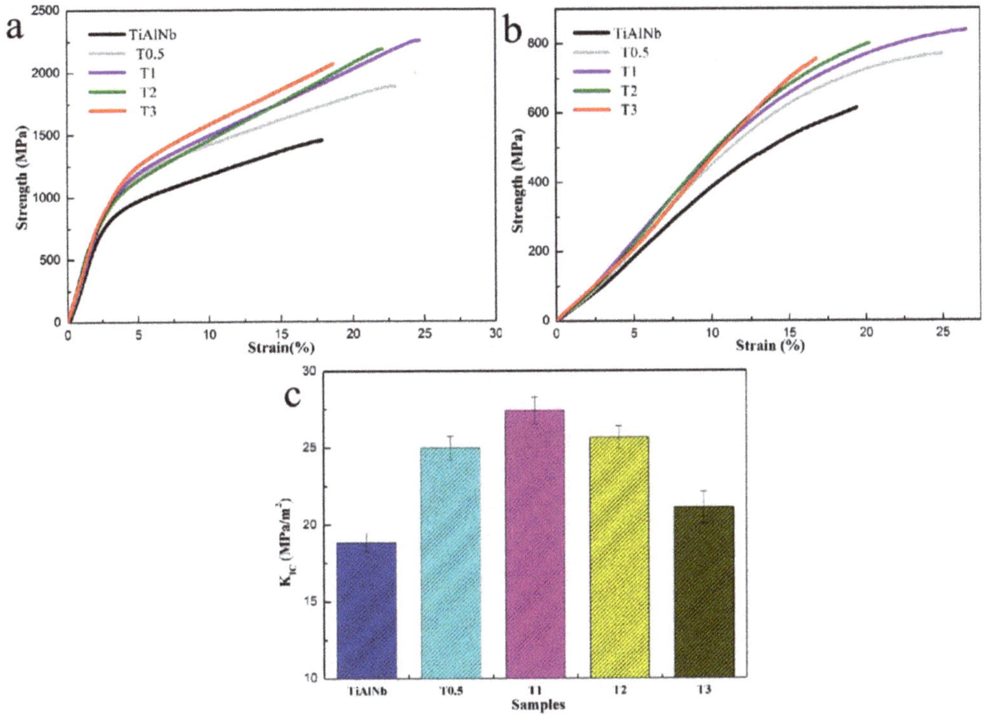

Figure 8. Comparison of mechanical properties between TiAlNb, T0.5, T1, T2 and T3. (**a**) Room-temperature compression curves. (**b**) High-temperature (1073 K) compression curves. (**c**) Fracture toughness results of different samples.

Table 1. Mechanical properties of different samples.

	Properties \ Samples	TiAlNb	T0.5	T1	T2	T3
Room Temperature	Compression Strain (%)	17.8	22.8	24.4	22.0	18.6
	Compression Stress (MPa)	1453	1887	2247	2183	2063
	Fracture toughness (MPa/m^2)	18.8	25.0	27.4	25.6	21.1
High Temperature (1073 K)	Compression Strain (%)	19.4	24.9	26.6	20.2	16.7
	Compression Stress (MPa)	612	769	836	798	753

Table 2. Microhardness of different samples.

Samples	TiAlNb	T0.5	T1	T2	T3
Microhardness (HV)	489	521	531	558	567

4. Discussion

4.1. The Refining Effects of Dual Reinforcements

Compared with the TiAlNb matrix, the grain size of the T1 sample was refined from 400 μm to less than 100 μm, and the lamellar spacing was refined from a few microns to less than 100 nm. The refining is attributed to the second phases, which act as the heterogeneous nuclei. The schematic diagrams are shown in Figure 9, which exhibits the evolution process of the second phases. Ex situ NbC has a melting point of 3500 °C, which can distribute stably in TiAlNb melt (Figure 9a). After that, as graphene reacts with Ti, TiC begins to generate in the melt (Figure 9b). Then, during the cooling process, TiC tends to react to transform into Ti_2AlC, which it does according to the transformation pathway L + TiC → L + Ti_2AlC [22]. Meanwhile, the Ti_2AlC can spontaneously grow into whiskers in the melt, as shown in Figure 9c. During the subsequent cooling process, NbC particles and Ti_2AlC whiskers will act as nucleation substrates and TiAlNb nucleates grow on the two second phases (Figure 9d). Abundant NbC particles, along with the Ti_2AlC whiskers, provide plenty of nucleation sites for the nucleation of the TiAl melt, which refines the grain size and lamellar width significantly. Due to the different sizes of the two reinforcements, the number of NbC particles is much more than that of Ti_2AlC whiskers. Therefore, NbC particles play a leading role in the refining effect of TiAlNb.

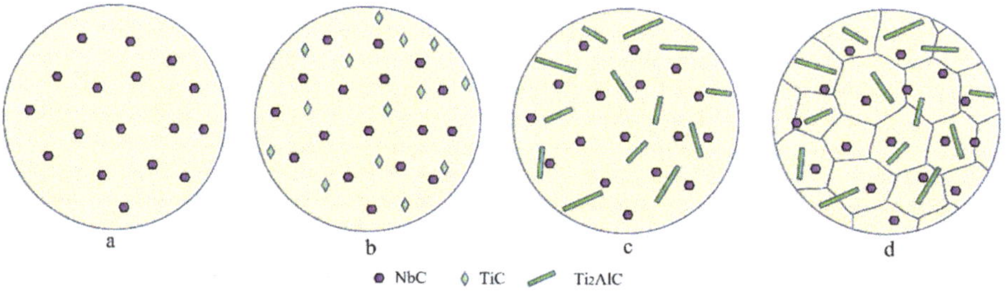

Figure 9. Schematic diagrams showing the formation of Ti_2AlC whiskers in TiAlNb. (**a**) NbC in TiAlNb melt. (**b**) TiC generates first in TiAlNb. (**c**) Ti_2AlC generates in TiAlNb. (**d**) TiAlNb nucleates and solidifies.

4.2. The Synergistic Strengthening Mechanisms

The increase in the strength of NTTC is attributed to the synergistic strengthening mechanisms brought by the hybrid NbC particles and Ti_2AlC whiskers, including fine grain strengthening, Orowan strengthening, Load Transferring (LT) strengthening and Coefficient of thermal expansion (CTE) mismatch strengthening.

The mechanical properties of the TiAl alloy are significantly influenced by the grain size and lamellar width. As the lamellar width is refined, the mechanical properties of TiAl are improved, such as the yield strength, ductility, fracture toughness and creep resistance. According to the Hall–Petch equation, yield strength σy of the alloy can be described as [23]:

$$\sigma_y = \sigma_0 + \frac{K_y}{\sqrt{d}} \quad (2)$$

where σ_0 is the lattice frictional resistance due to single dislocation movement, K_y is a material constant and d is the average grain size. As the grain size and lamellar width are both refined, the number of grain boundaries [24] and lamellar interfaces increases. As a result, the resistance to dislocation motion is greatly enhanced, thus increasing the strength of the TiAlNb alloy. According to the analysis of the refining effect, in the two reinforcements, NbC plays the main role in the fine grain strengthening mechanism.

When the dislocation moves near the second phase, due to the high strength of the second phase, the dislocation is unable to pass through the particle but bypasses them, leaving a dislocation loop around the particle, which is called the pinning effect on dislocation [25,26]. These bending dislocation lines will increase the lattice distortion energy in the affected region of dislocation and increase the resistance of the dislocation movement, thus resulting in an increase in material strength. The effect of the Orowan strengthening mechanism can be described by the following equation [27]:

$$\sigma_{Orowan} = \frac{MGb}{d_P}(\frac{6V_R}{\pi})^{1/3} \tag{3}$$

where M denotes the Taylor factor, G is the shearing modulus of TiAlNb matrix, b is the burger vector, d_p is the diameter of the second phase particles and V_R is the volume fraction of the second phases. It is easy to notice that for the second phase with a given volume fraction, the smaller the diameter, the stronger the effect of Orowan strengthening.

Ti_2AlC whiskers have a much larger scale than the NbC particles; hence, the Ti_2AlC/TiAl interfacial stability has a much greater influence on the mechanical properties of the composite, which should be analyzed. According to the HRTEM image and FFT patterns in Figure 7d,e, the parallel relationship between $(0006)_{Ti_2AlC}$ and $(111)_{\gamma-TiAl}$ has already been described. The planar mismatch value δ of $(0006)_{Ti2AlC}/(111)_{\gamma-TiAl}$ is calculated to be 2.59% ($d(0006)_{Ti2AlC}$ = 0.226 nm, $d(111)_{\gamma-TiAl}$ = 0.232 nm), which is smaller than 5%. Hence, the bonding mode of this interface belongs to the coherent interface. The coherent interface can effectively promote the transfer of load between the matrix and whisker. Thus, the excellent mechanical properties of whisker reinforcement can be fully utilized. In the service process, whiskers can effectively share the external load through the coherent interface so as to improve the comprehensive mechanical properties of materials and delay the fracture and failure of the materials.

Due to the difference in thermal expansion coefficient α between the matrix and the reinforcements (α_{TiAl} = 10.4~11.1 × 10^{-6} K^{-1}, α_{NbC} = 6.65 × 10^{-6} K^{-1}, α_{Ti2AlC} = 8.2 × 10^{-6} K^{-1}), the microscopic deformation of the matrix and reinforcements is different in the process of solidification, resulting in the arising of geometrically necessary dislocations and leading to the enhancement of material strength, which is named as the coefficient of thermal expansion (CTE) strengthening [28,29]. The contribution of CTE mismatch to the strength can be determined by [30]:

$$\sigma_{CTE} = \sqrt{3}\beta G\sqrt{\rho_{CTE}} \tag{4}$$

$$\rho_{CTE} = \frac{12V_R\Delta\alpha\Delta T}{1 - V_R - bd_R} \tag{5}$$

where β is the dislocation strengthening coefficient and ρ_{CTE} is the dislocation density, which is caused by the difference in thermal expansion coefficient between the matrix and the reinforcements. d_R is the particle size of the reinforcement. Δα is the CTE difference between the matrix and the nanoparticles and ΔT is the temperature change. Due to the existence of the mismatch of CTE, both NbC and Ti_2AlC contribute to the CTE strengthening.

The toughening of NTTC is mainly attributed to the grain refining effect and the effect of whiskers. By bearing the load which is transferred from the matrix, the fracture of the composite can be delayed. Meanwhile, when cracks start to extend, whiskers can cause the deflection of these cracks, thus increasing the distance of crack propagation and delaying the fracture.

In accordance with the microstructure and mechanical properties shown in Figures 3, 4 and 8, the amount of reinforcements influences the morphology and mechanical properties of MMCs. Especially when the content of reinforcements reaches a certain value, the ceramic phases will aggregate and deteriorate the properties. The aggregated bulk easily causes stress concentration, which causes the material to fracture under a small strain. Therefore, the total energy absorbed by the material from deformation to fracture is also reduced, which is reflected in the reduction of fracture toughness of the T3 sample compared with the T2 sample. This is a common phenomenon in MMCs containing particle reinforcement. Under the condition of the same volume fraction, whisker second phase has obvious excellent advantages compared with the agglomeration particles' second phase. Therefore, the dual scaled reinforcements in this study have the advantage of adjusting the morphology of second phases.

5. Conclusions

NbC particles and in situ Ti_2AlC whiskers are introduced into a TiAl alloy by a casting method; the mechanical properties of this composite are influenced by the content of the reinforcements. The key to this method is to control the morphology and distribution of the reinforcements. NTTC with 1 wt.% NbC/graphene powders exhibits the best mechanical properties, in which the second phases have ideal morphology and distribution. Particle reinforcements alone with the in situ Ti_2AlC whiskers can improve the comprehensive mechanical properties of the TiAl matrix by the synergistic strengthening mechanisms. The future development direction of TiAl composites in this work is mainly to reduce its casting defects by hot isostatic pressing (HIP) treatment to obtain high-temperature structural materials with low cost and high performance.

Supplementary Materials: The following supporting information can be downloaded at: https://www.mdpi.com/article/10.3390/ma16134661/s1, Table S1: Parallel test data of room temperature compression tests (group 1); Table S2: Parallel test data of room temperature compression tests (group 2); Figure S1: Macro photo of the sample T1.

Author Contributions: Conceptualization, S.C.; Methodology, S.C.; Software, S.C.; Supervision, C.C. and X.W.; Writing—original draft, S.C. All authors have read and agreed to the published version of the manuscript.

Funding: This work is supported by the National Natural Science Foundation of China, grant number 51871087. Natural Science Foundation of Hebei Province, China (No. E2021202140). Innovation Fund for Outstanding Youth (EERI_OY2022009), "S&T Program of Hebei" (225676163GH).

Institutional Review Board Statement: Not applicable.

Informed Consent Statement: Not applicable.

Data Availability Statement: All data included in this study are available upon request by contact with the corresponding author.

Conflicts of Interest: The authors declare that they have no known competing financial interest or personal relationships that could have appeared to influence the work reported in this paper.

References

1. Xu, R.R.; Li, M.Q.; Zhao, Y.H. A review of microstructure control and mechanical performance optimization of γ-TiAl alloys. *J. Alloys Compd.* **2023**, *932*, 167611. [CrossRef]
2. Avdeeva, V.; Bazhina, A.; Antipov, M. Relationship between Structure and Properties of Intermetallic Materials Based on γ-TiAl Hardened In Situ with Ti_3Al. *Metals* **2023**, *6*, 1002. [CrossRef]
3. Song, L.; Appel, F.; Wang, L. New insights into high-temperature deformation and phase transformation mechanisms of lamellar structures in high Nb-containing TiAl alloys. *Acta Mater.* **2020**, *186*, 575–586. [CrossRef]
4. Liu, Y.; Li, J.S.; Tang, B. The microstructure evolution and phase transformation behavior of a β-solidifying γ-TiAl alloy during creep. *Prog. Nat. Sci. Mater. Int.* **2023**, *33*, 193–202. [CrossRef]
5. Xu, W.C.; Shan, D.B.; Zhang, H. Effects of extrusion deformation on microstructure, mechanical properties and hot workability of β containing TiAl alloy. *Mater. Sci. Eng. A* **2013**, *571*, 199–206. [CrossRef]
6. Baradeswaran, A.; Elaya Perumal, A. Study on mechanical and wear properties of Al. 7075/Al_2O_3/graphite hybrid composites. *Compos. B Eng.* **2014**, *56*, 464–471. [CrossRef]
7. Sun, T.L.; Guo, Z.C.; Cao, J. Isothermal oxidation behavior of high-Nb-containing TiAl alloys doped with W, B, Y, and C/Si. *Corros. Sci.* **2023**, *213*, 110980. [CrossRef]
8. Sougavabar, M.A.; Niknam, S.A.; Davoodi, B. Experimental characterization of tool wear morphology in milling of Al520-MMC reinforced with SiC particles and additive elements Bi and Sn. *J. Mater. Res. Technol.* **2023**, *23*, 571–585. [CrossRef]
9. Wang, Y.M.; Zhu, M.; Dong, L.L. In-situ synthesized TiC/Ti-6Al-4V composites by elemental powder mixing and spark plasma sintering: Microstructural evolution and mechanical properties. *J. Alloys Compd.* **2023**, *947*, 169557. [CrossRef]
10. Li, J.; Zhao, K.; Ren, L. Microstructure evolution, mechanical properties, and strengthening mechanisms of heat-resistant Al-based composite reinforced by a combination of AlN and TiN particles. *J. Mater. Res. Technol.* **2023**, *24*, 5628–5641. [CrossRef]

11. Wang, F.C.; Zhang, J.F.; Sun, K.X. Synergistic reinforcement effect of Fe and in-situ synthesized MgAlB4 whiskers in Al matrix composites. *Compos. B Eng.* **2022**, *246*, 110267. [CrossRef]
12. Pourasiabi, H.; Gates, J.D. Effects of chromium carbide volume fraction on high-stress abrasion performance of NbC-bearing high chromium white cast irons. *Wear* **2022**, *498–499*, 204312. [CrossRef]
13. Wang, Z.J.; Zhou, M.J.; Jiang, Y.H. Effects of in situ NbC on the microstructure and high-temperature friction wear properties of 4Cr5MoSiV1 steel. *J. Mater. Res. Technol.* **2023**, *24*, 6159–6173. [CrossRef]
14. Yang, A.C.; Duan, Y.H.; Bao, L.K. Elastic properties, tensile strength, damage tolerance, electronic and thermal properties of TM_3AlC_2 (TM = Ti, Zr and Hf) MAX phases: A first-principles study. *Vacuum* **2022**, *206*, 111497. [CrossRef]
15. Chen, W.T.; Yu, W.B.; Ma, C.S. A review of novel ternary nano-layered MAX phases reinforced AZ91D magnesium composite. *J. Magnes. Alloy.* **2022**, *10*, 1457–1475. [CrossRef]
16. Salvo, C.; Chicardi, E.; Hernández-Saz, J. Microstructure, electrical and mechanical properties of Ti_2AlN MAX phase reinforced copper matrix composites processed by hot pressing. *Mater. Charact.* **2021**, *171*, 110812. [CrossRef]
17. Pei, X.; Yuan, M.N.; Wang, H.L. Revealing the interface properties of the $Ti_2AlC/TiAl$ composite from a first principles investigation. *Appl. Surf. Sci.* **2023**, *617*, 156491. [CrossRef]
18. Guo, Y.C.; Liang, Y.F.; Lin, J.P. In situ synthesis of nano/micron Ti_2AlC reinforced high-Nb TiAl composites: Microstructure and mechanical properties. *Intermetallics* **2023**, *159*, 107937. [CrossRef]
19. Chen, S.Y.; Tan, Y.M.; Wang, X. Comparing the role of Zr and Hf atoms on microstructure and mechanical properties optimization of Ti_2AlN reinforced Ti48Al0.5W composites. *J. Alloys Compd.* **2023**, *935*, 168159. [CrossRef]
20. Li, J.G.; Hu, R.; Zhou, M. High temperature micromechanical behavior of Ti_2AlN particle reinforced TiAl based composites investigated by in-situ high-energy X-ray diffraction. *Mater. Des.* **2021**, *212*, 110225. [CrossRef]
21. Cui, X.P.; Ding, H.; Zhang, Y.Y. Fabrication, microstructure characterization and fracture behavior of a unique micro-laminated TiB-TiAl composites. *J. Alloys Compd.* **2019**, *775*, 1057–1067. [CrossRef]
22. Lapin, J.; Kamyshnykova, K. Processing, microstructure and mechanical properties of in-situ Ti_3Al+TiAl matrix composite reinforced with Ti_2AlC particles prepared by centrifugal casting. *Intermetallics* **2018**, *98*, 34–44. [CrossRef]
23. Hall, E.O. The deformation and ageing of mild steel: III discussion of results. *Proc. Phys. Soc. Sect. B* **1951**, *64*, 747–752. [CrossRef]
24. Hu, P.H.; Song, H.W.; Chen, Y. Mechanisms for refining precipitation and improving strengthening of annealed Cu-2Ag alloy via La modification. *J. Alloys Compd.* **2021**, *883*, 160912. [CrossRef]
25. Lv, Y.T.; Zhang, Z.; Zhang, Q. Cracking inhibition behavior and the strengthening effect of TiC particles on the CM247LC superalloy prepared by selective laser melting. *Mater. Sci. Eng. A* **2022**, *858*, 144119. [CrossRef]
26. Wang, H.Y.; Lu, H.; Ren, Y.P. Strengthening mechanisms and tensile properties of 20Cr2Mn2Mo processed by laser shock peening and vacuum carbonitriding. *Surf. Coat. Technol.* **2022**, *439*, 128462.
27. Liu, G.; Zhang, G.; Jiang, F. Nanostructured high-strength molybdenum alloys with unprecedented tensile ductility. *Nat. Mater.* **2013**, *12*, 344–350. [CrossRef]
28. Mallmann, C.; Hannard, F.; Ferrié, E. Unveiling the impact of the effective particles distribution on strengthening mechanisms: A multiscale characterization of Mg+Y_2O_3 nanocomposites. *Mater. Sci. Eng. A* **2019**, *764*, 138170. [CrossRef]
29. Wu, Y.H.; Liu, B.X.; Kang, H.J. Ultrasound-assisted dispersion of TiB_2 nanoparticles in 7075 matrix hybrid composites. *Mater. Sci. Eng. A* **2022**, *840*, 142958. [CrossRef]
30. Goh, C.S.; Wei, J.; Lee, L.C. Properties and deformation behaviour of Mg-Y_2O_3 nanocomposites. *Acta Mater.* **2007**, *55*, 5115–5121. [CrossRef]

materials

Article

Effect of Nb on the Damping Property and Pseudoelasticity of a Porous Ni-Ti Shape Memory Alloy

Peng Sun [1], Qingzhou Wang [1,*], Jianhang Feng [1], Puguang Ji [1,2], Jianjun Zhang [1] and Fuxing Yin [3,*]

1 Tianjin Key Laboratory of Materials Laminating Fabrication and Interface Control Technology, School of Materials Science and Engineering, Hebei University of Technology, Tianjin 300130, China; sunptj@126.com (P.S.); jianhang_feng@126.com (J.F.); jipuguang@hebut.edu.cn (P.J.); zjjyyds888@163.com (J.Z.)
2 Xuzhou Jihua Metal Material Technology Co., Ltd., Xuzhou 221100, China
3 Institute of New Materials, Guangdong Academy of Sciences, Guangzhou 510651, China
* Correspondence: qzwang@hebut.edu.cn (Q.W.); fordjp@163.com (F.Y.)

Abstract: In order to develop novel high damping materials with excellent pseudoelasticity (PE) properties to meet the application requirements in aerospace, medical, military and other fields, porous $Ni_{50.8}Ti_{49.2}$ shape memory alloy (SMA) was prepared by the powder metallurgy method. Different contents of Nb element were added to regulate the microstructures. It was found that after adding the Nb element, the number of precipitates significantly decreased, and the Nb element was mainly distributed in the Ni-Ti matrix in the form of β-Nb blocks surrounded by Nb-rich layers. Property tests showed that with the increase in Nb content, the damping and PE increased first and then decreased. When the Nb content reached 9.0 at.%, the highest damping and the best PE could be achieved. Compared with the porous Ni-Ti SMA without Nb addition, the damping and PE increased by 60% and 35%, respectively. Correlated mechanisms were discussed.

Keywords: porous shape memory alloy; Ni-Ti; powder metallurgy; damping property; pseudoelasticity

1. Introduction

With the development of industry, problems such as vibration and noise are becoming increasingly serious. Damping materials can effectively convert mechanical energy into thermal energy, thus achieving the effect of vibration and noise reduction. Polymer materials exhibit excellent damping properties due to viscoelasticity, but owing to their low mechanical properties they cannot be directly used as structural components in engineering fields [1]. Although traditional metal materials have excellent mechanical properties, their applications are also limited due to their low damping properties [2,3]. SMAs exhibit high damping properties owing to their inherent thermoelastic martensitic transformation (MT) behavior and the hysteretic migration of a large number of interfaces such as the austenite/martensite interface, the twin interface as well as the martensite/martensite interface, so they have been widely used [4,5]. Compared with iron-based and copper-based SMAs, Ni-Ti SMAs with an almost equal atomic ratio have unique advantages because they have a better shape memory effect (SME) and PE besides high damping properties [6]. Moreover, the one-step reverse MT ($B19' \rightarrow B2$) of the Ni-Ti SMAs during the heating process can be easily changed to a two-step reverse MT ($B19' \rightarrow R \rightarrow B2$) by performing certain heat treatments [7]. Therefore, the regulation of its damping is more convenient than other SMAs.

With the development of science and technology, aerospace, medical, military and other fields have put forward higher requirements for the damping property of metal materials [2]. It has been shown that making Ni-Ti SMAs porous can further improve their damping properties. This is because the microplastic deformation of the pore wall as well as the motion of atoms and stress mode transformation around pores will dissipate additional mechanical energy [8,9]. Element addition is another effective method of improving the damping properties of Ni-Ti SMAs. For example, the addition of Cu can significantly improve the damping of Ni-Ti SMAs, but excess Cu will embrittle the alloy and reduce the machinability [10]. The addition of Hf and H elements can form a new damping peak in Ni-Ti SMAs, but the resultant alloys have low modulus and poor deformability in the martensitic state [11]. In recent years, the addition of Nb to Ni-Ti SMAs has attracted ever more attention because of the excellent comprehensive properties of the resultant alloys. By comparing the damping properties of the $Ni_{50}Ti_{50}$ SMA and the $Ni_{47}Ti_{44}Nb_9$ SMA, Cai et al. found that the addition of Nb contributed to an improvement in the damping property of the Ni-Ti SMA due to the formation of a high density of dislocations around the Nb phase [12]. Bao et al. fabricated a NiTiNb SMA through casting and hot rolling processes, and they found that the generation of the eutectic microstructure around the Nb phase could lead to an improvement in the damping [13].

Nevertheless, up to now, studies on the effect of Nb addition on damping properties have mostly focused on bulk Ni-Ti SMAs, and there are still few studies concerning porous Ni-Ti SMAs. In addition to damping properties, as mentioned above, the SME and PE are also important functional properties of Ni-Ti SMAs. In particular, the PE of porous Ni-Ti SMAs has shown an important application prospect in the field of high overload resistance related to high-speed launch in recent years [14]. Guo et al. demonstrated that Ni-Ti SMAs have good energy absorption ability and self-recovery capacity under dynamic loading, which can effectively improve the service life and material utilization of protective structures [15]. However, so far, there have been few studies on the PE property of porous Ni-Ti SMAs. In the present study, the effect of Nb addition on the microstructure, damping and PE properties of a porous NiTi SMA was systematically investigated. The obtained results can provide a theoretical basis for improving the comprehensive properties of the porous NiTi SMAs.

2. Materials and Methods

2.1. Fabrication of Porous Ni-Ti SMAs

Porous $Ni_{50.8}Ti_{49.2}$ SMAs with Nb addition were fabricated via the powder metallurgy process by using pure Ni, Ti and Nb powders (the purity was 99.9% and the average particle size was 75 μm) as raw materials. Powders were purchased from the Gripm Advanced Materials Co., Ltd. (Beijing, China). Ni, Ti powders and different contents (0, 3.0, 6.0, 9.0 and 12.0 at.%) of Nb powders were ball milled for 10 h on a planetary ball mill under the protection of argon (the weight ratio of ball to powder was 10:1, and the rotating speed of the ball mill was 200 rpm). Subsequently, the obtained Ni-Ti-Nb powders were homogeneously mixed with NaCl space-holders with an average particle size of 0.60 mm purchased from the China National Salt Industry Group Co., Ltd. (Beijing, China) in a V-type mixer. The mixed powders were then compressed into cylindrical green compacts under the pressure of 350 MPa on a computer numerical control (CNC) hydraulic machine purchased from the Henan Oukai Hydraulic Equipment Co., Ltd. (Xinxiang, China). The green compacts were sintered in a tube furnace purchased from the Tianjin Zhonghuan Electric Furnace Co., Ltd. (Tianjin, China) under the protection of high purity argon gas (the green compacts were firstly sintered at 790 °C for 2 h to densify the metal matrix, and then sintered at 1100 °C for 3 h). After natural cooling to room temperature, the specimens were taken out of the furnace and the residual NaCl in the pores of specimens was washed out with running water. In order to eliminate the precipitates in the Ni-Ti matrix, the porous Ni-Ti SMAs were subjected to solution treatment at 1000 °C for 60 min and then quenched in water. After that, the specimens were aged at 450 °C for 30 min. For convenience, the porous Ni-Ti SMAs with different contents (0, 3.0, 6.0, 9.0 and 12.0 at.%) of Nb were termed as Nb0, Nb3, Nb6, Nb9 and Nb12 specimens, respectively.

2.2. Characterization of Specimens

The porosity of porous Ni-Ti SMAs was calculated by using the following equation:

$$P = \left(1 - \frac{\rho^*}{\rho_0}\right) \times 100\% \quad (1)$$

where P and ρ^* denote the porosity and apparent density of the porous Ni-Ti SMAs, respectively, and ρ_0 denotes the theoretical density of the dense Ni-Ti SMA.

The phase composition of specimens was analyzed by X-ray diffraction (XRD, Bruker D8 Discover). The equipment was sourced from the Bruker Corporation (Billerica, MA, USA). The pore morphology and microstructure of specimens were observed by scanning electron microscopy (SEM, JSM-7100F) equipped with an energy dispersive spectrometer (EDS). Elements mapping was obtained by using electron probe microanalysis (EPMA, JEOL 8530F). Both SEM and EPMA were sourced from the JEOL Japan Electronics Co., Ltd. (Tokyo, Japan).

2.3. Property Tests

The damping property was characterized by the internal friction IF (Q^{-1}) and measured on a dynamic thermo mechanical analyzer (DMA, Q800) by using the method of forced vibration with a heating rate of 5°/min and a vibration frequency of 0.5 Hz. The DMA sourced from the TA Instruments (New Castle, DE, USA) is shown in Figure 1. The damping specimens had a dimension of $35 \times 10 \times 2$ mm^3. The PE of the specimens was tested at room temperature on a hot simulated testing machine (HSTM, Gleeble 3180) with a strain rate of 1.0×10^{-4} s^{-1}. The equipment was sourced from the Dynamic Systems Inc. (Poughkeepsie, NY, USA). Four pre-strains of 2.0%, 3.0%, 4.0% and 5.0% were used. The cylinder specimens had a dimension of $\varnothing 8 \times 12$ mm^3.

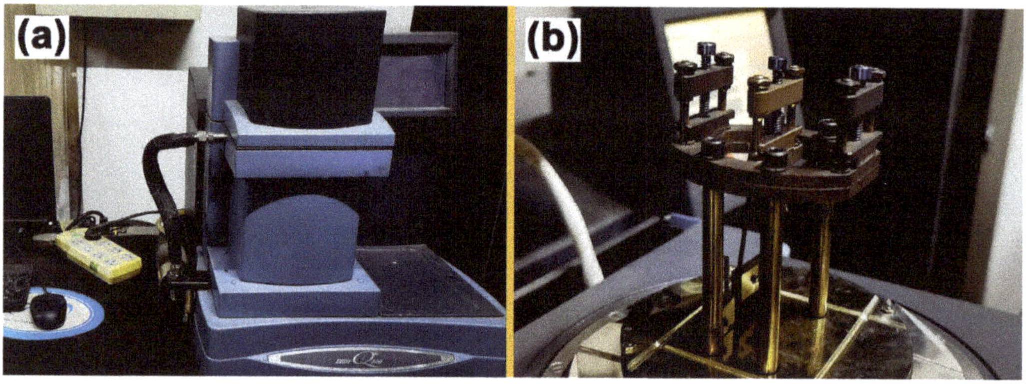

Figure 1. Images of (**a**) DMA, (**b**) sample fixture on the DMA.

3. Results and Discussion

3.1. Phase Composition and Microstructures of Specimens

The XRD patterns of Ni-Ti and Ni-Ti-Nb (Nb content: 9.0 at.%) powders as well as porous Ni-Ti SMAs are shown in Figure 2. The sharp characteristic diffraction peaks of Ni, Ti and Nb elements indicate that the ball-milled powders have an ordered lattice arrangement. The position of the peaks without deviation indicates that the mixed powder has no interatomic diffusion behavior. In Figure 2b, diffraction peaks of Na and Cl are not detected, indicating that the NaCl particles only play the role of space-holders and do not participate in the reaction between metal powders at a high sintering temperature. The porous Ni-Ti SMAs with Nb addition are mainly composed of NiTi(B2) matrix phase and a small amount of Ti-rich phase and β-Nb phase. The Ti-rich phase is probably $(Ti, Nb)_2Ni$ phase [16]. The existence of β-Nb phase indicates that the sintering temperature of

1100 °C could not provide enough energy for the Nb element to form an eutectic structure with the NiTi phase. Compared with the specimens with the Nb addition, a higher content of the Ni-rich phase appears in the Nb0 specimen. For the specimens with Ni content higher than 50.5 at.%, the solution treatment can make part of the Ni_3Ti phase re-dissolve in the NiTi phase [17,18]. However, in the subsequent aging treatment at 450 °C, the Ni-rich phase precipitates again as the Ni_4Ti_3 phase [19]. Interestingly, in this study the specimens with the Nb addition do not show the diffraction peaks of Ni_3Ti and Ni_4Ti_3 phases after heat treatment, indicating that the presence of the Nb element in the matrix can effectively hinder the precipitation of Ni-rich phases.

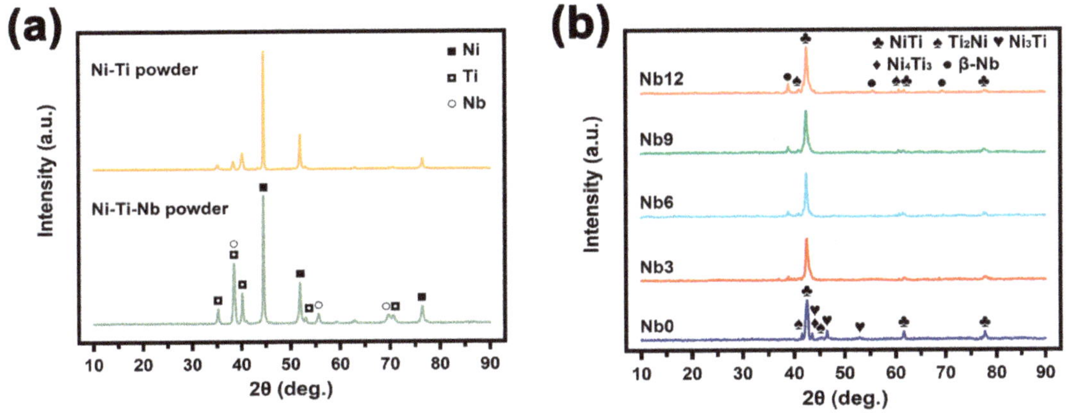

Figure 2. XRD patterns of (**a**) Ni-Ti and Ni-Ti-Nb powders, (**b**) porous Ni-Ti SMAs with Nb addition.

Figure 3a,b shows the morphologies of Ni-Ti and Ni-Ti-Nb (Nb content: 9at.%) powders, respectively. The two kinds of powders are irregularly shaped and have no obvious difference in the morphology. During the compressing process, the irregular shape gives a better meshing effect between the powders, which can effectively improve the strength and density of the compacts and can avoid the delamination and cracking of the compacts during the demolding process. Figure 3c shows the morphology of NaCl particles with an average particle size of 0.6 mm. Figure 3d–g shows the macro and micro morphologies of the Nb0 specimen, and Figure 3h–k shows the macro and micro morphologies of the Nb3 specimen. There is no significant difference in the macroscopic morphology between the two specimens. The pores are uniformly distributed in the Ni-Ti matrix and connected with each other, forming a three-dimensional network. The pores of the specimens retain the regular three-dimensional shape of the NaCl particles while losing their sharp-edged characteristics. This may be due to the fragmentation of the edges of the NaCl particles during the compressing process. The inner wall of pores is very rough, which is conducive to improving the ability of the material to absorb sound and reduce noise [20]. A 1100 °C sintering temperature provides enough energy for the formation and growth of the sintered neck. The powder particles exhibit a dense metallic framework due to good metallurgical bonding.

When we observe the micro-morphology of the pore wall at a higher magnification, it is clear that the Nb3 specimen has fewer micro pores than the Nb0 specimen, which can be attributed to the fact that the presence of Nb promotes the diffusion of atoms between Ni and Ti; this effectively reduces the vacancies generated by the Kirkendall effect during the high temperature sintering process [21]. Although Nb has a higher melting point than Ni and Ti, Nb and Ti belong to an infinite solid solution system, so Nb may still be dissolved in the Ni-Ti matrix [22]. Nb facilitates the sintering process by acting as a kind of channel for the diffusion and reaction between Ni and Ti. Thus, the migration of atoms in the powders is accelerated, which causes the distance between the powders of different elements to decrease continuously, resulting in the gradual disappearance of micro pores.

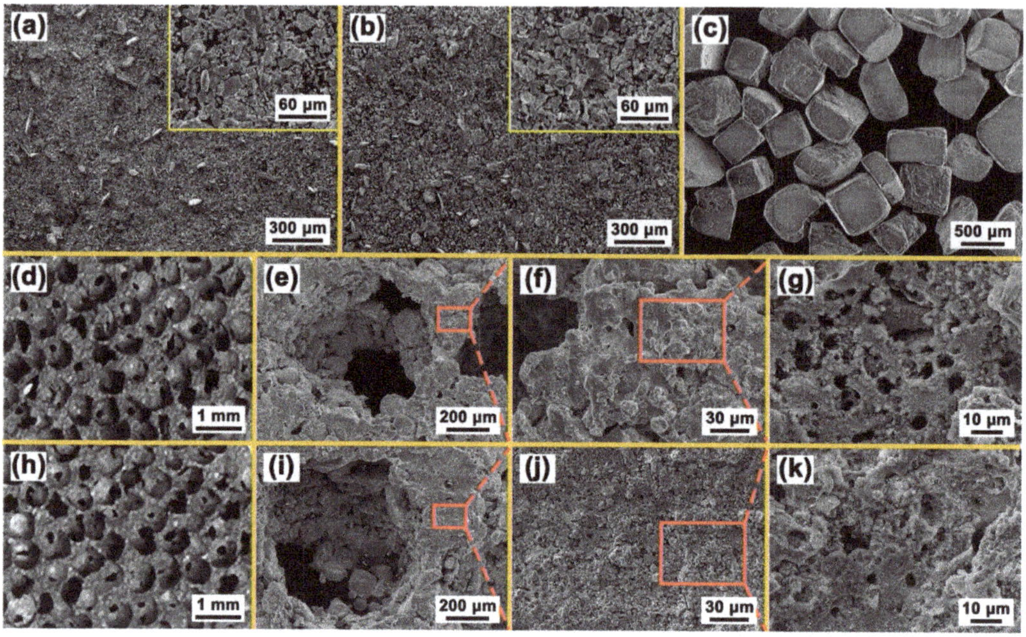

Figure 3. Morphologies of (**a**) Ni-Ti powders, (**b**) Ni-Ti-Nb powders, (**c**) NaCl particles, (**d**) Nb0 specimen, (**e**) pores in Nb0 specimen, (**f**) pore wall of Nb0 specimen, (**g**) micro pores in Nb0 specimen, (**h**) Nb3 specimen, (**i**) pores in Nb3 specimen, (**j**) pore wall of Nb3 specimen, (**k**) micro pores in Nb3 specimen.

Figure 4 shows the SEM images (backscattered electron images) of the porous Ni-Ti SMAs with different Nb contents. It can be seen from Figure 4a,b that the Nb0 specimen is mainly composed of a gray matrix and a small number of white precipitations. Combining the XRD analysis results shown in Figure 2b and the EDS analysis results of points A–D shown in Figure 5a, it can be determined that the gray matrix is the NiTi phase, whereas the white Ti-rich phase at point B, the flaky Ni-rich phase at point C, and the fine acicular Ni-rich phase at point D are the Ti_2Ni, Ni_3Ti and Ni_4Ti_3 phases, respectively [23,24]. From Figure 4b, it can be seen that the size of the Ni_4Ti_3 phase is 2–3 μm.

Figure 4c–g shows the backscattered electron images of the Nb3, Nb6, Nb9 and Nb12 specimens, respectively. In these specimens, white blocks appear, and with the increase in Nb content, the number of these white blocks increases. Figure 4d shows the image of the Nb3 specimen with higher magnification, and Figure 5b gives the EDS analysis results of points E–H. According to Figure 2b and the EDS result of point E, it can be determined that these white blocks are the β-Nb phase. In addition to the blocky β-Nb phase, a striped β-Nb phase is also found in Figure 4f,g, which may be related to the creep of the β-Nb phase during the sintering process [16]. From Figure 4d, it is apparent that the β-Nb phase is surrounded by a layer of Nb-rich phase. The clearly discernible transition region (the Nb-rich phase) increases the bonding effect between the β-Nb phase and the NiTi matrix. Moreover, Nb can also dissolve in NiTi and Ti_2Ni phases by replacing Ti and form (Ti, Nb)Ni and $(Ti, Nb)_2Ni$ phases, respectively. Because the sintering temperature (1100 °C) of porous Ni-Ti SMAs in this study is lower than the eutectic temperature (1150.7 °C) of the Ni-Ti-Nb alloy, the eutectic structure reported by Bao et al. is not found in Figure 4 [13,25,26].

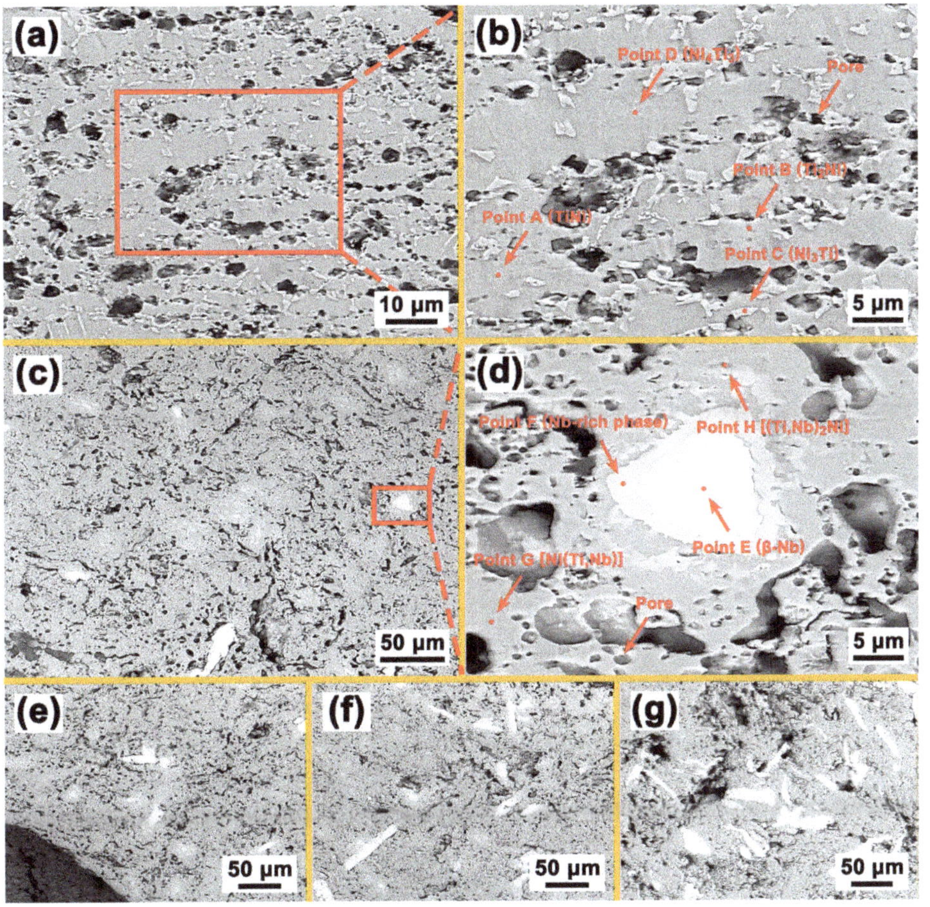

Figure 4. (**a**) Microstructure of Nb0 specimen; (**b**) precipitates in Nb0 specimen; (**c**) microstructure of Nb3 specimen; (**d**) precipitates in Nb3 specimen; (**e**) microstructure of Nb6 specimen; (**f**) microstructure of Nb9 specimen; (**g**) microstructure of Nb12 specimen.

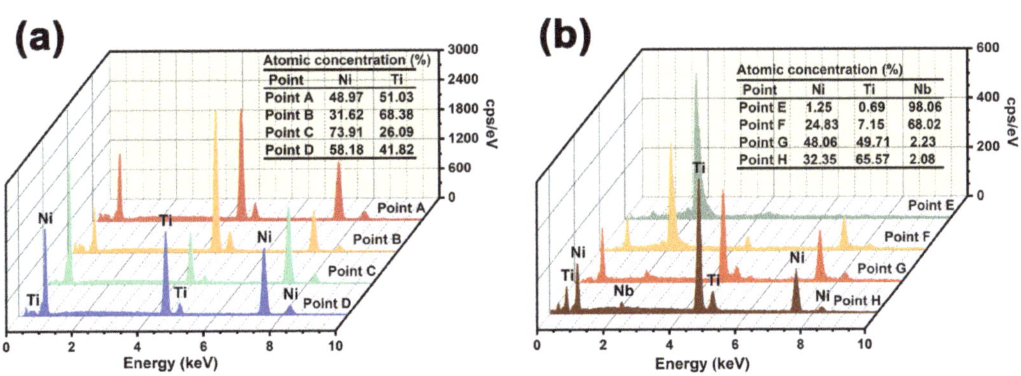

Figure 5. (**a**) EDS results of Nb0 specimen; (**b**) EDS results of Nb3 specimen.

Figure 6 shows the EPMA analysis results of the Nb0 specimen. Except for Ni and Ti, no other elements can be detected. Combining the EDS results shown in Figure 5a, it can be determined that the Ni-rich phase in Figure 6c is the Ni_3Ti phase, and the uniformly distributed Ti-rich phase in Figure 4d is the Ti_2Ni phase. The needle-like Ni_4Ti_3 phase, however, is not found in Figure 6 due to its small size. Figure 7 shows the EPMA analysis results of the Nb9 specimen. No other elements and Ni-rich phases can be detected except for the Ni, Ti, Nb elements and (Ti, Nb) Ni phase, which is in good accordance with the results shown in Figures 2b and 4c–g. The Ti-rich phase shown in Figure 4d cannot be detected due to its low content and the lower magnification of Figure 5 compared with that of Figure 4d. In Figure 7d, the layer of the Nb-rich phase shown in Figure 4d between the β-Nb phase and NiTi phase can still be clearly seen.

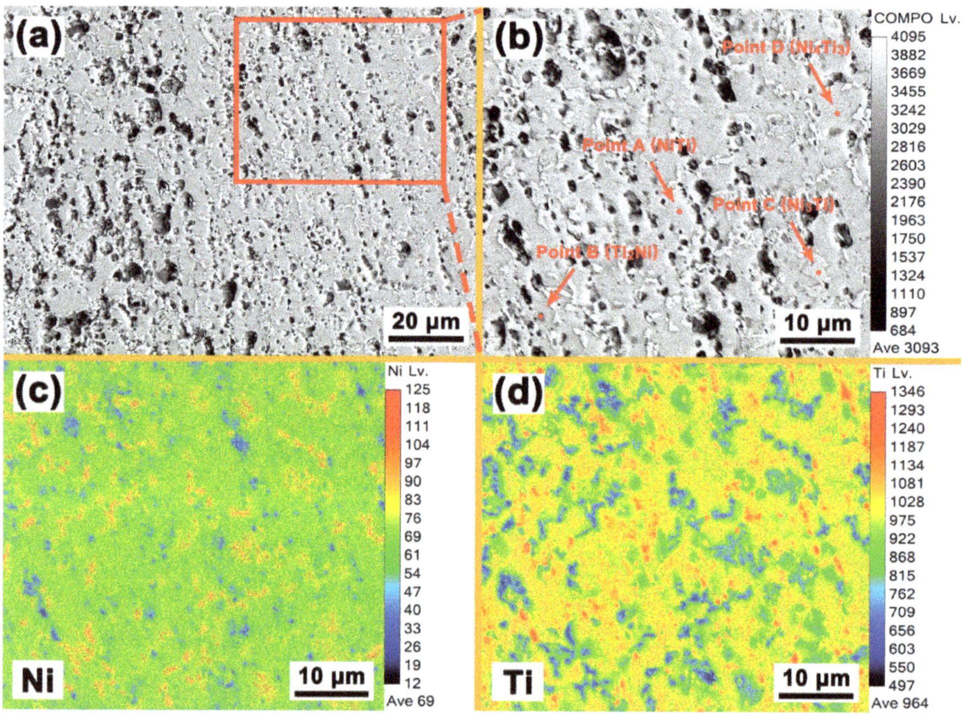

Figure 6. EPMA analysis results of Nb0 specimen: (**a**) microstructure; (**b**) high magnification image of the marked area in (**a**); (**c**) distribution of Ni element; (**d**) distribution of Ti element.

3.2. Effect of Nb on the Damping Property

Figure 8 shows the IF-temperature spectrum of the porous Ni-Ti SMAs with different contents of Nb. It can be seen that during the heating process two IF peaks arise at around −40 °C and 30 °C, respectively, which can be ascribed to the two steps transformation from the B19′ phase to the B2 phase (B19′ → R → B2) [7,27]. It was reported that the Ni_4Ti_3 phase precipitated after aging treatment could cause a stress field in the NiTi matrix and facilitate the transformation of the NiTi phase into the R phase; the two-step phase transformation of B19′ → R → B2 therefore occurs in the Nb0 specimen [7,28]. In the specimens containing Nb, however, the precipitation of the Ni_4Ti_3 phase is inhibited, so only the transformation of B19′ → B2 occurs (only one IF peak appears).

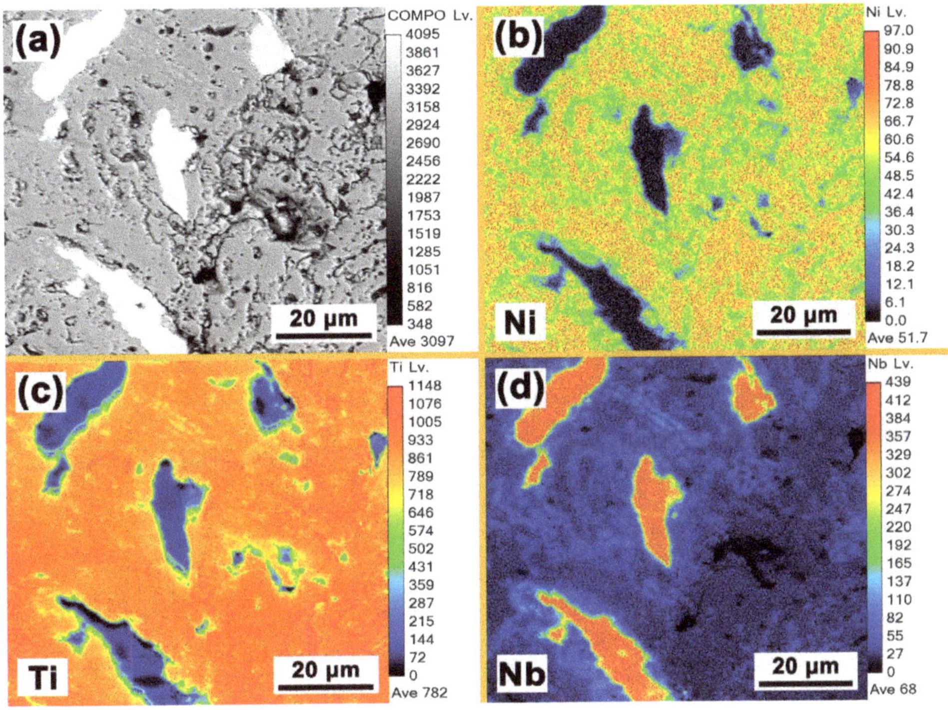

Figure 7. EPMA analysis results of Nb9 specimen: (**a**) microstructure; (**b**) distribution of Ni element; (**c**) distribution of Ti element; (**d**) distribution of Nb element.

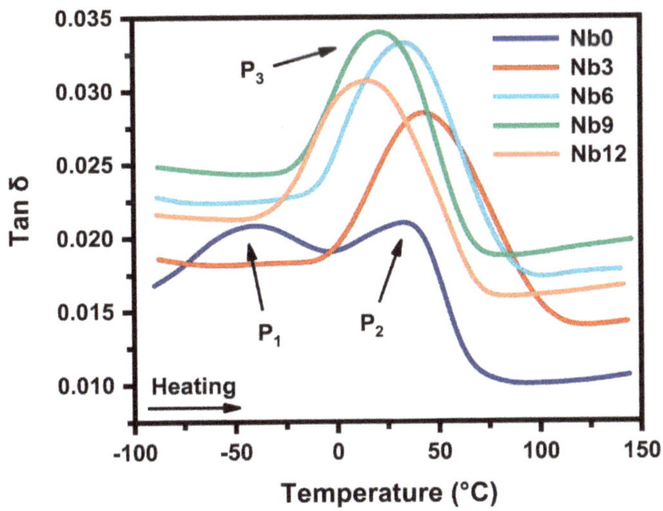

Figure 8. Effect of Nb on the damping property of the porous Ni-Ti SMAs.

From Figure 8, it is apparent that with the increase in Nb content, the damping (the level of the IF curve) of the porous Ni-Ti SMAs increases first and then decreases. When the Nb content reaches 9.0 at.%, the highest damping can be achieved. The damping capacity of the porous Ni-Ti SMAs has the following three influencing factors. One is the pore structure. Due to the existence of pores in the Ni-Ti matrix, the external stresses applied during IF measurements are not uniformly distributed in the specimens, making the strain significantly lag behind the stress and leading to the enhancement of the IF. In addition, uniformly distributed stresses will cause the pores to undergo the deformations of expansion and contraction, which are also conducive to the conversion of external mechanical energy into internal energy. However, in this study, all specimens have the same porosity. Therefore, the IF that originates from the pore structure is not the main influencing factor on the change in the IF with the Nb content. The other influencing factors are the damping sources in the matrix of the porous Ni-Ti SMAs. The hysteretic migration of interfaces in the Ni-Ti matrix is the main energy dissipation mechanism that leads to strain lagging behind stress and improvement in the damping [6]. Compared with the specimens containing the Nb element, the Nb0 specimen has a higher content of Ti_2Ni and Ni_3Ti phases without mobile interfaces, so the Nb0 specimen has lower damping. After adding the Nb element, as shown in Figure 4, the content of Ti_2Ni and Ni_3Ti phases significantly decreases. As a consequence, the damping increases. Moreover, from Figure 4d it is apparent that after adding the Nb element, a layer of Nb-rich phase appears around the β-Nb phase, forming a new interface of a β-Nb/Nb-rich layer/Ni-Ti matrix. This kind of interface is also mobile under applied stresses, so it acts as a new damping source of the porous Ni-Ti SMA [13]. It has been reported that there is a high density of dislocations in the regions adjacent to the β-Nb/Ni-Ti interface [12]. The slip of dislocations is also conductive to the improvement in the damping. Therefore, the addition of the Nb element not only decreases the content of precipitates, but also generates multiple damping sources by forming the β-Nb phase in the Ni-Ti matrix, and leads to the improvement in the damping. The third influencing factor of the damping is the existing form of the Nb element in the Ni-Ti matrix. In addition to the form of the β-Nb phase, a small number of Nb atoms dissolves into the NiTi phase. It has been shown that the dissolved Nb atoms can induce the movement of twins and effectively improve the intrinsic damping of martensite [13]. However, as the number of dissolved Nb atoms continues to increase, the damping of martensite will no longer increase, but will remain at a stable level. Therefore, dissolved Nb atoms are not the main reason for the change in damping with the Nb content. In addition, from Figure 8 it can also be seen that with the increase in the Nb content, the P_3 peak arising from the reverse MT ($B19' \rightarrow B2$) gradually shifts to a low temperature side. This can be attributed to the inhibiting effect of Nb on the precipitation of Ni-rich phases. The increase in the Ni/Ti atomic ratio leads to the decreased temperature of the reverse MT [29].

The porous Ni-Ti SMAs with Nb addition can be regarded as a two-phase structure consisting of the β-Nb phase and NiTi phase. The damping arising from the β-Nb phase consists of the intrinsic damping of the β-Nb phase and the damping of the β-Nb/Nb-rich layer interface and the Nb-rich layer/Ni-Ti interface. The intrinsic damping of the β-Nb phase is very low [21], so the damping of the β-Nb/Nb-rich layer interface and the Nb-rich layer/Ni-Ti interface is the main damping source. In addition, the Nb-rich layer/Ni-Ti interface provides nucleation sites for the formation of martensite during cooling, thus creating more interfaces, which further promote the improvement of the damping. With the increase in the Nb content, the volume fraction of the β-Nb phase increases, whereas the volume fraction of the NiTi phase decreases. According to Figure 8, the damping increases first and then decreases with the increase in the Nb content. When the Nb content reaches 9.0 at.%, the highest damping is obtained. When the increase in the interface damping can compensate for the decrease in the damping caused by the decrease in the volume fraction of the NiTi phase, the damping increases, whereas when the Nb content reaches 12.0 at.%, the level of the IF curve decreases, indicating that the increase in interface damping is no longer able to compensate for the decrease in damping caused by the decrease in the volume fraction of the NiTi phase.

3.3. Effect of Nb on the PE Property

The PE of the Ni-Ti matrix determines the shape recovery process of the porous Ni-Ti SMAs after stress unloading, whereas the pores can be regarded as the phase that does not reflect shape recovery. Figure 9 shows the effect of Nb on the shape recovery of the porous Ni-Ti SMAs at different pre-strains (2.0%, 3.0%, 4.0% and 5.0%). As can be seen from the figure, the strain of the compressed specimen after unloading is composed of residual strain (ε_R), superelastic reversion strain (ε_{SE}), and pure elastic reversion strain (ε_E). The reason for the high PE response of the specimens is the result of stress-induced MT and subsequent reverse MT, which is clearly reflected in the stress-strain curves during the loading–unloading processes. Taking the Nb9 specimen as an example, during the loading process, an inflection point appears on the stress-strain curve when the stress reaches 90 MPa, indicating that the stress-induced MT begins to occur. When the compressive strain reaches 5%, the stress is unloaded. Subsequently, the reverse MT begins to occur, and the strain decreases linearly to about 3.36%. After that, the strain continues to recover until reaching the ε_R of 1.77%. Then ε_E, ε_{SE} and the total shape reversion strain can be determined as 1.64%, 1.59% and 3.23%, respectively, so the total recovery rate can reach 64.6% when the pre-strain is 5%.

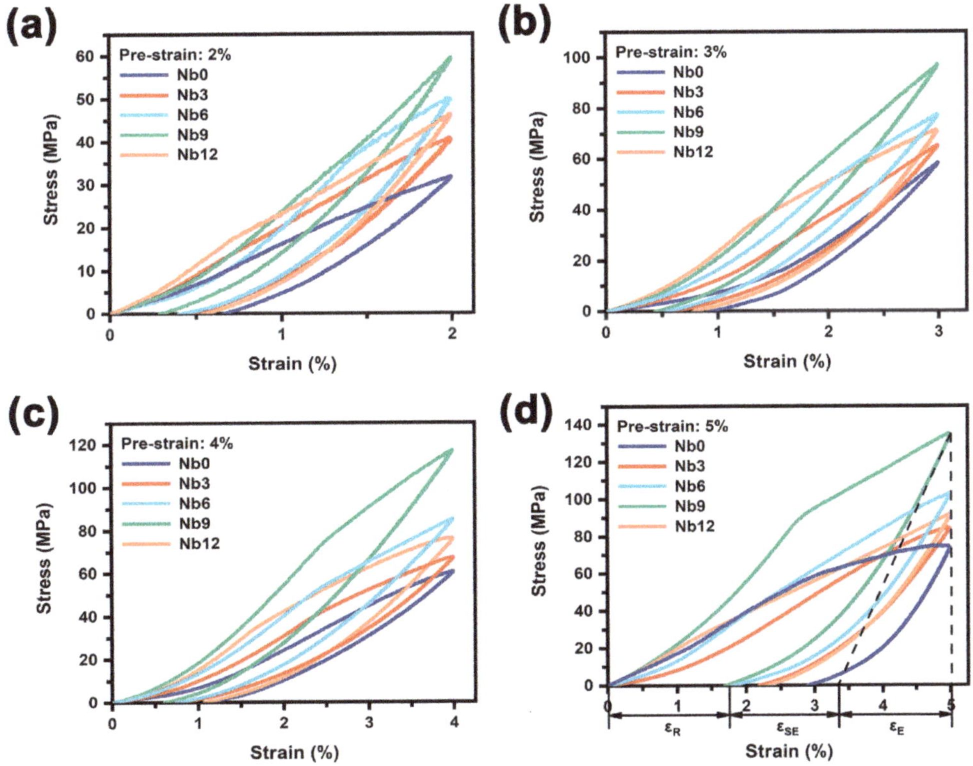

Figure 9. Effect of Nb on the shape recovery of the porous Ni-Ti SMAs at different pre-strains: (**a**) 2%; (**b**) 3%; (**c**) 4%; (**d**) 5%.

The superelastic recovery strain of different specimens after unloading under the four test conditions is summarized in Figure 10. It can be seen that with the increase in the Nb content, the superelastic recovery strain increases first and then decreases. The Nb9 specimen has the highest superelastic recovery strain, indicating that Nb has a positive contribution to the PE. In addition, the Ti_2Ni and Ni_3Ti phases in the Nb0 specimen have no PE, so the reduction in the number of these phases also contributes to the improvement in the PE.

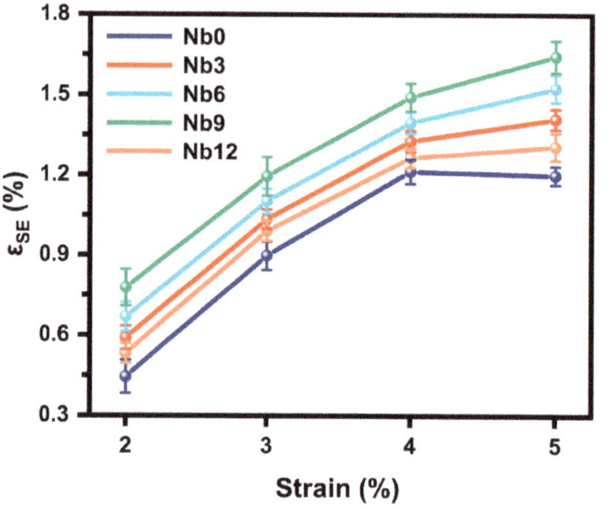

Figure 10. Effect of Nb content on the PE of the porous Ni-Ti SMAs.

It has been reported that in Ni-Ti SMAs with the Nb addition, there is a large number of twinned martensites at the interface of the Nb-rich phase/Ni-Ti matrix [29], and this structure is considered necessary for the self-accommodation of martensite variants [30]. During loading and unloading processes, with the increase in the strain, reorientation occurs between martensite variants when the stress exceeds the yield stress of martensites. The reorientation and de-twinning of twinned martensites facilitate the hyperelastic recovery of the matrix [31]. However, the β-Nb phase has no PE, and its plastic deformation is not conducive to shape recovery. Therefore, as the content of the β-Nb phase continues to increase and the volume fraction of the Ni-Ti matrix phase continues to decrease, the PE of the porous Ni-Ti SMAs will decrease. Moreover, from Figure 9 it can be seen that the level of stress-strain curves increases first and then decreases with the increase in the Nb content. The increase in stress-strain curves can be attributed to the solid solution strengthening effect of Nb atoms as well as the dislocation strengthening effect due to the thermal mismatch between the β-Nb phase and the Ni-Ti matrix [29]. However, an excessive soft β-Nb phase will inevitably degrade the strength of the porous Ni-Ti SMAs. Therefore, the Nb9 sample has the highest level of stress-strain curve. Clearly, the strengthening of the porous Ni-Ti alloy contributes to the improvement in PE.

4. Conclusions

- Porous Ni-Ti SMAs with different Nb contents were fabricated. The pores were uniformly distributed in the Ni-Ti matrix and connected with each other, forming a three-dimensional network. The addition of Nb can effectively reduce the numbers of micro pores and precipitates in the Ni-Ti matrix.
- With the increase in Nb content, the IF peak of the porous Ni-Ti SMAs arising from the reverse MT gradually shifts towards the low temperature side. This can be attributed to the hindering effect of Nb on the precipitation of Ni-rich phases as well as the replacement of Nb atoms on Ti atoms.
- With the increase in Nb content, both the damping and the PE of the porous Ni-Ti SMAs increase first and then decrease. The Nb9 specimen has the highest damping and the best PE. The improvement in damping can be ascribed to the formation of mobile interfaces around β-Nb phase, whereas the improvement in PE is related to

the formation of twinned martensites and the strengthening of the NiTi matrix. In addition, the decrease in the content of low damping Ni-rich phases without PE is also conductive to the improvements in damping and PE. The decrease in damping and PE is caused by the decrease in the NiTi content.

Author Contributions: Conceptualization, Q.W. and F.Y.; methodology, Q.W. and J.F.; investigation, P.S., P.J. and J.Z.; data curation, P.S. and P.J.; writing—original draft preparation, P.S.; writing—review and editing, F.Y.; project administration, Q.W.; funding acquisition, Q.W. and F.Y. All authors have read and agreed to the published version of the manuscript.

Funding: This research was funded by the Natural Science Foundation of Hebei Province (No. E2021202017), the Guangdong Academy of Sciences (No. 2021GDASYL-20210102002), and the Jiangsu Provincial Policy Guidance Program (Special Project for the Introduction of Foreign Talents) Talent Introduction Program Project (No. BX2021024).

Institutional Review Board Statement: Not applicable.

Informed Consent Statement: Not applicable.

Data Availability Statement: Not applicable.

Conflicts of Interest: The authors declare no conflict of interest.

References

1. Stanley, J.; Kumar, S.A.; Ariraman, M.; Mandhakini, M. Role of amines in microstructure tuning for synergistic damping and mechanical properties in epoxy resins. *Prog. Org. Coat.* **2021**, *150*, 105896. [CrossRef]
2. Poddaeva, O.; Fedosova, A. Damping capacity of materials and its effect on the dynamic behavior of structures. Review. *Energy Rep.* **2021**, *7*, 299–307. [CrossRef]
3. Yi, X.Y.; Huang, B.W.; Gao, W.H.; Sun, B.; Feng, X.X.; Cao, X.J.; Guo, Y.L.; Meng, X.L.; Gao, Z.Y.; Wang, H.Z. Damping behaviors and strain recovery characteristics of Hf-modified TiNb-based shape memory alloys. *Mater. Res. Bull.* **2023**, *158*, 112084. [CrossRef]
4. He, Y.J.; Yin, H.; Zhou, R.H.; Sun, Q.P. Ambient effect on damping peak of NiTi shape memory alloy. *Mater. Lett.* **2010**, *64*, 1483–1486. [CrossRef]
5. Nair, V.S.; Nachimuthu, R. The role of NiTi shape memory alloys in quality of life improvement through medical advancements: A comprehensive review. *J. Eng. Med.* **2022**, *236*, 923–950. [CrossRef]
6. Safavi, M.S.; Bordbar-Khiabani, A.; Walsh, F.C.; Mozafari, M.; Khalil-Allafi, J. Surface modified NiTi smart biomaterials: Surface engineering and biological compatibility. *Curr. Opin. Biomed. Eng.* **2023**, *25*, 100429. [CrossRef]
7. Otsuka, K.; Ren, X. Physical metallurgy of Ti-Ni-based shape memory alloys. *Prog. Mater. Sci.* **2005**, *50*, 511–678. [CrossRef]
8. Kohl, M.; Bram, M.; Moser, A.; Buchkremer, H.P.; Beck, T.; Stover, D. Characterization of porous, net-shaped NiTi alloy regarding its damping and energy absorbing capacity. *Mater. Sci. Eng. A* **2011**, *528*, 2454–2462. [CrossRef]
9. Zhang, X.X.; Hou, H.W.; Wei, L.S.; Chen, Z.X.; Wei, W.T.; Geng, L. High damping capacity in porous NiTi alloy with bimodal pore architecture. *J. Alloys Compd.* **2013**, *550*, 297–301. [CrossRef]
10. Villa, F.; Nespoli, A.; Passaretti, F.; Villa, E. Microstructural and Thermo-Mechanical Characterization of Cast $NiTiCu_{20}$ Shape Memory Alloy. *Materials* **2021**, *14*, 3770. [CrossRef]
11. Shuitcev, A.; Li, L.; Markova, G.V.; Golovin, I.S.; Tong, Y.X. Internal friction in $Ti_{29.7}Ni_{50.3}Hf_{20}$ alloy with high temperature shape memory effect. *Mater. Lett.* **2020**, *262*, 127025. [CrossRef]
12. Cai, W.; Lu, X.L.; Zhao, L.C. Damping behavior of TiNi-based shape memory alloys. *Mater. Sci. Eng. A* **2005**, *394*, 78–82. [CrossRef]
13. Bao, Z.Z.; Guo, S.; Xiao, F.; Zhao, X.Q. Development of NiTiNb in-situ composite with high damping capacity and high yield strength. *Prog. Nat. Sci.* **2011**, *21*, 293–300. [CrossRef]
14. Pant, D.C.; Pal, S. Phase transformation and energy dissipation of porous shape memory alloy structure under blast loading. *Mech. Mater.* **2019**, *132*, 31–46. [CrossRef]
15. Guo, X.; Dong, X.J.; Yu, Z.L.; Zhang, Z.H.; Xie, X.Y.; Wang, X.B.; Xin, R.L.; Yan, W. Study on the mechanical properties of bionic protection and self-recovery structures. *Materials* **2020**, *13*, 389. [CrossRef]
16. Zheng, H.X.; Mentz, J.; Brarn, M.; Buchkremer, H.P.; Stover, D. Powder metallurgical production of TiNiNb and TiNiCu shape memory alloys by combination of pre-alloyed and elemental powders. *J. Alloys Compd.* **2008**, *463*, 250–256. [CrossRef]
17. Li, B.; Zhang, F.; Zheng, L.; Zhang, H. The role of Sc on the microstructures and mechanical properties of the Ni-rich NiTi alloy. *Mater. Today Commun.* **2022**, *31*, 103556. [CrossRef]
18. Oezerim, G.; Anlas, G.; Moumni, Z. The effect of heat treatment on pseudoelastic behavior of spark plasma sintered NiTi. *Mater. Today Commun.* **2022**, *31*, 11. [CrossRef]

19. Sun, B.; Fu, M.W.; Lin, J.P.; Ning, Y.Q. Effect of low temperature aging treatment on thermally and stress induced phase transformations of nanocrystalline and coarse grained NiTi wires. *Mater. Des.* **2017**, *131*, 49–59. [CrossRef]
20. Cao, L.T.; Fu, Q.X.; Si, Y.; Ding, B.; Yu, J.Y. Porous materials for sound absorption. *Compos. Commun.* **2018**, *10*, 25–35. [CrossRef]
21. Guo, W.; Kato, H. Submicron-porous NiTi and NiTiNb shape memory alloys with high damping capacity fabricated by a new top-down process. *Mater. Des.* **2015**, *78*, 74–79. [CrossRef]
22. Fan, Q.C.; Sun, M.Y.; Zhang, Y.H.; Wang, Y.Y.; Zhang, Y.; Peng, H.B.; Sun, K.H.; Fan, X.H.; Huang, S.K.; Wen, Y.H. Influence of precipitation on phase transformation and mechanical properties of Ni-rich NiTiNb alloys. *Mater. Charact.* **2019**, *154*, 148–160. [CrossRef]
23. Fan, G.; Zhou, Y.; Chen, W.; Yang, S.; Ren, X.; Otsuka, K. Precipitation kinetics of Ti_3Ni_4 in polycrystalline Ni-rich TiNi alloys and its relation to abnormal multi-stage transformation behavior. *Mater. Sci. Eng. A* **2006**, *438*, 622–626. [CrossRef]
24. Fan, Q.C.; Zhang, Y.H.; Wang, Y.Y.; Sun, M.Y.; Meng, Y.T.; Huang, S.K.; Wen, Y.H. Influences of transformation behavior and precipitates on the deformation behavior of Ni-rich NiTi alloys. *Mater. Sci. Eng. A* **2017**, *700*, 269–280. [CrossRef]
25. Chen, Y.; Jiang, H.C.; Rong, L.J.; Xiao, L.; Zhao, X.Q. Mechanical behavior in NiTiNb shape memory alloys with low Nb content. *Intermetallics* **2011**, *19*, 217–220. [CrossRef]
26. Wang, L.Q.; Wang, C.; Dunand, D.C. Microstructure and strength of NiTi-Nb eutectic braze joining NiTi wires. *Met. Mater. Trans. A* **2015**, *46A*, 1433–1436. [CrossRef]
27. Wang, X.B.; Verlinden, B.; Van Humbeeck, J. R-phase transformation in NiTi alloys. *Mater. Sci. Technol.* **2014**, *30*, 1517–1529. [CrossRef]
28. Zeng, J.M.; Jiang, H.C.; Liu, S.W.; Rong, L.J. Damping behavior of $Ti_{50.1}Ni_{49.9}$ alloy in reverse martensitic transformation region. *Sci. China Technol. Sci.* **2012**, *55*, 470–474. [CrossRef]
29. Shi, H.; Frenzel, J.; Martinez, G.T.; Van Rompaey, S.; Bakulin, A.; Kulkova, S.; Van Aert, S.; Schryvers, D. Site occupation of Nb atoms in ternary Ni-Ti-Nb shape memory alloys. *Acta Mater.* **2014**, *74*, 85–95. [CrossRef]
30. Uchida, K.; Shigenaka, N.; Sakuma, T.; Sutou, Y.; Yamauchi, K. Effect of Nb content on martensitic transformation temperatures and mechanical properties of Ti-Ni-Nb shape memory alloys for pipe joint applications. *Mater. Trans.* **2007**, *48*, 445–450. [CrossRef]
31. Wang, L.Q.; Xie, L.C.; Zhang, L.C.; Chen, L.Y.; Ding, Z.H.; Lv, Y.T.; Zhang, W.; Lu, W.J.; Zhang, D. Microstructure evolution and superelasticity of layer-like NiTiNb porous metal prepared by eutectic reaction. *Acta Mater.* **2018**, *143*, 214–226. [CrossRef]

materials

MDPI

Article

Preparation of Al_3Ti-Al_2O_3/Al Inoculant and Its Inoculation Effect on Al-Cu-Mn Alloy

Jinhua Ding [1], Chao Wang [1], Cheng Lu [2,*], Guangming Zhu [1], Nana Guo [1], Xujie Gao [1], Xin Wang [3,*] and Chunxiang Cui [3]

1 School of Mechanical Engineering, Shandong University of Technology, Zibo 255000, China
2 School of Physics and Optoelectronic Engineering, Shandong University of Technology, Zibo 255000, China
3 Hebei Key Laboratory of New Functional Materials, School of Material Science and Engineering, Hebei University of Technology, Tianjin 300401, China
* Correspondence: luchengcg@163.com (C.L.); ahaxin@hebut.edu.cn (X.W.); Tel.: +86-0533-2786-815 (C.L.); +86-022-6020-4125 (X.W.)

Abstract: The grain size plays a pivotal role in determining the properties of the alloy. The grain size can be significantly decreased by adding inoculants. Aiming to address the shortcomings of existing inoculants, the Al_3Ti-Al_2O_3/Al inoculant was successfully prepared using Al-Ti master alloy and Al_2O_3 whiskers as raw materials. With the aid of ultrasonic energy, the Al_2O_3 whiskers were uniformly dispersed within the inoculants. Under the combined action of ultrasonic and titanium, the Al_2O_3 whiskers were broken into small particles at high temperature. To enhance the morphology of Al_3Ti and achieve even particle dispersion throughout the matrix, vacuum rapid quenching treatment was applied to the inoculant. The SEM test results indicated a significant reduction in particle size after vacuum rapid quenching. The Al_3Ti-Al_2O_3/Al inoculants exhibited excellent grain refinement effects on the weldable Al-Cu-Mn alloy. Crystallographic calculations and HRTEM analysis revealed that Al_2O_3 and Al have orientation relationships, indicating their potential as effective heterogeneous nucleation sites. The mechanical properties of the Al-Cu-Mn alloy were obviously improved after the Al_3Ti-Al_2O_3/Al inoculant was added.

Keywords: inoculant; Al_2O_3 whiskers; grain refinement; microstructure; property

1. Introduction

The refinement of grain is widely acknowledged as a straightforward and efficient approach to enhance the microstructure and properties of aluminum alloys [1–5]. Grain refinement can be achieved through adding inoculants, increasing melt cooling rate, stirring, ultrasonic treatment, plastic deformation and heat treatment [6–10]. Among these refining methods, adding inoculants to aluminum alloy melt is the most simple and effective method. The inoculants of aluminum alloy mainly include some alloying elements and ceramic particles [11–14]. Alloying elements usually play a role in heterogeneous nucleation of Al matrix by forming intermetallic compounds with Al. The main requirements for ceramic inoculants are high melting points, excellent physical and chemical stability, and low lattice misfit with the substrates [15–20].

For most aluminum alloys, the Al-5Ti-B refiner can play a role in grain refinement. However, in the Al-5Ti-B ingot, TiB_2 particles tend to aggregate and form clusters, and the Al_3Ti phase is usually needle-like, which are detrimental to the refinement effect. Moreover, the presence of elements such as Zr, Si, Cr, and V in aluminum alloy can lead to a "poisoning" effect on Al-5Ti-B [21–24]. For instance, even 0.2 wt.% of Zr can rapidly diminish the refining capability of Al-5Ti-B. The reason is the formation of $(Ti_{1-x}Zr_x)Al_3$ and $(Ti_{1-x}Zr_x)B_2$, which leads to the weakening of heterogeneous nucleation ability of TiB_2 and Al_3Ti particles. Additionally, the formation of $(Ti_{1-x}Zr_x)Al_3$ weakens the constraining effect of free Ti on grain growth, while Zr disrupts the two-dimensional monatomic layer of Al_3Ti on the surface of TiB_2, resulting in a loss of grain refinement effectiveness [25–27].

Then, Al-Ti-B-RE [28,29] was developed to further improve the refining effect of Al-Ti-B refiner. The addition of rare earth can not only change the morphology of Al_3Ti particles but also improve the dispersibility of TiB_2 and Al_3Ti particles and increase the degree of supercooling of aluminum melt. Compared with the traditional Al-Ti-B refiner, Al-Ti-B-RE refiner has a better refining effect, but its optimization effect is limited, the process is complicated and the cost is high.

A1-Ti-C intermediate alloy is also an aluminum alloy refiner widely used in industrial production. In crystallography, there exists a specific orientation relationship between TiC particles and the α-Al matrix, and so, it has heterogeneous nucleation potential for α-Al. Li et al. [30] found that the dispersion of TiC in aluminum matrix is obviously better than that of TiB_2 particles, and the particle size of TiC is much smaller than that of TiB_2. Therefore, for two kinds of particles with the same mass fraction, the number of TiC particles is more than that of TiB_2, so that more nucleation sites can be provided for α-Al, thus achieving a better refining effect. However, the major disadvantage of A1-Ti-C is that TiC and Al will react to form Al_4C_3 brittle phase at a certain temperature: $13Al\ (l) + 3TiC\ (s) = 3TiAl_3\ (s) + Al_4C_3\ (s)$ [31], resulting in poor stability of TiC and fading behaviors of the refinement effect. In addition, the preparation of Al-Ti-C refiner is usually based on graphite powder, carbon tetrachloride and other carbon sources, which are added to the Al-Ti intermediate alloy melt at high temperature and then reacted with each other. However, due to the poor wettability of carbon in aluminum, the low utilization rate of C and the high reaction temperature, the required preparation equipment is complicated and the cost is high. Therefore, Al-Ti-C refiner cannot be used on a large scale in industrial production [32].

In order to improve the refining effect of Al-Ti-C refiner, Zhao, Ding and Xu et al. [33,34] prepared Al-Ti-C-RE refiner. After the addition of rare earth elements, the wettability of C and Al melt is improved, which promotes the formation of TiC particles. In addition, rare earth can improve the morphology of Al_3Ti particles, reduce the size of Al_3Ti particles, improve the dispersion of TiC and Al_3Ti particles and improve the refining effect and property of anti-degeneration of refiners. However, the development of Al-Ti-C-RE refiner is the same as that of Al-Ti-B-RE refiner, with limited optimization effect, cumbersome process and high cost [35–37]. Therefore, it is necessary to develop new inoculants to overcome the shortcomings of existing inoculants.

α-Al_2O_3 belongs to the trigonal system and is composed of hexagonal tightly packed crystals. The lattice constant of α-Al_2O_3 is a = b = 0.4759 nm, c = 1.3 nm. According to the edge-to-edge model proposed by Zhang et al. [17], the orientation relationships between α-Al and α-Al_2O_3 are as follows: $(200)_{Al}\,||\,(006)_{Al_2O_3}$, $[011]_{Al}\,||\,[\bar{1}20]$; $(200)_{Al}\,||\,(006)_{Al_2O_3}$, $[011]_{Al}\,||\,[\bar{1}10]_{Al_2O_3}$; $(111)_{Al}\,||\,(011)_{Al_2O_3}$, $[01\bar{1}]_{Al}\,||\,[11\bar{1}]_{Al_2O_3}$. In addition, α-Al_2O_3 has the advantages of high strength, high melting point, high elastic modulus and good stability [38–42]; as a result, it can serve as heterogeneous substrates for α-Al and effectively refine aluminum alloys. In this paper, we desired to replace TiB_2 with Al_2O_3 to overcome the shortcomings of the existing inoculants.

The wettability between α-Al_2O_3 particles and liquid aluminum is poor, and so, it is easy to aggregate into clusters on the surface of liquid aluminum. In this paper, the Al_2O_3 whiskers instead of Al_2O_3 particles were added to the liquid Al-Ti master alloy at high temperature, and ultrasonic treatment was used to improve the dispersion and wettability.

2. Materials and Methods

The primary equipment utilized in the preparation of Al-Cu-Mn alloy was predominantly the MZG series high-frequency induction melting furnace. First, commercial pure aluminum (99 wt.%), commercial pure copper (99 wt.%), commercial pure zinc (99 wt.%), Al-Mn, Al-Ti, Al-Zr and Al-V were proportionally weighed using an electronic scale. The weighed pure aluminum and pure copper were placed in the graphite crucible and heated in the MZG series high-frequency induction melting furnace. After these two metals were melted, the weighed commercial pure zinc (99 wt.%), Al-Mn, Al-Ti, Al-Zr and Al-V were quickly put in and fully stirred with graphite rods. After all the raw materials were melted and fully stirred, the dross on the surface of the molten liquid was removed with a scraping spoon and was then poured into the preheated steel mold to produce alloy sample rods with Φ20 mm × 100 mm.

The raw materials for preparation of Al_3Ti-Al_2O_3/Al inoculant were Al-Ti master alloy and alumina whiskers. The alumina whiskers were provided by the Shenzhen Research Institute of Tsinghua University (Shenzhen, China). The Al-Ti master alloy was melted in a high frequency induction furnace, and then the alumina whiskers were added to the melt in a certain proportion. To make the whisker evenly dispersed in the melt, ultrasonic vibration treatment was applied to the melt, and the metal liquid was then cast into the steel mold. To achieve a uniform dispersion of the inoculant in the Al-Cu-Mn matrix alloy, the inoculant ingots were remelted and transformed into ribbons through vacuum rapid quenching, that is, by pouring liquid metal on a spinning cold copper wheel.

The Al-Cu-Mn alloy was melted at 800 °C in the crucible resistance furnace. After the Al-Cu-Mn alloy was melted, the inoculant ribbons were added to the melt and thoroughly stirred with a graphite rod and were then cast into the steel mold of Φ20 mm × 100 mm. The Al-Cu-Mn alloys underwent T6 heat treatment, which involved solution treatment at 520 °C for 24 h followed by water quenching and artificial aging treatment at 165 °C for 14 h.

The phase composition of the inoculants was determined via X-ray diffraction (XRD). The metallographic microscope was utilized to observe the changes in grain morphology and size. The microstructures and element composition of the alloy and inoculant were analyzed by the scanning electron microscopy (SEM) equipped with energy dispersive X-ray spectroscopy (EDS). To enable observation of the sample under both optical and scanning electron microscopes, it underwent a series of preparation steps including sandpaper polishing with varying roughness, mechanical polishing using a specialized machine and, finally, corrosion treatment utilizing Keller reagent. The heterogeneous nucleation particles in alloy were characterized using JEM-2100 F (JEOL, Tokyo, Japan) transmission electron microscope (TEM). The specimens for TEM observation were prepared by ion milling. The mechanical properties of the alloy was measured using a universal testing machine.

3. Results and Discussion

The XRD pattern of Al_3Ti-Al_2O_3/Al inoculant is presented in Figure 1, revealing that the phase composition of the inoculant primarily comprised α-Al, α-Al_2O_3 and Al_3Ti.

The SEM images of Al_3Ti-Al_2O_3/Al inoculants and the EDS patterns of the second phases are presented in Figure 2. The matrix contained distributed rod-like and granular second phases, as illustrated in Figure 2a. In Figure 2b, it can be seen that the granular second phases were nanometer and submicron in size. The rod-like second phases were alumina whiskers, and it can be found that there were craters on the surface of the alumina whiskers and the whisker length was obviously reduced. At high temperature (1600~1700 °C), Al_2O_3 will decompose and release O atoms, the O atoms can easily combine with the Ti atoms in the melt to form TiO_2 or Al_2TiO_5 [43–45]. However, no diffraction peaks corresponding to other phases were observed in the XRD pattern shown in Figure 1, probably because too few reaction products were generated to be detected. Because Al_2O_3 can react with titanium at high temperatures, titanium in the Al-Ti master alloy will corrode the whisker and break it. The EDS pattern of point B indicates that the particles in Figure 2b were the chips dropped from the alumina whiskers after being fragmented. The gray phases in Figure 2a,b are Al_3Ti, and it can be seen that before the vacuum rapid quenching treatment, its morphology was short rod. The refining effect of large particles was less than ideal when compared to that of fine particles. Therefore, to achieve optimal refining results, the inoculant ingot was rapidly quenched under vacuum conditions. The SEM image of the Al_3Ti-Al_2O_3/Al ribbon in Figure 2f reveals a disappearance of large rod-like Al_3Ti particles and an emergence of numerous gray nanoparticles. In the process of remelting, the Al_3Ti

particles were dissolved in liquid aluminum. In the subsequent solidification stage, due to the rapid cooling rate, these particles did not have sufficient time to grow and, thus, resulted in the formation of numerous nanoparticles.

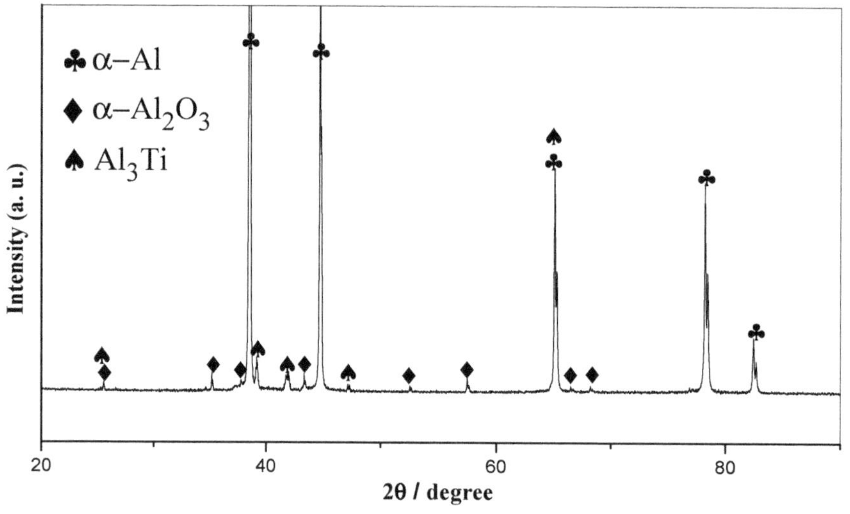

Figure 1. XRD pattern of Al_3Ti-Al_2O_3/Al inoculant.

To evaluate the refining effect of Al_3Ti-Al_2O_3/Al inoculant on Al-Cu-Mn alloy, 1 wt.% of the inoculant was introduced into the alloy. The metallographic images of the Al-Cu-Mn alloy, both pre- and post-inoculation, are depicted in Figure 3. The addition of the inoculant resulted in a significant reduction in grain size for the Al-Cu-Mn alloy. The main reasons for the decrease in the grain size after adding Al_3Ti-Al_2O_3/Al inoculant were as follows: first, Al_3Ti can serve as a heterogeneous nucleation site for α-Al; second, Ti can limit the growth of aluminum grains; third, the calculation results show that Al_2O_3 and Al also had crystal orientation relationships and, therefore, it can serve as a site for heterogeneous nucleation of α-Al during the solidification process of the alloy.

The (111), (200) and (220) planes of Al corresponded to the first, second and third close-packed planes, respectively, with interplanar spacings of 0.234 nm, 0.202 nm and 0.143 nm. The close-packed crystallographic planes of Al_2O_3 were (012), (104), (110) and (006) with corresponding interplanar spacings of 0.348 nm, 0.255 nm, 0.238 nm and 0.217 nm, respectively. By calculating, Al and Al_2O_3 can form the following orientation relationships: $(111)_{Al}//(110)_{Al_2O_3}$, $[011]_{Al}//[111]_{Al_2O_3}$; $(200)_{Al}//(006)_{Al_2O_3}$, $[011]_{Al}//[1\bar{2}0]_{Al_2O_3}$; $(200)_{Al}//(006)_{Al_2O_3}$, $[011]_{Al}//[1\bar{1}0]_{Al_2O_3}$. The Materials Studio (MS 7.0) software was utilized to generate a schematic diagram illustrating the lattice matching between Al_2O_3 and Al, as depicted in Figure 4. In Figure 4(a_1,a_2), it can be seen the crystal plane spacing of $(111)_{Al}$ was close to that of $(110)_{Al_2O_3}$, the atomic spacing of $[011]_{Al}$ was close to that of $[111]_{Al_2O_3}$; therefore, they can form a good interface combination in theory. In Figure 4(b_1–b_3), it can be seen that when the $(200)_{Al}$ was combined with the $(006)_{Al_2O_3}$, the plane spacing between them and the atomic spacing of the corresponding crystal orientation were very close, and so, $(200)_{Al}$ and $(006)_{Al_2O_3}$ also can form a good interface combination in theory.

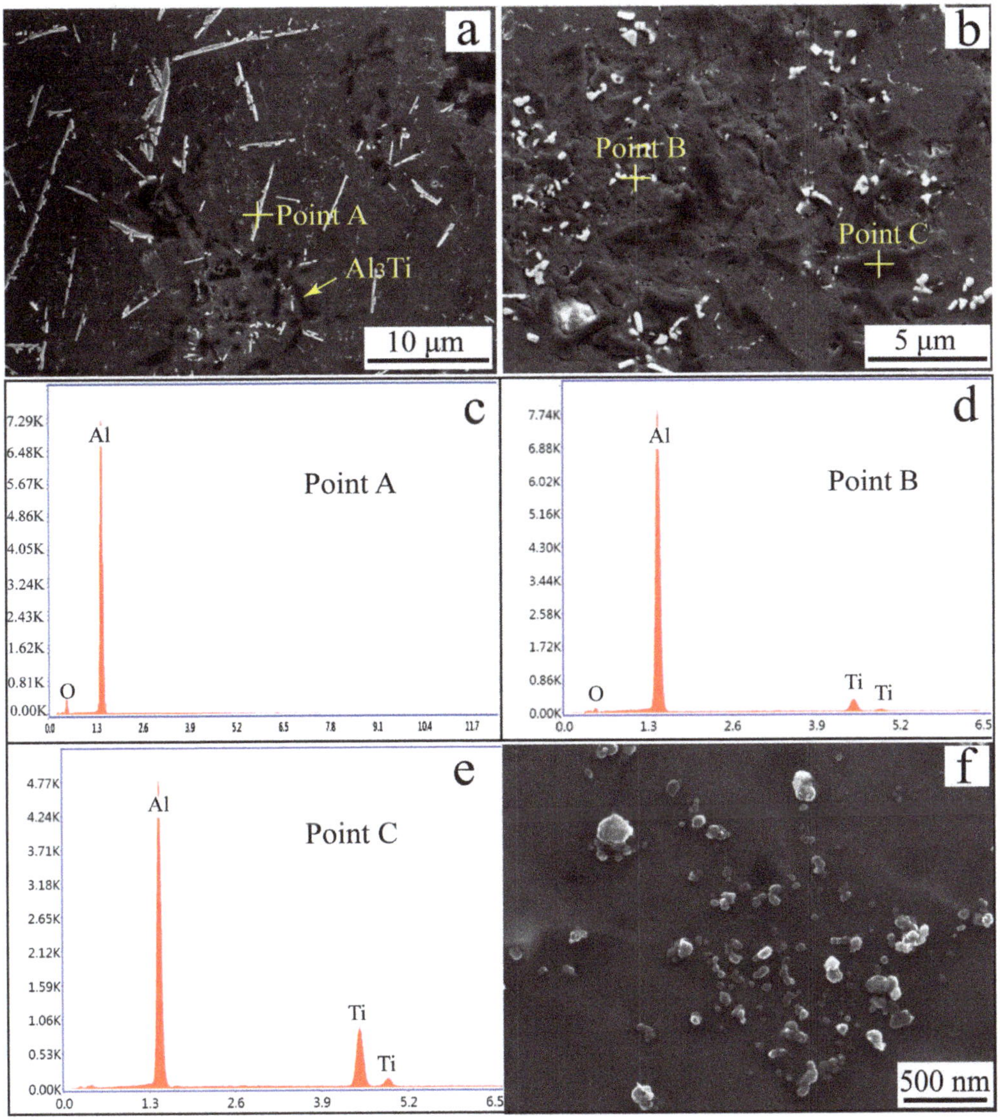

Figure 2. (**a**,**b**) SEM images of Al_3Ti-Al_2O_3/Al inoculants, (**c**) EDS pattern of Point A, (**d**) EDS pattern of Point B, (**e**) EDS pattern of Point C, (**f**) SEM image of Al_3Ti-Al_2O_3/Al inoculant after vacuum rapid quenching.

The HRTEM image of the Al-Cu-Mn alloy inoculated with the Al_3Ti-Al_2O_3/Al inoculant is presented in Figure 5c, revealing a slight variation in atomic arrangement within the circular box region compared to its surrounding area. The measurement results show that the interplanar spacing d_1 was 0.2134 nm and d_2 was 0.2505 nm, and the angle between the two crystallographic planes was 38.54°. Figure 5a is the simulation of Al_2O_3 lattice built by MS 7.0 software. Figure 5b is the simulation of (104) and (006) planes cut from Al_2O_3 lattice, the plane spacing of (104) and (006) was 0.255 nm and 0.217 nm, respectively, and the angle between them was 38.2°. It indicates that the phase of the round frame region corresponded to the Al_2O_3, and so, the round frame area can be identified as Al_2O_3. Figure 5d–f depict the fast Fourier transform (FFT) patterns of zones A, B and C. As evidenced by the HRTEM image and FFT pattern, the $(006)_{Al_2O_3}$ plane was parallel to $(200)_{Al}$ with closely matched plane spacing. The analysis results indicate that a perfect interface can be formed between Al_2O_3 and Al, thus enabling Al_2O_3 to serve as the heterogeneous nucleation site for Al. These test findings are consistent with the aforementioned theoretical analysis.

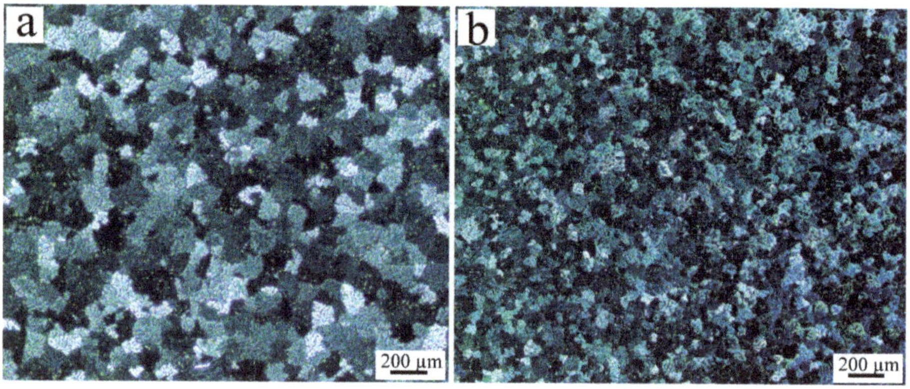

Figure 3. Metallographic image of (**a**) Al-Cu-Mn alloy, (**b**) Al-Cu-Mn alloy inoculated by Al_3Ti-Al_2O_3/Al.

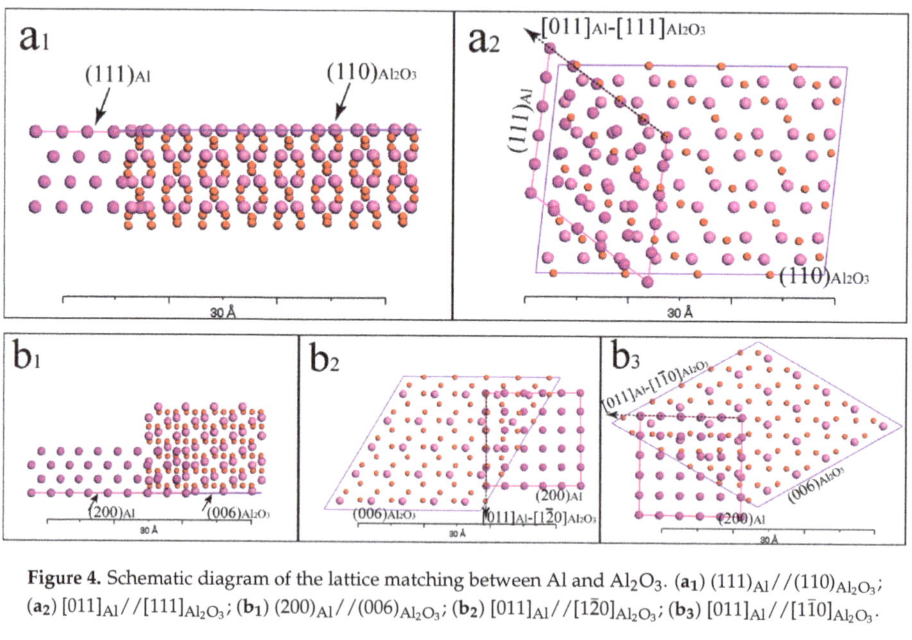

Figure 4. Schematic diagram of the lattice matching between Al and Al_2O_3. (**a$_1$**) $(111)_{Al}//(110)_{Al_2O_3}$; (**a$_2$**) $[011]_{Al}//[111]_{Al_2O_3}$; (**b$_1$**) $(200)_{Al}//(006)_{Al_2O_3}$; (**b$_2$**) $[011]_{Al}//[1\bar{2}0]_{Al_2O_3}$; (**b$_3$**) $[011]_{Al}//[1\bar{1}0]_{Al_2O_3}$.

Figure 6a illustrates the stress–strain behavior of Al-Cu-Mn alloy and Al-Cu-Mn alloy treated with the Al_3Ti-Al_2O_3/Al inoculant. The enhancement of the ultimate tensile strength and a slight increase in elongation were evidently observed in the alloy. The Al-Cu-Mn alloy exhibited an initial tensile strength of 428 MPa and an elongation of 8.2%. Upon inoculation with the Al_3Ti-Al_2O_3/Al inoculant, the ultimate tensile strength was enhanced to 492 MPa, while the elongation increased to 8.8%. According to Figure 3, the addition of Al_3Ti-Al_2O_3/Al inoculants resulted in a significant reduction in the average grain size of Al-Cu-Mn alloy, leading to an increase in grain boundaries. During plastic deformation of the alloy, the presence of grain boundaries impeded dislocation movement, thereby increasing the tensile strength of the alloy. In addition, the Al_3Ti and Al_2O_3 particles can also act as the obstacles to the dislocation motion, thus leading to higher tensile strength. The alloy could maintain good plasticity while the tensile strength was increased, on the one hand because the grain refinement made the applied stress more evenly dispersed in each grain and, thus, not making it easy to form stress concentration. On the other hand, unreacted Al_2O_3 whiskers can serve as a bridging agent in the matrix (as illustrated in Figure 6b), thereby contributing positively to the plasticity of the alloy.

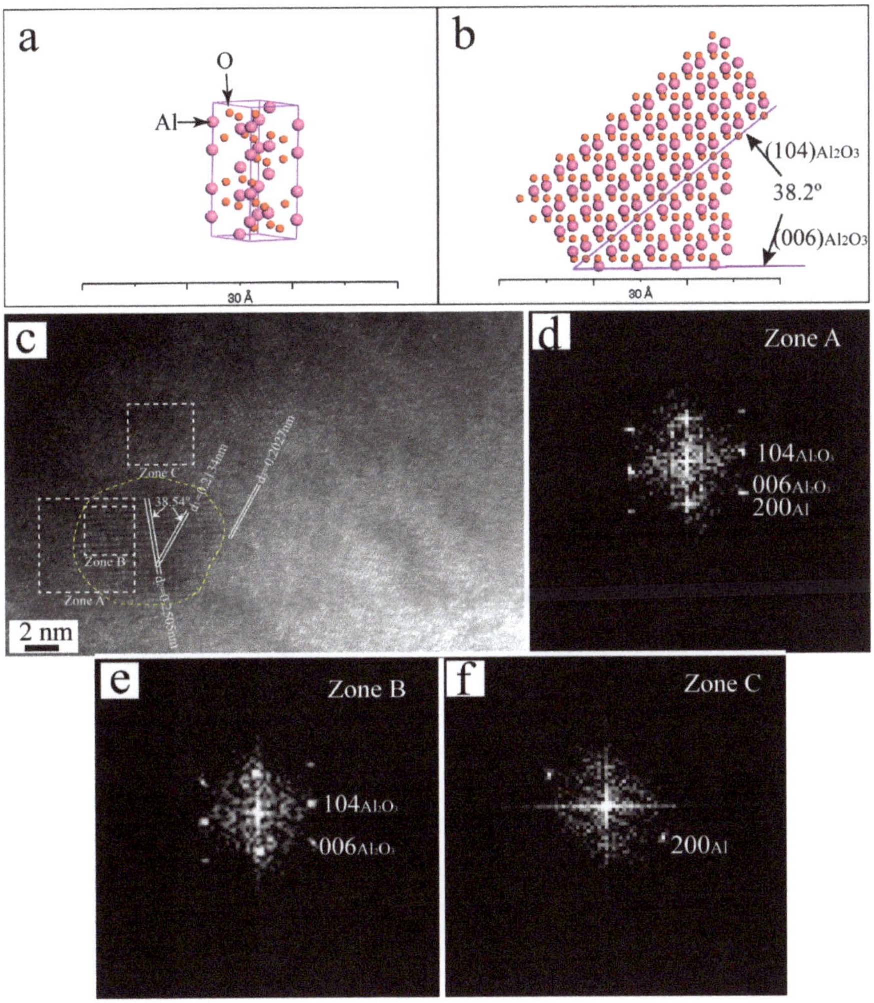

Figure 5. **(a,b)** Al_2O_3 lattice and crystal plane simulation diagrams, **(c)** HRTEM image of Al_2O_3 particle, **(d–f)** fast Fourier transform (FFT) patterns of zone A, zone B and Zone C.

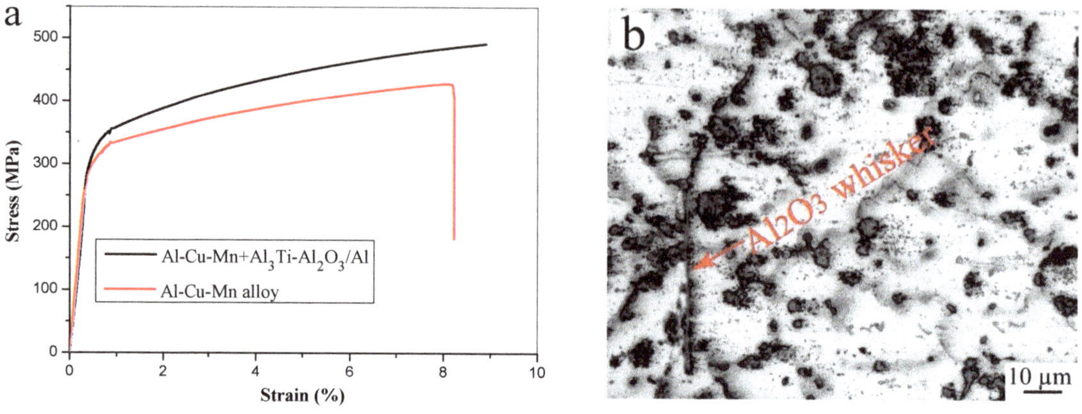

Figure 6. (**a**) Stress–strain curves of Al-Cu-Mn alloy and Al-Cu-Mn alloy + Al_3Ti-Al_2O_3/Al inoculant, (**b**) metallographic image of Al-Cu-Mn alloy inoculated by Al_3Ti-Al_2O_3/Al.

4. Conclusions

In this study, the Al_3Ti-Al_2O_3/Al inoculant was prepared with Al-Ti master alloy and Al_2O_3 whiskers as raw materials. The Al_3Ti-Al_2O_3/Al inoculants were used to inoculate the Al-Cu-Mn alloy and showed good grain refining effect. The key findings can be succinctly summarized as follows:

(1) Under the combined action of high temperature, ultrasonic and Ti element, parts of the Al_2O_3 whiskers were fragmented into granular form. After the vacuum-quenching treatment, the large sizes Al_3Ti were replaced by small size particles.

(2) There exist orientation relationships between α-Al_2O_3 particles and the aluminum matrix, whereby the former can serve as heterogeneous nucleation substrates for the latter. The Al_3Ti-Al_2O_3/Al inoculants exhibited a significant grain refining effect on the Al-Cu-Mn alloy.

(3) The tensile strength of the Al-Cu-Mn alloy was improved from 428 MPa to 492 MPa and the elongation was enhanced from 8.2% to 8.8% after being inoculated by the Al_3Ti-Al_2O_3/Al inoculants.

Author Contributions: Conceptualization, J.D.; methodology, C.L.; software, J.D.; validation, J.D., N.G. and X.G.; formal analysis, N.G.; investigation, C.W.; resources, G.Z.; data curation, J.D.; writing—original draft preparation, C.W. and X.W.; writing—review and editing, J.D. and G.Z.; visualization, X.W.; supervision, C.C.; project administration, G.Z. and C.C; funding acquisition, C.L. All authors have read and agreed to the published version of the manuscript.

Funding: This research was funded by the Natural Science Foundation of Shandong Province, grant number ZR2021QE190 and No. ZR2021QE178.

Institutional Review Board Statement: Not applicable.

Informed Consent Statement: Not applicable.

Data Availability Statement: Any further detailed data may be obtained from the authors upon a reasonable request.

Conflicts of Interest: The authors declare no conflict of interest.

References

1. Zhang, L.L.; Jiang, H.X.; He, J.; Zhao, J.Z. Improved grain refinement in aluminium alloys by re-precipitated TiB_2 particles. *Mater. Lett.* **2022**, *312*, 131657. [CrossRef]
2. Li, J.H.; Hage, F.S.; Ramasse, Q.M.; Schumacher, P. The nucleation sequence of α-Al on TiB_2 particles in Al-Cu alloys. *Acta Mater.* **2021**, *206*, 116652. [CrossRef]

3. Huang, B.; Liu, Y.Z.; Zhou, Z.G.; Cheng, W.; Liu, X.H. Selective laser melting of 7075 aluminum alloy inoculated by Al–Ti–B: Grain refinement and superior mechanical properties. *Vacuum* **2022**, *200*, 111030. [CrossRef]
4. Liu, X.H.; Liu, Y.Z.; Zhou, Z.G.; Wang, K.D.; Zhan, Q.K.; Xiao, X.J. Grain refinement and crack inhibition of selective laser melted AA2024 aluminum alloy via inoculation with TiC–TiH_2. *Mater. Sci. Eng. A* **2021**, *813*, 141171. [CrossRef]
5. Guo, Y.W.; Wei, W.; Shi, W.; Zhang, B.; Zhou, X.R.; Wen, S.P.; Wu, X.L.; Gao, K.Y.; Rong, L.; Huang, H.; et al. Effect of Er and Zr additions and aging treatment on grain refinement of aluminum alloy fabricated by laser powder bed fusion. *J. Alloys Compd.* **2022**, *912*, 165237. [CrossRef]
6. Sigworth, G.K.; Kuhn, T.A. Grain Refinement of Aluminum Casting Alloys. *Int. J. Metalcast.* **2007**, *1*, 31–40. [CrossRef]
7. Samuel, F.H.; Champier, G. Grain Refinement in Aluminum Alloys by Rapid Solidification. *Mater. Sci. Forum* **1992**, *94–96*, 747–752. [CrossRef]
8. Ghadimi, H.; Hossein, S.; Nedjhda, B. Eghbali, Enhanced grain refinement of cast aluminum alloy by thermal and mechanical treatment of Al-5Ti-B master alloy. *Trans. Nonferrous Met. Soc.* **2013**, *23*, 1563–1569. [CrossRef]
9. Qiu, K.; Wang, R.C.; Peng, C.Q.; Wang, N.G.; Cai, Z.Y.; Zhang, C. Effects of Mn and Sn on microstructure of Al-7Si-Mg alloy modified by Sr and Al-5Ti-B. *Trans. Nonferrous Met. Soc.* **2015**, *25*, 3546–3552. [CrossRef]
10. Kai, X.Z.; Tian, K.L.; Wang, C.M.; Jiao, L.; Chen, G.; Zhao, Y.T. Effects of ultrasonic vibration on the microstructure and tensile properties of the nano ZrB_2/2024Al composites synthesized by direct melt reaction. *J. Alloys Compd.* **2016**, *668*, 121–127. [CrossRef]
11. Guo, Y.W.; Wei, W.; Shi, W.; Xue, D.; Zhou, X.R.; Wen, S.P.; Wu, X.L.; Gao, K.Y.; Huang, H.; Nie, Z.R. Selective laser melting of Er modified AlSi7Mg alloy: Effect of processing parameters on forming quality, microstructure and mechanical properties. *Mater. Sci. Eng. A* **2022**, *842*, 143085. [CrossRef]
12. Zhang, X.; Xiao, Z.; Yu, W.; Chua, C.K.; Zhu, L.; Wang, Z.; Xue, P.; Tan, S.; Wu, Y.; Zheng, H. Influence of erbium addition on the defects of selective laser-melted 7075 aluminium alloy. *Virtual Phys. Prototyp.* **2022**, *17*, 406–418. [CrossRef]
13. Geng, Y.; Jia, C.; Xu, J.; Zhang, Z.; Ju, H.; Wang, D.; Yu, L. Selective laser melting of a novel high-strength Er- and Zr-modified Al-Mn-Mg alloy. *Mater. Lett.* **2022**, *313*, 131762. [CrossRef]
14. Wang, Z.; Lin, X.; Kang, N.; Chen, J.; Tang, Y.; Tan, H.; Yu, X.; Yang, H.; Huang, W. Directed energy deposition additive manufacturing of a Sc/Zr-modified Al-Mg alloy: Effect of thermal history on microstructural evolution and mechanical properties. *Mater. Sci. Eng. A* **2021**, *802*, 140606. [CrossRef]
15. Dong, Y.; Wang, M.J.; Zhang, G.W.; Xu, H. Influence of Ti/C mass ratio on the microstructure of Al-Ti-C master alloy and refinement effect on pure aluminum. *Results Phys.* **2021**, *23*, 104000. [CrossRef]
16. Zhang, M.X.; Kelly, P.M.; Easton, M.A.; Taylor, J.A. Crystallographic study of grain refinement in aluminum alloys using the edge-to-edge matching model. *Acta Mater.* **2005**, *53*, 1427–1438. [CrossRef]
17. Kai, X.Z.; Wang, Y.H.; Chen, R.K.; Peng, Y.J.; Shi, A.J.; Tao, R.; Liang, X.F.; Li, G.R.; Chen, G.; Xu, X.J.; et al. Effects of in-situ ZrB_2 nanoparticles and scandium on microstructure and mechanical property of 7N01 aluminum alloy. *J. Rare Earths* **2023**. [CrossRef]
18. Liu, T.; Hu, M.L.; Li, X.J.; Piao, J.J. Synthesis of Al-5Ti-1B-1Ce alloy from remelted chips and its refinement effect. *Mater. Lett.* **2023**, *349*, 134798. [CrossRef]
19. Liao, H.B.; Mo, L.L.; Li, C.B.; Zhan, M.Y.; Du, J. Grain refinement of Mg-Al binary alloys inoculated by in-situ oxidation. *Trans. Nonferrous Met. Soc.* **2022**, *32*, 3212–3221. [CrossRef]
20. Wang, F.; Qiu, D.; Liu, Z.L.; Taylor, J.A.; Easton, M.A.; Zhang, X.M. The grain refinement mechanism of cast aluminium by zirconium. *Acta Mater.* **2013**, *61*, 5636–5645. [CrossRef]
21. Limmaneevichitr, C.; Eidhed, W. Fading mechanism of grain refinement of aluminium-silicon alloy with Al-Ti-B grain refiners. *Mater. Sci. Eng. A* **2003**, *349*, 197–206. [CrossRef]
22. Qiu, D.; Taylor, J.A.; Zhang, M.X.; Kelly, P.M. A mechanism for the poisoning effect of silicon on the grain refinement of Al-Si alloys. *Acta Mater.* **2007**, *55*, 1447–1456. [CrossRef]
23. Ding, H.M.; Liu, X.F.; Yu, L.N. Influence of zirconium on grain refining efficiency of Al-Ti-C master alloys. *J. Mater. Sci.* **2007**, *42*, 9817–9821. [CrossRef]
24. Lee, Y.C.; Dahle, A.K.; StJohn, D.H.; Hutt, J.E.C. The effect of grain refinement and silicon content on grain formation in hypoeutectic Al-Si alloys. *Mater. Sci. Eng. A* **1999**, *259*, 43–52. [CrossRef]
25. Zaid, A.I.O.; Al-Qawabah, S.M.A. Effect of Zirconium Addition on the Grain Size and Mechanical Behavior of Aluminum Grain Refined by Titanium Plus Boron (Ti+B) in the as Cast and Cold Extruded Conditions. *Key Eng. Mater.* **2012**, *510–511*, 241–247. [CrossRef]
26. Johnsson, M. Influence of Zirconium on the grain refinement of Aluminum. *Z. Met.* **1994**, *85*, 786–789.
27. Hamid, A.; Ahmend, A. Effect of other element on the grain refinement of aluminum by titanium or titanium and boron. Part 2. Effect of the refractory metals vanadium, molybdenum, zirconium, tantalum. *Z. Fuer Met.* **1989**, *80*, 643–647.
28. Wang, Z.J.; Si, N.C. Synthesis and Refinement Performance of the Novel Al-Ti-B-RE Master Alloy Grain Refiner. *Rare Met. Mater. Eng.* **2015**, *44*, 2970–2975.
29. Wang, K.; Cui, C.X.; Wang, Q.; Zhao, L.C.; Hu, Y. Microstructure of Al-5Ti-1B-1RE nanoribbon and its refining efficiency on as-cast A356 alloys. *J. Rare Earths* **2013**, *31*, 313–320. [CrossRef]
30. Li, P.T.; Ma, X.G.; Li, Y.G.; Nie, J.F.; Liu, X.F. Effects of trace C addition on the microstructure and refining efficiency of Al-Ti-B master alloy. *J. Alloys Compd.* **2010**, *503*, 286–290. [CrossRef]
31. Svendsen, L.; Jarfor, A. Al-Ti-C phase diagram. *Mater. Sci. Technol.* **1993**, *9*, 948–957. [CrossRef]

32. Kearns, M.A.; Cooper, P.S. Effects of Solutes on Grain Refinement of Selected Wrought Aluminum Alloys. *Mater. Sci. Technol.* **1997**, *13*, 650–654. [CrossRef]
33. Zhao, H.L.; Song, Y.; Li, M.; Guan, S. Grain refining efficiency and microstructure of Al-Ti-C-RE master alloy. *J. Alloys Compd.* **2010**, *508*, 206–211. [CrossRef]
34. Ma, T.F.; Chen, Z.Y.; Nie, Z.R.; Huang, H. Microstructure of Al-Ti-B-Er refiner and its grain refining performance. *J. Rare Earths* **2013**, *31*, 622–627. [CrossRef]
35. Xu, C.X.; Liang, L.P.; Lu, B.F.; Zhang, J.S.; Liang, W. Effect of La on microstructure and grain-refining performance of Al-Ti-C grain refiner. *J. Rare Earths* **2006**, *24*, 596–601. [CrossRef]
36. Lai, J.; Zhang, Z.; Chen, X.G. Effect of Sc, Zr, and Ti on the interfacial reactions of the B_4C/Al system. *J. Mater. Sci.* **2010**, *46*, 451–459. [CrossRef]
37. Cong, X.U.; Xiao, W.L.; Zhao, W.T.; Wang, W.; Shuji, H.; Hiroshi, Y.; Ma, C. Microstructure and formation mechanism of grain-refining particles in Al-Ti-C-RE grain refiners. *J. Rare Earths* **2015**, *33*, 553–561.
38. Corrochano, J.; Cerecedo, C.; Valcárcel, V.; Lieblich, M.; Guitián, F. Whiskers of Al_2O_3 as reinforcement of a powder metallurgical 6061 aluminium matrix composite. *Mater. Lett.* **2008**, *62*, 103–105. [CrossRef]
39. Shen, Y.F.; Zou, Z.G.; Xiao, Z.G.; Liu, K.; Long, F.; Wu, Y. Properties and electronic structures of titanium aluminides–alumina composites from in-situ SHS process. *Mater. Sci. Eng. A* **2011**, *528*, 2100–2105. [CrossRef]
40. Ana, A.; Maria, J.A.; Rosalıa, N. Alumina Whisker Selection to Design Fiber-Reinforced Polyethylene Composites. *Polym.-Plast. Technol.* **2014**, *53*, 1119–1130.
41. Sumalatha, C.; Chandra Sekhar Rao, P.V.; Subba Rao, V.V. Synthesis of nickel coated graphene and analysis of the efficacy of its reinforcement on the hyper eutectic AlSi alloys that are grain modified and refined. *Diam. Relat. Mater.* **2022**, *130*, 109519. [CrossRef]
42. Carmen, C.; Victor, V.; Monica, G.; Francisco, G. Production of Chromium-Doped a-Al_2O_3 Whiskers Via Vapor Liquid Solid Deposition. *J. Am. Ceram. Soc.* **2006**, *89*, 323–327.
43. Cui, S.; Cui, C.X.; Lv, J.; Chen, S.Y.; Xie, J.Q.; Liu, S.J. Fabrication, microstructure and mechanical properties of Al_2O_3 whiskers reinforced Ti-46Al-4Nb alloy. *Mater. Lett.* **2020**, *259*, 126902. [CrossRef]
44. Sabuan, N.A.; Asmelash, M.; Azhari, A.; Mulubrhan, F.; Alemu, T. Investigation on the electrical conductivity of Al_2O_3-Ti ceramic composites using a pressureless sintering process. *Mater. Today Proc.* **2021**, *46*, 1718–1723. [CrossRef]
45. Shi, S.F.; Cho, S.H.; Goto, T.; Sekino, T. The effects of sintering temperature on mechanical and electrical properties of Al_2O_3/Ti composites. *Mater. Today Commun.* **2020**, *25*, 101522. [CrossRef]

materials

MDPI

Article

Design Optimization and Mechanical Properties of SiC Particle Reinforced Ti-Based Metallic Glass Matrix Composite

Huawei Liu [1], Jing Li [1], Jingyao Zhang [1], Pan Gong [2], Wei Yang [1], Lichen Zhao [1] and Xin Wang [1,*]

1 Hebei Key Laboratory of New Functional Materials, School of Material Science and Engineering, Hebei University of Technology, No. 5340, Xiping Road, Beichen District, Tianjin 300401, China

2 State Key Laboratory of Materials Processing and Die & Mould Technology, School of Materials Science and Engineering, Huazhong University of Science and Technology, Wuhan 430074, China

* Correspondence: ahaxin@hebut.edu.cn; Tel.: +86-022-6020-4125

Abstract: Ti-based bulk metallic glass (BMG) alloys have attracted widespread attention due to their strong glass forming ability, high specific strength, and good corrosion resistance. However, the poor plasticity of BMGs limits their further application in the aerospace and aircraft fields, as well as others. We optimized the composition of SiC-reinforced, Ti-based metallic glass matrix composites (MGMCs) through finite element modeling (FEM). FEM of MGMCs containing irregularly shaped SiC particles with different contents was conducted. Stress and strain analyses were conducted to evaluate the effect of the particle volume fraction on the mechanical behavior of MGMCs, and an optimization value of 30% was obtained, which is conducive to plasticity improvement. Arc melting copper mold injection casting was used to verify the optimized SiC content. The results show that the electroless nickel plating treatment effectively improves the wettability between SiC particles and the amorphous matrix, enabling the successful preparation of SiC/MGMC with a volume fraction of 29.5% through traditional injection casting. The volume fraction of SiC plays a crucial role in the transition of fracture mode from splitting to shear in MGMCs. After adding lightweight SiC particles, the yield strength, plasticity, modulus, and specific strength were improved by 25%, 1471%, 46%, and 33%, indicating that the use of nickel-plated SiC particles in MGMCs is an effective strengthening and toughening method for BMGs.

Keywords: metallic glass matrix composite; titanium; silicon carbide; finite element simulation; mechanical properties

1. Introduction

Titanium is an abundant element in reserves on Earth, and is considered one of the most important material resources required for the development of modern technology. Benefiting from the essential properties of Ti atoms, Ti-based alloys and composites generally have high specific strength, high specific modulus, wear resistance, corrosion resistance, and good high-temperature oxidation resistance, making them widely used in many fields such as in the aerospace field, chemical engineering, biomedical materials, and in constructing ships [1–3]. Recently, with the development of glass-forming technology and theory, it was found that some titanium alloys can be prepared into bulk metallic glasses (BMGs) at lower cooling rates, which can further enhance their specific strength [4]. Benefiting from their disordered atomic arrangement, BMGs usually have a highly uniform chemical composition and isotropy, and lack the typical crystallographic defects such as dislocations and grain boundaries, generally resulting in a higher strength, corrosion resistance, and wear resistance than crystalline alloys with the same composition [5,6]. It should be emphasized that BMG combines two important properties, high strength, and high elasticity, with an elastic limit that is several to several tens of times that of a typical crystalline alloy, and an elastic strain of up to ~2% [7,8]. However, due to the difficulty in extending the plastic behavior that originates from the local range (limited to the shear band) to the macroscopic overall range, BMGs typically lack the ability of macroscopic room temperature plastic deformation and exhibit significant brittleness, which greatly limits their widespread application as structural materials [9,10].

1.1. Progress in SiC Particle Reinforced MGMCs

In order to improve the macroscopic plasticity of BMGs, one of the most effective strategies is to prepare metallic glass matrix composites (MGMCs) by introducing a crystalline second phase as a reinforcement system to promote the shear band nucleation and suppress the shear band propagation [11–13]. The reinforcement of MGMCs is generally designed as a crystalline second phase, including in situ-precipitated crystalline phases and ex situ-added classical reinforcements such as carbon fibers [14–17], W fibers [18–22], ceramic particles [23–25], etc. As a common reinforcing particle in crystalline matrix composites, silicon carbide (SiC) particles are also used in MGMCs through different preparation methods and metallic glass (MG) matrices.

In order to obtain the macroscopic properties of composite materials, various preparation processes have been used for the preparation of SiC-reinforced MGMCs. Li et al., used induction melting and copper mold injection casting to prepare SiC particle-reinforced, Mg-based MGMCs and found that cracks can penetrate, terminate, or bypass SiC particles when the Mg-based MGMC fractures, indicating that the presence of SiC particles is effective in preventing rapid crack propagation [26]. Lin et al., prepared SiC particle-reinforced $Ti_{50}Cu_{28}Ni_{15}Sn_7$ MGMC through mechanical alloying and hot-pressing sintering methods, and they found that partial crystallization of the glassy matrix occurred during the hot-pressing process [27]. It was also found that the addition of SiC particles improved the hardness and the fracture strength of the composite, but the sharp boundaries of the SiC particles led to brittle failure of the composite material [27,28]. Xie et al., prepared SiC particle-reinforced $Ni_{52.5}Nb_{10}Zr_{15}Ti_{15}Pt_{7.5}$ MGMC through a spark plasma sintering (SPS) process and found that no crystallization of the matrix glass alloy occurred during the SPS process, and no interfacial reaction layer was observed [29]. They also prepared Fe-based MGMC without interfacial reaction using the SPS process and found that the addition of SiC particles is beneficial for improving high-frequency soft magnetic properties [30]. Yue et al., prepared a SiC particle-reinforced $Zr_{65}Al_{7.5}Ni_{10}Cu_{17.5}$ MGMC coating using laser cladding technology. It was found that an interface reaction layer containing ZrC was formed between the SiC particles and the glass phase matrix, but there was no significant negative impact on thermal stability, wear resistance, or corrosion resistance [31]. This interface reaction introduces ZrC and SiC phases into the MG matrix, which can further improve the hardness and wear resistance of the coating through the presence of a large number of ceramic phases [32]. Chen et al., prepared porous SiC-reinforced, zirconium-based MGMC using pressure infiltration and water quenching methods, and found that the SiC volume fraction had a significant impact on the deformation and fracture behavior [33,34]. When the SiC volume fraction increased to a certain extent, the compression fracture mode of the composite material changed from a shear mode to an axial splitting mode [33]. Wang et al., found that cracks in porous SiC/Ti-based MGMCs typically originate at the interface between SiC and MG matrix or within SiC particles [35]. During dynamic compression, the initiation and propagation of cracks are faster, resulting in the dynamic compressive fracture strength of the MGMC material being much lower than that of the quasi-static compression [35]. They also found that the deformation behavior under dynamic loading is more severe than static, and that the dynamic hardness of composite materials is significantly higher than their static hardness [36].

From the above document review, it can be seen that one of the challenges for MGMCs containing SiC is the control of interface reactions. Obviously, using low heating temperature processing technology (i.e., SPS) and appropriate matrix materials (i.e., Mg-based or Ni-based) is beneficial for avoiding interfacial reactions. Zhang et al., studied the wetting behavior between Zr-based alloy melt and SiC, and found that, although the wettability increased with an increasing temperature, interfacial reactions occurred at the Zr-based alloy/SiC interface [37]. They also found that the interface reaction between Zr-based MG matrix and SiC depends on the content of Zr element [38]. In our early work, we used electroless copper plating treatment on SiC particles to improve the interfacial wettability and plastic deformation ability of the composite material [39]. However, due to the large brittleness of the Mg-based MG alloy matrix, the addition of SiC particles did not cause significant macroscopic plasticity [39].

1.2. Application of Finite Element Analysis in Particle Reinforced MGMCs

The deformation behavior of MGMCs is limited by complex stress states, making it difficult to conduct quantitative analysis and direct experimental observation. Therefore, finite element numerical analysis based on computer simulation is widely used. Finite element modeling (FEM) can not only be used to analyze the stress and strain changes in BMG during deformation [40–44], but also to analyze the interfacial stress conduction in MGMCs.

The spherical second phase model was first used to analyze the stress and strain states of MGMCs under different external forces. Zhu et al., established a two-dimensional finite element model to analyze the deformation behavior of spherical Ta particle-reinforced, Zr-based MGMC. They believe that the presence of Ta particles can cause the maximum shear stress direction to deviate from the 45° plane towards the nearest particle, so the shear band can be diffused between particles [45]. Ding et al., obtained similar results in the FEA of a high-entropy, alloy-reinforced, and Zr-based MGMC [46]. They proposed that the strain hardening of particles played an important role in the bifurcation of shear bands. Ott et al., built a two-dimensional plane-strain FEM. Due to the lateral constraint imposed by the elastic matrix on particle deformation, the Ta particles did not undergo significant hardening. The nucleation of shear bands is not caused by differences in elastic properties between the matrix and particles, but by stress concentration around the particle that is induced by mismatched plastic strains [47]. They believe that, for small volume-fraction MGMCs, the presence of particles has a significant impact on the yield of MGMCs, but has almost no effect on the fracture behavior, which largely depends on the properties and behavior of the MG matrix [48]. In another FEM study of porous tungsten/Zr-based MGMC, the plastic mismatch strain of the two phases was also considered the main cause of fracture, which resulted in large stress concentration near the interface, leading to the failure of the tungsten phase or the separation of the two phases at the interface [49].

FEM analysis method has been used to analyze the influence of particle shape on the deformation behavior of MGMCs. Jiang et al., conducted FEM analysis on the deformation behavior of particle-reinforced MGMC based on a two-dimensional, eight-node plane strain element, and found that the back stress in crystal particles affects the evolution of shear bands [50]. They pointed out that the work hardening ability of particles has a significant impact on both the motion strain hardening and the strength of MGMC [50]. They also studied the influence of particle strength, shape, and volume fraction on the deformation behavior of MGMCs, and found that improving the aspect ratio and the dispersion randomness of particles would be beneficial to inhibiting the formation of main shear bands and improving the ductility of MGMCs [51,52]. Jia et al., used the FEM method to investigate the influence of different crystal geometries, including cubes/spheres/dendrites, on the deformation behavior of MGMCs [53]. They found that a complex geometric shape and favorable dendritic orientation (45°) introduce more stress concentration, thereby improving ductility. The volume fraction of the second phase should be higher than the percolation limit to block and deflect the propagation of shear bands. Bian et al., established a two-dimensional micromechanical FEM for elliptical Ta particle-reinforced MGMCs and found that, the larger the angle between the axis of charge and particle orientation, the better the toughness of MGMCs. The shape of the particles (respect ratio) has a significant impact on the plasticity improvement of MGMCs, but plays a small role in the ultimate strength [54].

To sum up, the FEM study of particle-reinforced MGMCs is mainly based on the second phase with simple shapes such as spherical, elliptical, and fibrous shapes, which do not match the actual three-dimensional morphology of SiC particles. In this work, we used linear polygonal particles as the basis for FEM analysis to study the effect of SiC volume fraction on the deformation behavior of Ti-based MGMCs, optimizing the design of an appropriate SiC volume fraction. We also used the copper mold injection casting method to prepare the optimized SiC particle-reinforced Ti-based MGMC. The mechanical properties and the deformation mechanism were investigated for validation.

2. Materials and Methods

2.1. FEM Analysis

The typical ANSYS Workbench software was used to establish a two-dimensional finite element model, as shown in Figure 1. The necessary material parameters required for the calculation of FEM simulation are shown in Table 1. The height of the sample is 7 mm, the width is 3.5 mm, and the shape of the particles is a random linear quadrilateral with a size of approximately 0.5 mm. In order to ensure that the computer used in this work would not crash, the size of SiC particles was set to about 10 times the actual nominal size. We estimated the volume fraction of particles in the sample by calculating the ratio of total particle area to sample area, and drew two-dimensional models with volume fractions of 10%, 20%, 30%, and 40%, respectively. The grid division mode adopts automatic random allocation of four points, with a size of approximately 0.1 mm. This size is the minimum value obtained through repeated testing to ensure that the computer would not crash (as the grid size has a significant impact on computational accuracy). The upper and lower surfaces of the model were set as fixed constraints (nondeformable). To observe the stress state inside the composite material sample during the loading changes from small to large enough, a displacement of 0.2–5% was applied from top to bottom to obtain the equivalent stress cloud map and equivalent plastic strain cloud map of the sample. Here, 0.2% has no special meaning, it is a small value we randomly selected; 5% is a value selected to ensure the fracture of the composite. In addition, it should be noted that we improved the wettability of particles and titanium alloy melt by electroless nickel plating, where the nickel coating melts and disappears. Therefore, no coating was established on the surface of SiC particles in the FEM model.

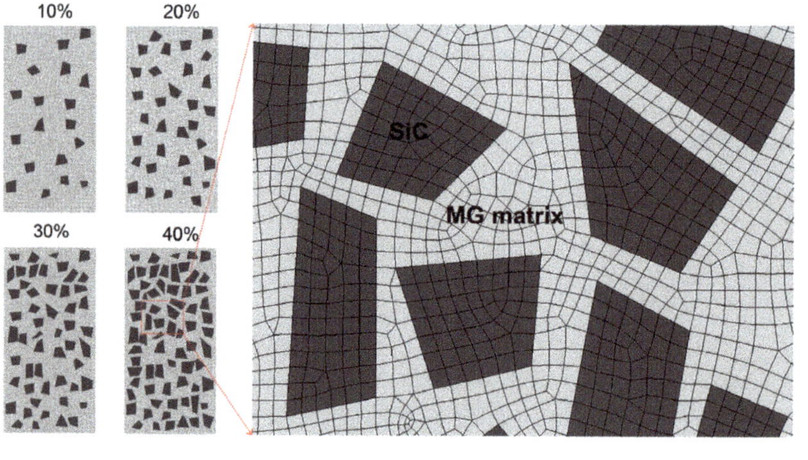

Figure 1. FEM models of Ti-based MGMC reinforced with different volume fractions of SiC particles, with an enlarged image showing the details of mesh division.

Table 1. Property data of the matrix and SiC reinforcement phase used for designing MGMC during ANASYS simulation.

Materials	Properties		Reference
MG matrix	density	5.36 g·cm^{-3}	[55]
	Compression strength	1955 MPa	This work
	Young's modulus	100 GPa	[56]
	Poisson's ratio	0.35	[56]
SiC particle	density	3.2 g·cm^{-3}	[57]
	Compression strength	15 GPa	[58]
	Young's modulus	500 GPa	[57]
	Poisson's ratio	0.2	[59]

2.2. Fabrication of MGMC

Commercial SiC particles (Guangzhou Hongwu Material Technology Co., Ltd., Guangzhou, China) with 600 mesh were used as the reinforcement of MGMC, and the particle size was approximately 10–50 μm according to the commodity specification. In order to improve the interfacial wettability, SiC particles were pretreated by common electroless nickel plating. The plating process was as follows: (1) soak in acetone for 1 h and ultrasonic cleaning for 15 min to remove organic pollutants from the surface of SiC particles; (2) soak the dried SiC particles in concentrated HF acid for 1 h to increase the surface roughness of the particles through corrosion for roughening treatment; (3) soak in HCl solution of $SnCl_2$ for 30 min for sensitization, and activate the particles with HCl solution of $PdCl_2$ after cleaning, the treated SiC particles must be reduced with sodium hypophosphite solution after activation; (4) carry out nickel plating treatment using nickel plating solutions prepared with $NiSO_4 \cdot 6H_2O$, $NaC_6H_5O_7 \cdot 2H_2O$, $NaH_2PO_2 \cdot H_2O$, CH_3COONa, NH_4Cl, and NaOH at a plating temperature of 60 °C.

Fragments of Ti, Zr, Be, Fe, and Cu with a purity of more than 99.9% were used as raw materials. Analytical balance with an accuracy of 0.0001 g (Mettler, ME204E, Columbus, OH, USA) was used to weigh the alloy ingredient of $(Ti_{41}Zr_{25}Be_{28}Fe_6)_{98}Cu_2$ [55]. Under the protection of high-purity argon gas, the MG master alloy ingot for matrix alloy of MGMC was prepared using a vacuum arc furnace (Shanghai MTINST, DHL-600, Shanghai, China). In order to ensure the uniformity of the composition, special attention should be paid to the following aspects in the experimental operation: firstly, ensure that the vacuum degree reaches 1×10^{-3} Pa or above to reduce the oxygen content in the furnace body; secondly, high-purity Ar with a purity of 99.9999% should be used to avoid introducing impurity gas molecules; thirdly, it is necessary to frequently replace pure titanium ingots used for oxygen absorption to avoid weakening their oxygen absorption effect; fourthly, during the re-melting process, the ingot needs to be flipped up and down 5 times to avoid gravity segregation; fifthly, the weight of the obtained ingots must be carefully weighed, and ingots with a mass change of no more than 0.5% before and after melting are considered qualified samples for subsequent MGMC preparation (ingots exceeding 0.5% are considered unqualified samples and discarded). Then, the nickel-plated SiC particles and the $(Ti_{41}Zr_{25}Be_{28}Fe_6)_{98}Cu_2$ ingot fragments were placed in the vacuum arc furnace (Shanghai MTINST, DHL-600, Shanghai, China) to prepare the SiC/$(Ti_{41}Zr_{25}Be_{28}Fe_6)_{98}Cu_2$ composite master alloy material. Finally, a round rod-shaped specimen with a diameter of 2 mm and a length of about 70 mm was prepared using the copper mold injection casting method [60] by an injection casting furnace (Shanghai MTINST, VIMC-1000, Shanghai, China). The vacuum degree of the furnace chamber is -1.0×10^{-2} Pa, and the pressure of the injection-cast Ar was about 2 MPa.

2.3. Microstructural Characterization

An X-ray diffractometer (XRD, Bruker D8 Advance, Cu Kα radiation, Billerica, MA, USA) was used to examine the phase composition of SiC particle samples and MGMC rod samples with a scanning speed of 10°/min and a scanning angle range of 20–80°. A differential scanning calorimeter (DSC, Setaram Themys ONE, Cranbury, NJ, USA) was used to measure the thermal stability of the as-prepared samples under N_2 protection at a heating rate of 20 K/min. A scanning electron microscope (SEM, Hitachi S-4800, Tokyo, Japan) was used to examine the microstructure of the plated SiC particles and the rod MGMC samples. An EDS detector assembled on the S-4800 SEM was used to determine the chemical composition of the matrix of the composite material.

2.4. *Mechanical Property Testing*

A Vickers micro hardness tester (SHIMADZU, HMV-2TE, Kyoto, Japan) was used to test the micro hardness of composite materials, with a loading load of 980 mN and a loading time of 15 s. Metallographic microscope (Olympus, BX41RF, Tokyo, Japan) was used to take photos of hardness indentation. A slow diamond cutting machine (Kejing, SYJ-150, Shenyang, China) was used to cut a circular rod with a diameter of 2 mm into a cylindrical sample with a length of 4 mm for compression testing. A computer-controlled electro-hydraulic servo universal testing machine (Guanteng, PA200, Changchun, China) was used to test the compressive mechanical properties of the MGMC and BMG samples with a strain rate of 0.5 mm/min. To ensure the reliability of the data, each sample was tested at least 3 times, and the mechanical property was taken from the average value of the samples. The density of different as cast samples was tested based on the Archimedes drainage method, and the specific strength was evaluated by dividing the yield strength obtained from the compression experiment by the density.

3. Results

3.1. *Optimization of SiC Particle Volume Fraction*

Figure 2 shows the FEM analysis results of the Ti-based MGMC samples reinforced with different volume fractions (V_f) of SiC particles under a total deformation of 5%. From the equivalent stress analysis shown in Figure 2a–d, it can be seen that the maximum compressive stress in the sample occured within the SiC particles, and there is a gradual increasing trend with the increase in V_f. For the stress distribution on the matrix, when the V_f value is less than 20%, the maximum stress (green area) is limited to the area between two adjacent particles arranged vertically, indicating that yielding or fracture tends to occur along the axial direction (splitting fracture mode). When the V_f value reaches over 30%, the maximum stress range exhibits transverse crosslinking, and the higher the V_f, the stronger the transverse crosslinking. As a result, the MGMC sample with a V_f value of 40% exhibited the most stress concentration zones in the 45° direction (maximum shear stress direction), which clearly facilitated the transition of the fracture mode from the splitting mode to the shear mode.

Figure 2e–g shows the FEM analysis results for the equivalent plastic strain, where the low strain zone is essentially located within the SiC particles. This can be attributed to the fact that the Young's modulus of a SiC particle (~400–550 GPa [57,58,61,62]) is about six times that of MG matrix alloy (~100 GPa [56,63,64]), resulting in the smaller deformation of SiC particles under the same stress-bearing conditions. However, as V_f increases, the low strain region in the SiC particles gradually decreases, which can be attributed to the increasing effect of SiC V_f on the maximum stress under the same deformation amount, as shown in Figure 2a–d. In addition, when the V_f value is less than 20%, the distribution of the maximum strain zone (green and yellow areas) is consistent with the equivalent stress, and also tends to be vertically distributed. When the V_f value reaches over 30%, the maximum equivalent plastic strain zone exhibits multiple 45° distributions, which means that the plastic deformation zone in the sample can extend along the main shear direction to the entire sample. In other words, this stress distribution contributes to the formation of the multiple shear banding mode.

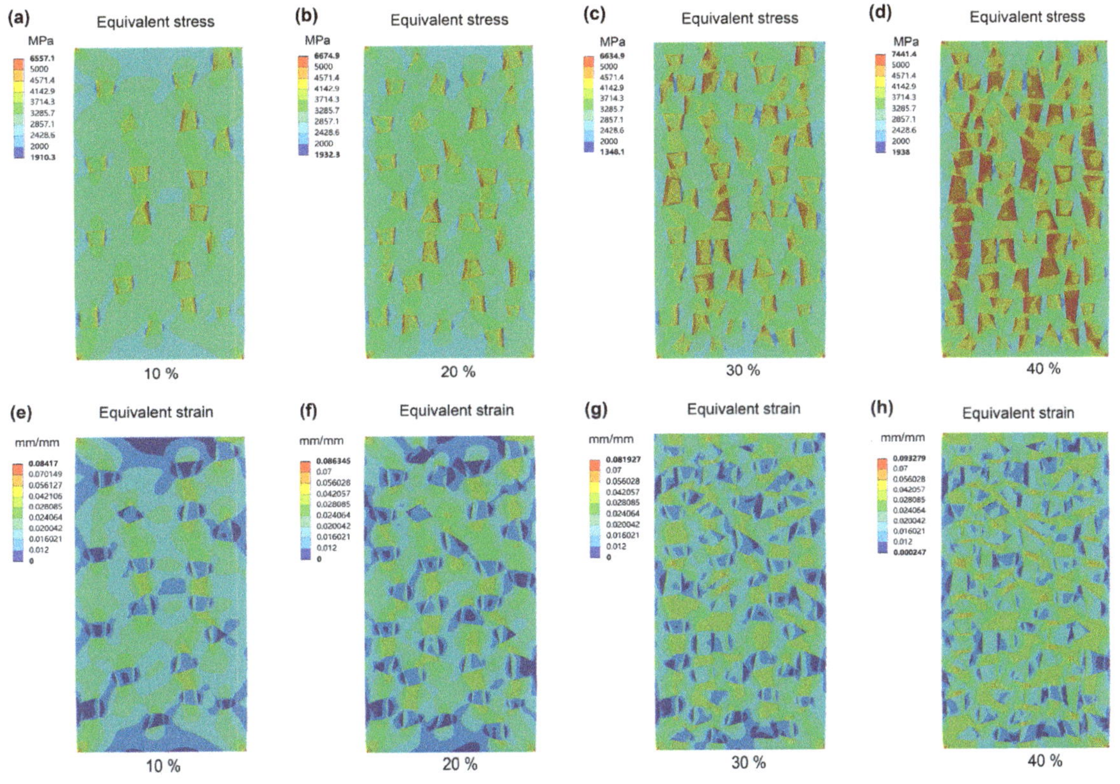

Figure 2. Contour maps of equivalent stress analysis at 5% deformation for MGMC samples with different SiC volume fractions of (**a**) 10%, (**b**) 20%, (**c**) 30%, and (**d**) 40%, and the corresponding equivalent plastic strain analysis of samples with different SiC content of (**e**), 10%; (**f**), 20%; (**g**), 30; and (**h**), 40%.

The FEM analysis results for equivalent stress and equivalent plastic strain both support that V_f is a key parameter which determines the deformation fracture mode of SiC-reinforced MGMCs. From the results in Figure 2, it can be seen that the ideal SiC V_f value for achieving good ductility should be above 30–40%. Considering that, if V_f is too large, SiC particles are prone to agglomerate and form defects during the fabrication process, which is not conducive to injection molding, the V_f should not be too large.

3.2. Fabrication and Microstructure of Composite Materials

Figure 3 shows the XRD patterns of the original SiC particles and the SiC particles after electroless Ni plating. The sharp diffraction peaks in the XRD pattern of the as-received SiC particles correspond to SiC crystals, indicating that the tested original powder sample contains a single SiC phase and its purity is high. For the XRD pattern of the Ni-coated SiC particles, in addition to the weak peaks of crystalline SiC, there is also a group of diffraction peaks that have undergone significant broadening, with peak angles consistent with those of Ni crystals. We know that the products of alkaline electroless nickel plating are P-rich, Ni-based amorphous alloy or nanocrystalline alloy, which may be the main reason for the broadening of the Ni peaks. It should be noted that the height of the Ni peak is significantly higher than the main peak of SiC, indicating that the surface of SiC particles has been successfully coated with Ni after plating. The existence of weak peaks in SiC can be attributed to the fact that the electroless Ni coating is very thin, thus, the conductive X-ray can penetrate the coating. In addition, some SiC particles were missed during electroless plating, resulting in their failure to form a surface coating and exposing a certain amount of surface to be detected.

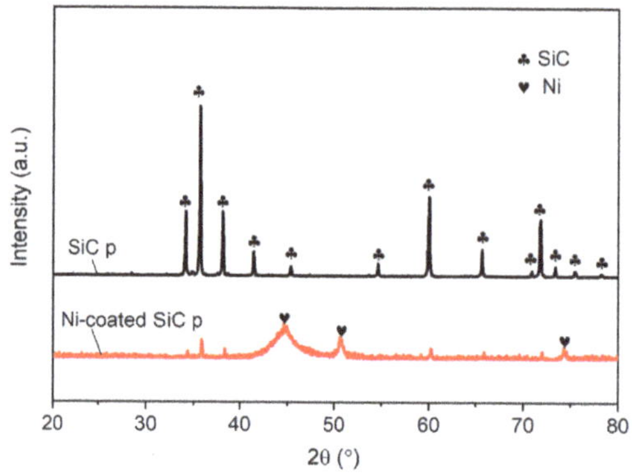

Figure 3. XRD patterns of SiC particles before and after electroless Ni plating.

Figure 4 compares the microstructures of the SiC particles before and after electroless nickel plating. Figure 4a shows the microstructure of commercial SiC, indicating that the size of these particles ranges from ~10 μm–~40 μm. Most particles were in the form of single particles, with sharp edges and smooth surfaces, as shown in the inset in Figure 4a, which is not conducive to wettability between the Ti-based MG alloy melt and the SiC particles. Meanwhile, some particles showed a status of loose clusters with rough surfaces—in our opinion, such particles may result in the formation of cast defects which would induce a low strength and Young's modulus of MGMCs due to the surface status. Therefore, it is necessary to carry out electroless nickel plating to modify the surface structure of SiC particles. Figure 4b shows the microstructure of SiC particles after electroless nickel plating. It can be seen that, after electroless nickel plating, the surface of the SiC particles is covered with a nickel coating as shown in the inset of Figure 4b. It is worth noting that, as the size of the SiC particle increases, the coverage of the Ni coating plated on the particle surface becomes better, while small particles are prone to be partially missing plating.

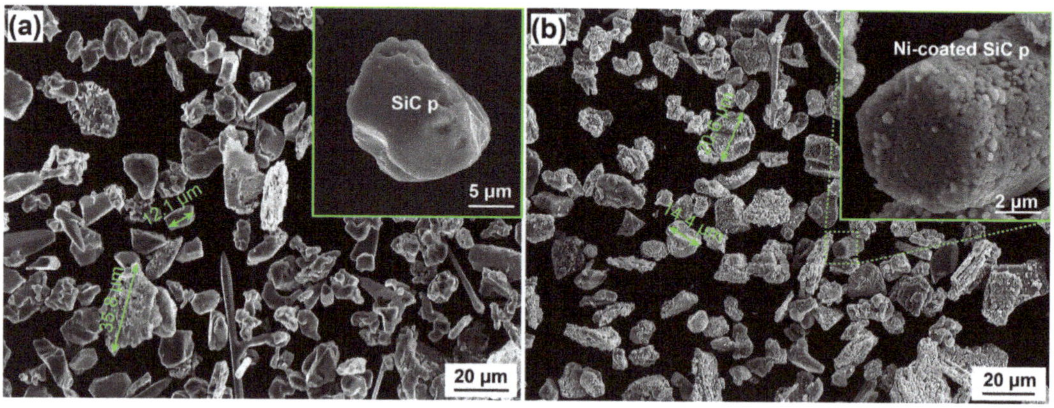

Figure 4. SEM images of SiC particle samples before electroless nickel plating (**a**) and after electroless nickel plating (**b**).

Figure 5 shows an optical photo of the MGMC rod prepared by an injection casting system. The diameter of the cast sample is 2 mm and the length is 70 mm, which is consistent with the size of the inner cavity of the copper mold used, indicating that the composite material melt could completely fill the mold. We know that solid–liquid mixtures containing a large number of solid particles typically have poor flowability, which is not conducive to the formation of castings. In this work, a mixture of $(Ti_{41}Zr_{25}Be_{28}Fe_6)_{98}Cu_2$ melt and SiC particles was injected into a copper mold using high-pressure Ar gas up to 2 MPa, which relatively weakened the flowability requirements for the mixed liquid. Under high-pressure injection casting conditions, the forming process is mainly influenced by the heating temperature (which affects viscosity and fluidity) and the content of solid SiC particles. Figure 5 shows that, when the addition amount of SiC is ~30%, its flowability meets the requirements of spray casting molding. We also attempted to achieve a 50% addition of SiC, but it is impossible to obtain cast rods for mechanical performance testing due to the poor formability. Therefore, it is not advisable to add higher volume fractions of solid particles (i.e., over 50%) for injection casting. This is consistent with the limit value calculated by the granular soil model (47%) [65], indicating that the permeation behavior of metal melt between solid particles is similar to that of water in soil.

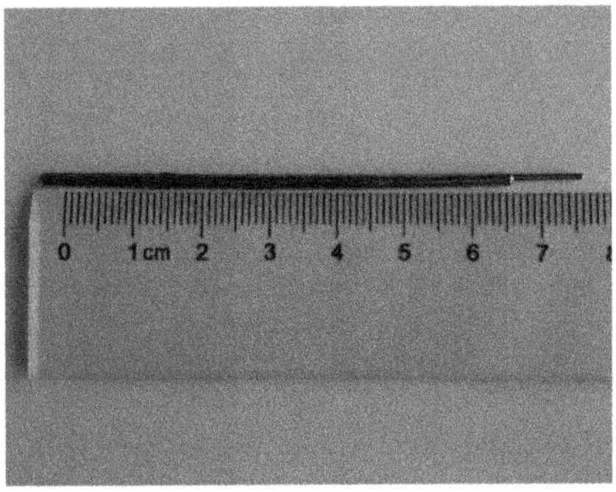

Figure 5. Typical macroscopic optical image of as-cast $SiC/(Ti_{41}Zr_{25}Be_{28}Fe_6)_{98}Cu_2$ MGMC rod.

Figure 6 shows the XRD patterns of the MGMC sample and the $(Ti_{41}Zr_{25}Be_{28}Fe_6)_{98}Cu_2$ BMG sample, comparatively. It is known that, due to the lack of long-range ordering in amorphous structures, typical XRD patterns typically exhibit diffuse peaks within a specific angle range. Obviously, both patterns in Figure 6 have this typical feature, indicating that both samples contained a certain amount of amorphous phase. For the MGMC sample, in addition to the diffuse peaks belonging to the amorphous phase, there are also some sharp diffraction peaks corresponding to the crystalline Be_5Zr and SiC, respectively. This shows that, after adding Ni-plated SiC particles, a certain amount of crystalline phase precipitates onto the amorphous matrix. It is interesting that the precipitated crystalline phase is not an intermediate phase containing Si and C, and this indicates that SiC did not undergo significant decomposition during the preparation process. Here, the formation of crystalline phases may be related to the change in matrix composition caused by the introduction of Ni, C, and Si atoms, leading to a decrease in the glass forming ability (GFA) of the matrix alloy and resulting part crystallization. In addition, by comparing the XRD patterns of the MGMC sample in Figure 6, it can be seen that the height of the main diffraction peaks of SiC is significantly higher than that of Be_5Zr, indicating that the amount of SiC is much higher and that SiC is the main second phase in the MGMC sample.

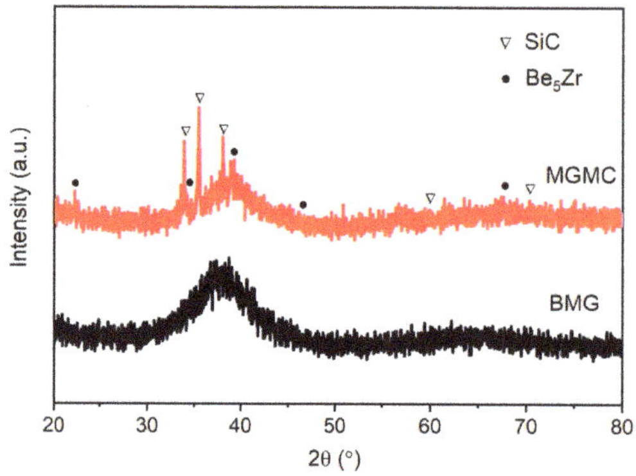

Figure 6. XRD patterns of $SiC/(Ti_{41}Zr_{25}Be_{28}Fe_6)_{98}Cu_2$ MGMC and $(Ti_{41}Zr_{25}Be_{28}Fe_6)_{98}Cu_2$ BMG sample.

To further confirm the presence of amorphous phases, the DSC curves of the MGMC sample and the BMG sample were tested, as shown in Figure 7. For BMG, a glass transition occurs when it is heated to approximately 609 K, and crystallization begins at 724 K, forming a clear exothermic peak with the peak temperature T_{p1} being 735.7 K. The first crystallization is followed by a second crystallization, forming a second exothermic crystallization peak, with a peak temperature T_{p2} of 788.9 K. This is a typical behavior of amorphous alloys during heating, which is consistent with the results in the literature [55]. For the MGMC sample, the DSC curve exhibited similar heat absorption and release behavior to the BMG sample, but the glass transition temperature T_g, the onset crystallization temperature T_x, the first crystallization peak temperature T_{p1}, and the second crystallization peak temperature T_{p2} were all lower than that of the BMG sample (as shown in Table 2). It was demonstrated that the MGMC samples contain amorphous phases, but their thermal stability was reduced compared to the BMG samples. In addition, the difference between T_g and T_x is referred to as the width of the supercooled liquid phase zone ΔT ($T_g - T_x$), which can also be regarded as an indicator of the thermal stability of the sample. The ΔT value of the MGMC sample was decreased by up to 30 K compared to the BMG sample, which also proves that its thermal stability was deteriorated. This can be attributed to the presence of SiC particles in the MGMC sample, which leads to the preferential nucleation of crystalline precipitates on the SiC surface, requiring lower energy and resulting in the formation of a lower T_x.

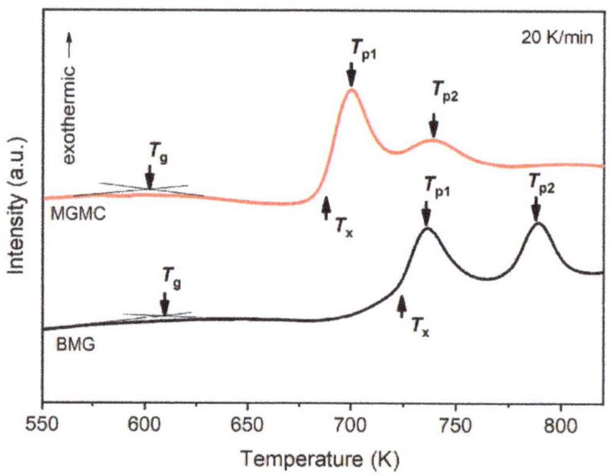

Figure 7. DSC curves of $SiC/(Ti_{41}Zr_{25}Be_{28}Fe_6)_{98}Cu_2$ MGMC and $(Ti_{41}Zr_{25}Be_{28}Fe_6)_{98}Cu_2$ BMG samples.

Table 2. Thermal properties measured from the DSC curves of SiC/$(Ti_{41}Zr_{25}Be_{28}Fe_6)_{98}Cu_2$ MGMC and $(Ti_{41}Zr_{25}Be_{28}Fe_6)_{98}Cu_2$ BMG samples.

Sample	T_g (K)	T_x (K)	T_{p1} (K)	T_{p2} (K)	ΔT (K)
BMG	609.1	724.2	735.7	788.9	115.1
MGMC	601.7	687.3	700.1	739.1	85.6

Figure 8 shows typical SEM images of $(Ti_{41}Zr_{25}Be_{28}Fe_6)_{98}Cu_2$ BMG and SiC/$(Ti_{41}Zr_{25}Be_{28}Fe_6)_{98}Cu_2$ MGMC. Figure 8a shows the microstructure of the BMG sample, which clearly shows that there are no other phases present on the gray matrix, indicating that the BMG sample prepared in this work had a full amorphous structure. This is consistent with the results in Figures 6 and 7. Figure 8b shows the SEM image of the MGMC sample. There are many dark gray irregular particles evenly distributed on the light gray matrix. From the particle morphology, it can be inferred that these particles are SiC particles. To analyze the content of the particles, we stained dark gray particles (Figure 8c) and analyzed the size distribution of the particles using Image Pro Plus 6.0 software, as shown in Figure 8d. The average size of the particles was 14.8 μm and the particle area fraction was 29.5%, indicating that the prepared MGMC sample reached the preset SiC volume fraction value (30%) according to the FEM analysis results.

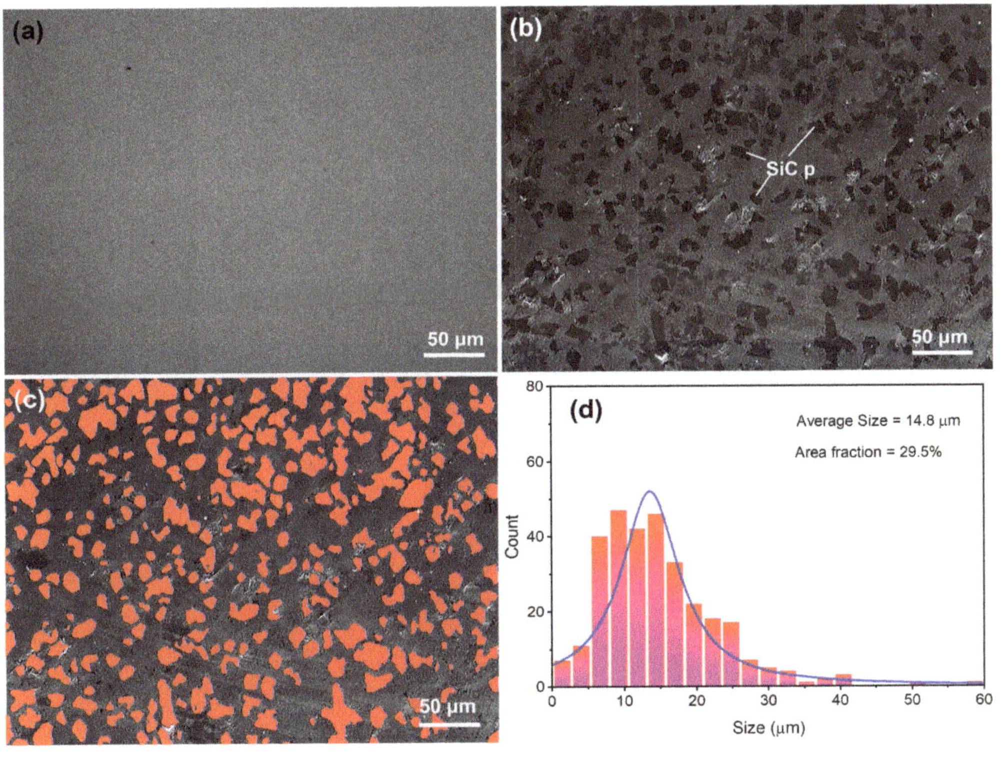

Figure 8. Typical SEM images of (**a**) $(Ti_{41}Zr_{25}Be_{28}Fe_6)_{98}Cu_2$ BMG and (**b**) SiC/$(Ti_{41}Zr_{25}Be_{28}Fe_6)_{98}Cu_2$ MGMC, (**c**) partial rendering image of (**b**), and (**d**) histogram of particle size distribution and area fraction measurement results.

Figure 9a shows a high magnification SEM image of as-cast SiC/$(Ti_{41}Zr_{25}Be_{28}Fe_6)_{98}Cu_2$ MGMC. The dark gray irregularly shaped particles are the SiC phase, and there is no obvious phenomenon of Ni coating on the surface of the SiC particles, indicating that the electroless Ni coating melted and might have entered the matrix during the preparation process of the MGMC. The continuous area between the dark gray SiC particles is the matrix of the MGMC. Under high magnification SEM, the matrix exhibited a pitted shape formed by mixing small particles of one darker color and another lighter color, indicating the presence of multiple phases in the matrix. Based on the XRD analysis results of the MGMC sample in Figure 6, we believe that the matrix phase should actually consist of two phases: the metallic glass phase and the Be_5Zr crystalline phase. In order to analyze the whereabouts of the Ni layer, the chemical composition of the matrix was determined by EDS, and the results are shown in Figure 9b. Since EDS cannot measure elements with an atomic number less than 12, the values for the contents of C and Be are inaccurate. At the same time, the presence of SiC particles below the sample testing area may have also caused some interference in the testing. Therefore, the composition analysis based on EDS should exclude the contents of C, Si, and Be elements. In this way, the content ratio of the remaining elements in the nominal composition is Ti:Zr:Fe:Cu = 55.4:33.8:8.1:2.8. For the MGMC samples, the proportion of other elements after excluding the interfering elements is approximately Ti:Zr:Fe:Cu:Ni = 52.8:29.5:8.1:4.6:5.0. It can be confirmed that Ni atoms in the SiC surface coating entered the matrix, and due to the introduction of Ni, the relative content of Ti and Zr in the matrix decreased while the Cu content increased, ultimately resulting in the matrix composition deviating from the nominal composition. This will alter the GFA and thermal stability of the matrix, which is consistent with the XRD and DSC test results shown in Figures 6 and 7.

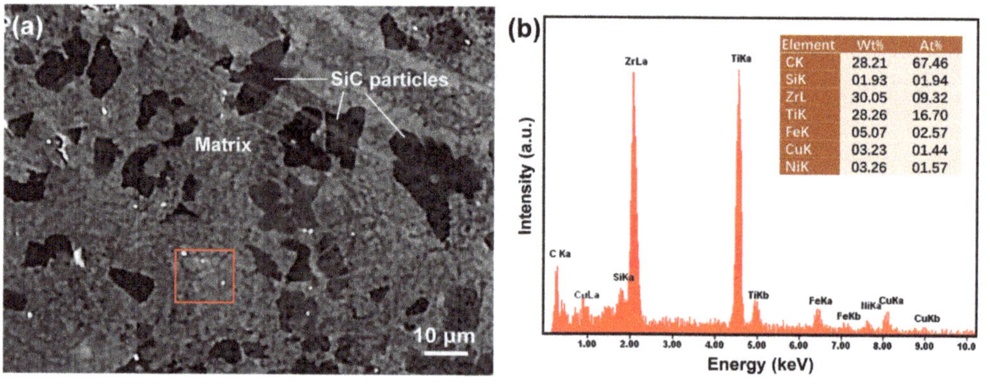

Figure 9. (**a**) High magnification SEM image of SiC/$(Ti_{41}Zr_{25}Be_{28}Fe_6)_{98}Cu_2$ MGMC and (**b**) EDS analysis results of the matrix area marked by the red box in (**a**).

3.3. Mechanical Properties

Figure 10 shows the quasi-static compression stress–strain curve of Ti-based BMG alloy and SiC/MGMC, and Table 3 shows the calculation results of the conditional yield strength $\sigma_{0.2}$, fracture strength σ_f, elastic strain ε_e, plastic strain ε_p, elastic modulus E, and specific strength σ_{sp} measured from the stress–strain curves. The compressive fracture strength σ_f of the BMG sample is 1955 MPa, and the plastic strain ε_p is almost "0," indicating that its mechanical behavior exhibits brittle characteristics. The compressive fracture strength σ_f of the MGMC sample after adding SiC is 2453 MPa, and the plastic strain ε_p is 1.1%, indicating that the compressive strength of the MGMC was enhanced and the plastic deformation ability was significantly improved after adding Ni-plated SiC particles. It is worth noting that the slope of the elastic deformation stage on the stress–strain curve of MGMC is larger than that of BMG sample, indicating a significant increase in its elastic modulus E. This is because the E value of SiC particle is higher than that of MG matrix, which increases the modulus of the composite material. In addition, the decrease in density and increase in strength of the MGMC resulted in a further increase in its specific strength.

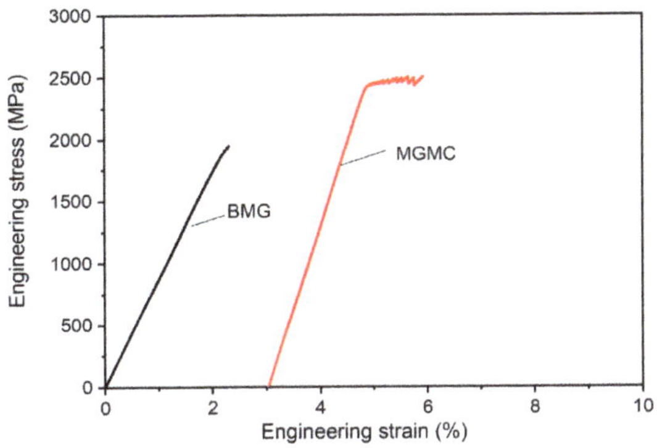

Figure 10. Compression engineering stress–strain curves of BMG and MGMC.

Table 3. Mechanical properties of BMG and MGMC samples measured from compression engineering stress–strain curves.

Sample	ρ (g/cm^3)	$\sigma_{0.2}$ (MPa)	σ_f (MPa)	ε_e (%)	ε_p (%)	E (GPa)	σ_{sp} (Nm/Kg)
BMG	5.36	1955	1955	2.1	0.07	93	3.6×10^5
MGMC	5.04	2453	2502	1.8	1.1	136	4.8×10^5
Improvement rate	−6%	25%	28%	−14%	1471%	46%	33%

Figure 11 shows the results of Vickers microhardness analysis for BMG and MGMC samples. The hardness of the BMG sample was 553 ± 36 kg·cm^{-2}, and the hardness of the MGMC sample was 612 ± 58 kg·cm^{-2}. It can be seen that the hardness of the MGMC sample was significantly higher than that of the BMG sample, which is consistent with the yield strength results listed in Table 3. The increase in the average hardness of the MGMC can be attributed to the introduction of high hardness-enhancing phase SiC. In addition, typical optical photos of hardness indentations on the two samples are also shown in comparison in Figure 11. The indentation on the BMG sample has a typical diamond shape, and the boundary of the indentation is very clear. The indentation of the MGMC sample is slightly flat, and there are circular stripes stacked around it, which may be due to the presence of SiC particles around the indentation, affecting the indentation behavior of the hardness tester's indenter. In addition, the hardness indentation pattern of circular shear band stacking is a typical feature of ductile BMG, indicating that the toughness of the MGMC samples was better.

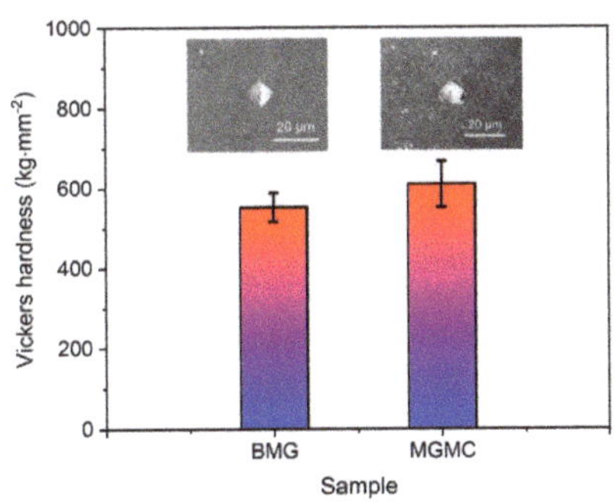

Figure 11. Vickers hardness analysis of the BMG and MGMC sample with typical optical photos of indentations.

4. Discussion

The toughening mechanism of BMGs using the MGMC strategy is mainly related to the properties of the second phase [13]. Numerous studies have shown that adding plastic metallic particles is beneficial for toughening, as plastic particles can undergo plastic deformation earlier than the MG matrix, as the applied load increases, and absorb the deformation energy of the sample through the deformation of the particles themselves [66–70]. However, the strength of the plastic phase is generally low, leading to an improvement in MGMC plasticity accompanied by an obvious strength degradation, inevitably weakening the MGMCs competitive advantage of high strength. The use of high-strength ceramic particles to prepare MGMC has not achieved significant toughening effects in many research cases [26–36]. Most scholars attribute this to the poor plastic deformation capability of high hard second-phase particles [25,26,71], which easily induce crack initiation and form catastrophic fractures in series with each other. However, our calculation results indicate that, when the V_f value of SiC particles increases to over 30%, internal stress conditions that are more favorable for shear deformation can be obtained under large deformation conditions.

Figure 12 shows the Contour plots of the equivalent stress, equivalent elastic strain, and equivalent plastic strain of the 30%-SiC-MGMC samples under different deformation amounts. Obviously, under the condition of 2% strain, SiC particles in MGMC can induce cross stress distribution, which is beneficial for the uniform loading of different parts of the sample. SiC particles have an irregular shape, which may result in significant stress concentration at local edges and corners, which, in turn, may lead to local failure. However, our calculation results indicate that, due to the large elastic deformation of the MG matrix, this very limited stress concentration phenomenon is not as large as expected when the deformation is less than 2%. Figure 12b shows that the stress on the matrix is mainly distributed between particles. In two adjacent particles arranged vertically, this phenomenon is especially obvious. It is already known that the maximum stress distribution line between circular particles is very straight. In contrast, the regularity of the maximum stress distribution line between irregularly shaped particles is worse, which is beneficial for the random deflection of shear bands. The stress near circular particles is mainly concentrated in four directions: up, down, left, and right, while the stress concentration points of irregular particles are more random. When the number of particles is large, this randomness is beneficial for dispersing stress and avoiding the maximum stress overlapping with the main shear direction.

From the distribution of the equivalent plastic strain (Figure 12g–i), when the deformation amount is about 2%, plastic deformation occurs almost everywhere in the MGMC sample, which means that the nucleation sites of shear bands are more numerous and dispersed, which is conducive to the formation of multiple shear bands. However, this requires the matrix to have sufficient intrinsic toughness, which means that the matrix should have sufficient resistance to early local shear band nucleation before the formation of multiple shear bands. Therefore, a matrix with poorer intrinsic toughness (i.e., Mg-based MG [39]) often undergoes catastrophic fracturing before SiC plays a role in deflecting shear bands.

In addition to the V_f of SiC, mentioned earlier, other factors such as casting defects and crystalline precipitates in the matrix/interface can also affect the mechanical properties of MGMC. In this work, the mechanical properties, such as Young's modulus E, yield strength $\sigma_{0.2}$, and fracture strength of composite materials σ_f, do not meet the composite law describing the relationship between the volume fraction and performance of reinforcements in the field of composite materials. For example, the yield strength σ_C of the MGMC should follow $\sigma_C = \sigma_M \cdot (1 - V_f) + \sigma_{SiC} \cdot V_f$, where σ_M and σ_{SiC} refer to the yield strength of the matrix alloy and SiC particle, respectively. However, the test value 2453 MPa is lower than the calculated ones 5800–7300 MPa (theoretical strength value of polycrystalline SiC, 15–20 GPa [58]). This can be attributed to the following aspects: firstly, the casting defect caused by the high content of SiC particles in the composite, including pores, insufficient infiltration, inclusions, etc., will cause the decline of the nominal strength of the MGMC due to the inability to share the external load (weakening the strengthening effect); secondly, the possible interface reactions at the interface may result in unfavorable interface segregation, including the formation of small amounts of carbides, silicide, etc., through the local decomposition of SiC, which will cause interface cracking through stress concentration; thirdly, the dissolution of the electroless nickel plating layer leads to changes in the composition of the matrix, deviating from the optimal GFA range and causing the precipitation of certain nanoscale crystalline phases (Figure 6), which will increase the brittleness of the MGMC matrix.

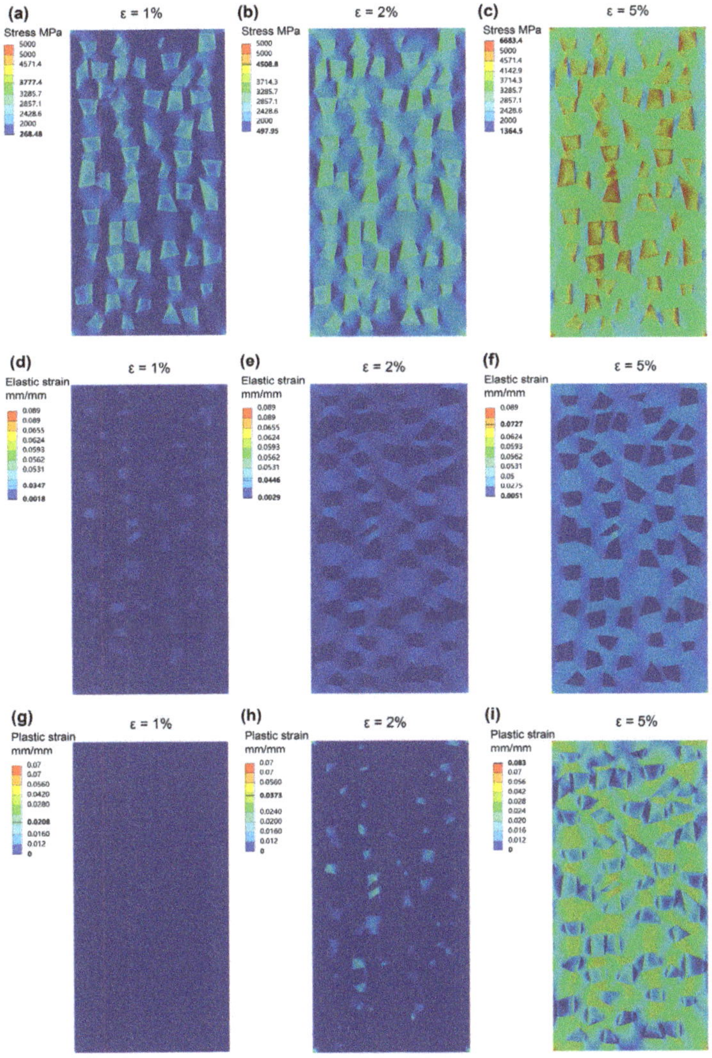

Figure 12. The Contour maps of equivalent stress (**a–c**), equivalent elastic strain (**d–f**), and equivalent plastic strain (**g–i**) calculated by FEM based on the 30%-SiC-MGMC samples under different strains of 1%, 2%, and 5%, respectively.

In order to improve the strengthening effect of SiC particles on MGMC, future work should focus on modifying the interface microstructure of composite materials. In short, the interface reaction on the SiC surface can be reduced through appropriate matrix alloy selection, optimization of the composite material preparation process, and interface structure regulation technology. Our present work has confirmed that the electroless nickel plating treatment on the surface of SiC particles can improve the wettability between the matrix alloy and SiC particles, which also shows the potential for regulating the interface microstructure of MGMC. In the future, the structure and composition of the electroless Ni coating should be optimized to reduce the drastic changes in the matrix composition caused by the melting of the Ni layer. For example, the content of P and O in the coating can be reduced by changing the pH value and the composition of the plating solution. The addition of refining agents can also refine Ni grains to improve the coverage rate of the coating on the SiC particles.

5. Conclusions

This article uses the FEM method to study the micro mechanical behavior of SiC particle-reinforced, Ti-based MGMC, and obtains the optimal volume fraction based on the calculation results. Subsequently, 30%-volume-fraction, nickel-plated $SiC/(Ti_{41}Zr_{25}Be_{28}Fe_6)_{98}Cu_2$ MGMC was prepared using an arc melting copper mold spray casting method, and its mechanical properties were tested. The main conclusions drawn are as follows:

(1) The FEM analysis based on irregular SiC particles shows that, when the volume fraction is below 20%, the maximum stress and strain tend to be distributed parallel to the force axis, which may lead to splitting fracture; when the volume fraction is higher than 30%, the maximum stress and strain tend to form a cross network, which is conducive to shear fracture;

(2) Electroless nickel plating on the surface of SiC particles can significantly improve the interfacial wettability of composite materials. As a result, MGMC with a SiC volume fraction of 29.5% was successfully prepared using copper mold injection casting. The Ni-coated SiC particles exhibited dispersed distribution in the amorphous matrix, but resulted in the precipitation of a small amount of Be_5Zr phase in the amorphous matrix, indicating the GFA degradation of the matrix alloy and the occurrence of partial crystallization;

(3) The glass transition temperature and the crystallization temperature of the MGMC sample significantly decreased, indicating a reduction in the thermal stability of the composite. This can be attributed to the presence of Ni-coated SiC, which changes the composition of the MG matrix, resulting in reduced thermal properties of the matrix. In addition, the presence of SiC particles reduces the free energy barrier of crystalline-phase precipitation due to heterogeneous nucleation, thereby lowering the crystallization temperature;

(4) The compression test showed that the compressive strength and hardness of the prepared MGMC were significantly increased compared to the matrix MG alloy, reaching over 2500 MPa and 612 $kg \cdot cm^{-2}$, and the plastic strain prior to fracture increased from almost "0" to about 1.1%. The use of lightweight reinforcements also resulted in a high specific strength of 4.8×10^5 Nm/kg;

(5) The introduction of electroless Ni-plated SiC particles play multiple roles in weight lightening, strengthening, and toughening, providing a basis for the design and development of MGMCs. This work sheds light on the manufacturing strategy of high specific strength and wear-resistant small parts needed in the fields of consumer electronics, unmanned aerial vehicles, and robots.

Author Contributions: Investigation, software, writing—original draft preparation, H.L.; data curation, formal analysis, J.L.; Investigation, validation, J.Z.; formal analysis, L.Z.; methodology, visualization, W.Y.; resources, P.G.; supervision, writing—review and editing, project administration, funding acquisition, X.W. All authors have read and agreed to the published version of the manuscript.

Funding: This research was funded by the "Natural Science Foundation of Hebei Province, China, grant number E2022202055", the "Scientific Research Projects of Colleges and Universities in Hebei Province, China, grant number ZD2022099", the "Hebei Province funding project for introducing overseas persons, grant number C20220314"and the "Open Fund of State Key Laboratory of Materials Processing and Die & Mould Technology, Huazhong University of Science and Technology, grant number P2023-023".

Institutional Review Board Statement: Not applicable.

Informed Consent Statement: Not applicable.

Data Availability Statement: Any further detailed data may be obtained from the authors upon a reasonable request.

Conflicts of Interest: The authors declare no conflict of interest.

References

1. Cui, C.; Hu, B.; Zhao, L.; Liu, S. Titanium alloy production technology, market prospects and industry development. *Mater. Des.* **2011**, *32*, 1684–1691. [CrossRef]
2. Gupta, M.K.; El Etri, H.; Korkmaz, M.E.; Ross, N.S.; Krolczyk, G.M.; Gawlik, J.; Yasar, N.; Pimenov, D.Y. Tribological and surface morphological characteristics of titanium alloys: A review. *Arch. Civ. Mech. Eng.* **2022**, *22*, 72. [CrossRef]
3. Shen, Y.; Jin, Z.; Ling, M.; Sun, Z.; Feng, M.; Xu, C.; Liu, S. Advances in Research on Titanium and Titanium Alloys with Antibacterial Functionality for Medical Use—A Review. *J. Biomater. Tiss. Eng.* **2023**, *13*, 1–17. [CrossRef]
4. Gong, P.; Deng, L.; Jin, J.; Wang, S.; Wang, X.; Yao, K. Review on the Research and Development of Ti-Based Bulk Metallic Glasses. *Metals* **2016**, *6*, 264. [CrossRef]
5. Eckert, J.; Das, J.; Pauly, S.; Duhamel, C. Mechanical properties of bulk metallic glasses and composites. *J. Mater. Res.* **2007**, *22*, 285–301. [CrossRef]
6. Louzguine-Luzgin, D.V.; Louzguina-Luzgina, L.V.; Churyumov, A.Y. Mechanical Properties and Deformation Behavior of Bulk Metallic Glasses. *Metals* **2013**, *3*, 1–22. [CrossRef]
7. Gu, X.J.; Poon, S.J.; Shiflet, G.J.; Lewandowski, J.J. Compressive plasticity and toughness of a Ti-based bulk metallic glass. *Acta Mater.* **2010**, *58*, 1708–1720. [CrossRef]
8. Wang, X.; Gong, P.; Yao, K.-F. Mechanical behavior of bulk metallic glass prepared by copper mold casting with reversed pressure. *J. Mater. Process. Tech.* **2016**, *237*, 270–276. [CrossRef]
9. Xi, X.K.; Zhao, D.Q.; Pan, M.X.; Wang, W.H.; Wu, Y.; Lewandowski, J.J. Fracture of brittle metallic glasses: Brittleness or plasticity. *Phys. Rev. Lett.* **2005**, *94*, 125510. [CrossRef] [PubMed]
10. Zhao, J.X.; Qu, R.T.; Wu, F.F.; Zhang, Z.F.; Shen, B.L.; Stoica, M.; Eckert, J. Fracture mechanism of some brittle metallic glasses. *J. Appl. Phys.* **2009**, *105*, 103519. [CrossRef]
11. Hofmann, D.C.; Suh, J.-Y.; Wiest, A.; Duan, G.; Lind, M.-L.; Demetriou, M.D.; Johnson, W.L. Designing metallic glass matrix composites with high toughness and tensile ductility. *Nature* **2008**, *451*, 1085-U3. [CrossRef] [PubMed]
12. Qiao, J. In-Situ Dendrite/Metallic Glass Matrix Composites: A Review. *J. Mater. Sci. Technol.* **2013**, *29*, 685–701. [CrossRef]
13. Qiao, J.; Jia, H.; Liaw, P.K. Metallic glass matrix composites. *Mater. Sci. Eng. R* **2016**, *100*, 1–69. [CrossRef]
14. Yin, H.L.; Yang, W.; Zhao, L.C.; Hu, X.M.; Liu, S.Q.; Cui, C.X.; Wang, X. Fabrication and mechanical property of three-dimensional carbon fiber reinforced Mg-based bulk metallic glass matrix composite. *Mater. Sci. Eng. A* **2022**, *839*, 142853. [CrossRef]
15. Yin, H.L.; Liu, S.Q.; Zhao, L.C.; Cui, C.X.; Wang, X. Vacuum infiltration molding and mechanical property of short carbon fiber reinforced Ti-based metallic glass matrix composite. *J. Mater. Process. Techol.* **2021**, *295*, 117151. [CrossRef]
16. Liu, J.; Zhang, H.; Yuan, X.; Fu, H.; Hu, Z. Synthesis and Properties of Carbon Short Fiber Reinforced ZrCuNiAl Metallic Glass Matrix Composite. *Mater. Trans.* **2011**, *52*, 412–415. [CrossRef]
17. Shamlaye, K.F.; Laws, K.J.; Ferry, M. Supercooled liquid fusion of carbon fibre-bulk metallic glass composites with superplastic forming properties. *Scr. Mater.* **2016**, *111*, 127–130. [CrossRef]
18. Chen, S.; Li, W.Q.; Zhang, L.; Fu, H.M.; Li, Z.K.; Zhu, Z.W.; Li, H.; Zhang, H.W.; Wang, A.M.; Wang, Y.D.; et al. Dynamic compressive mechanical properties of the spiral tungsten wire reinforced Zr-based bulk metallic glass composites. *Compos. Part B* **2020**, *199*, 108219. [CrossRef]
19. Zhou, F.; Du, C.; Cheng, C.; Xu, L.; Du, Z.; Gao, G.; Liu, A.; Fu, H. Penetration performance and fragmentation mechanism behind target of tungsten fibre/zirconium-based bulk metallic glass matrix composite rod. *Int. J. Refract. Met. Hard Mater.* **2023**, *112*, 106160. [CrossRef]
20. Zou, M.; Guo, M.; Ye, X. Penetrating Power of High-Density Tungsten Fiber/Zr-Based Metallic Glass Matrix Composite. *Adv. Mater. Sci. Eng.* **2022**, *2022*, 8087525. [CrossRef]
21. Qiu, K.Q.; Wang, A.M.; Zhang, H.F.; Ding, B.Z.; Hu, Z.Q. Mechanical properties of tungsten fiber reinforced ZrAlNiCuSi metallic glass matrix composite. *Intermetallics* **2002**, *10*, 1283–1288. [CrossRef]
22. Zhang, H.F.; Li, H.; Wang, A.M.; Fu, H.M.; Ding, B.Z.; Hu, Z.Q. Synthesis and characteristics of 80 vol.% tungsten (W) fibre/Zr based metallic glass composite. *Intermetallics* **2009**, *17*, 1070–1077. [CrossRef]
23. Moelle, C.; Lu, I.R.; Sagel, A.; Wunderlich, R.K.; Perepezko, J.H.; Fecht, H.J. Formation of ceramic/metallic glass composite by mechanical alloying. In *Mechanically Alloyed, Metastable and Nanocrystalline Materials, Part 1*; Baro, M.D., Surinach, S., Eds.; Trans Tech Publications, Ltd.: Bäch, Switzerland, 1998; Volume 269–272, pp. 47–52.

24. Rezaei, M.R.; Albooyeh, A.; Chachei, R.; Malahi, P. Effect of the spark plasma sintering temperature on the microstructure and mechanical properties of a ceramic/metallic glass reinforced hybrid composite. *J. Compos. Mater.* **2022**, *56*, 2779–2788. [CrossRef]
25. Xu, Y.K.; Xu, J. Ceramics particulate reinforced $Mg_{65}Cu_{20}Zn_5Y_{10}$ bulk metallic glass composites. *Acta Metall. Sin.* **2004**, *40*, 726–730. [CrossRef]
26. Li, J.Q.; Wang, L.; Cheng, H.W.; Zhang, H.F.; Hu, Z.Q.; Cai, H.N. Fracture surface morphology of Mg-based bulk metallic glass and composite during quasi-static and dynamic compressive deformation. *J. Alloys Compd.* **2009**, *478*, 827–830. [CrossRef]
27. Lin, H.-M.; Jeng, R.-R.; Lee, P.-Y. Microstructure and mechanical properties of vacuum hot-pressing SiC/Ti-Cu-Ni-Sn bulk metallic glass composites. *Mater. Sci. Eng. A* **2008**, *493*, 246–250. [CrossRef]
28. Lin, H.-M.; Chen, G.-S.; Lee, P.-Y. Microstructure and properties of vacuum hot-pressing SiC/Ti-Cu-Ni-Sn bulk metallic glass composites. In Proceedings of the 5th China Cross-Strait Conference on Composite Materials, Shanghai, China, 22–26 October 2006; pp. 26–30. [CrossRef]
29. Xie, G.; Louzguine-Luzgin, D.V.; Inoue, A. Characterization of interface between the particles in NiNbZrTiPt metallic glassy matrix composite containing SiC fabricated by spark plasma sintering. *J. Alloys Compd.* **2009**, *483*, 239–242. [CrossRef]
30. Xie, G.; Kimura, H.; Louzguine-Luzgin, D.V.; Men, H.; Inoue, A. SiC dispersed Fe-based glassy composite cores produced by spark plasma sintering and their high frequency magnetic properties. *Intermetallics* **2012**, *20*, 76–81. [CrossRef]
31. Yue, T.M.; Su, Y.P. Laser cladding of SiC reinforced $Zr_{65}Al_{7.5}Ni_{10}Cu_{17.5}$ amorphous coating on magnesium substrate. *Appl. Surf. Sci.* **2008**, *255*, 1692–1698. [CrossRef]
32. Qian, Y.; Zhang, D.; Hong, J.; Zhang, L.; Jiang, M.; Huang, H.; Yan, J. Microstructure and mechanical properties of SiC particle reinforced Zr-based metallic glass surface composite layers produced by laser alloying. *Surf. Coat. Technol.* **2022**, *446*, 128784. [CrossRef]
33. Chen, Y.; Wang, A.; Fu, H.; Zhu, Z.; Zhang, H.; Hu, Z.; Wang, L.; Cheng, H. Preparation, microstructure and deformation behavior of Zr-based metallic glass/porous SiC interpenetrating phase composites. *Mater. Sci. Eng. A* **2011**, *530*, 15–20. [CrossRef]
34. Chen, Y.; Wang, A.; Zhang, H.; Hu, Z. Preparation and characterization of amorphous alloy/porous SiC bi-continuous structure composite. *Int. J. Modern Phys. B* **2009**, *23*, 1294–1299. [CrossRef]
35. Wang, B.P.; Wang, L.; Xue, Y.F.; Wang, S.Y.; Wang, Y.Q.; Zhang, H.F.; Fu, H.M. Strain rate-dependent compressive deformation and failure behavior of porous SiC/Ti-based metallic glass composite. *Mater. Sci. Eng. A* **2014**, *609*, 53–59. [CrossRef]
36. Wang, B.-p.; Wang, L.; Xue, Y.-f.; Wang, Y.-w.; Zhang, H.-f.; Fu, H.-m. Dynamic indentation response of porous SiC/Ti-based metallic glass composite. *T. Nonferr. Metal. Soc.* **2016**, *26*, 3154–3160. [CrossRef]
37. Zhang, B.; Li, W.; Li, H.; Zhang, H.-f. Spontaneous infiltration and wetting behaviors of a Zr-based alloy melt on a porous SiC substrate. *Int. J. Min. Met. Mater.* **2018**, *25*, 817–823. [CrossRef]
38. Zhang, H.; Li, S.; Liu, Z.; Li, H.; Geng, T.; Zhang, J.; Jiao, D.; Zeng, S.; Zhang, H.; Zhu, Z. Simultaneously enhancing strength and fracture toughness of bulk metallic glass composites containing SiC scaffolds with nacre-like lamellar architectures. *Mater. Sci. Eng. A* **2022**, *840*, 143000. [CrossRef]
39. Wang, X.; Zhao, L.; Hu, X.; Cheng, Y.; Liu, S.; Chen, P.; Cui, C. Fabrication and Mechanical Behavior of Ex Situ Mg-Based Bulk Metallic Glass Matrix Composite Reinforced with Electroless Cu-Coated SiC Particles. *Materials* **2017**, *10*, 1371. [CrossRef]
40. Li, J.-F.; Wang, X.; Yang, G.-N.; Chen, N.; Liu, X.; Yao, K.-F. Enhanced plasticity of a Fe-based bulk amorphous alloy by thin Ni coating. *Mater. Sci. Eng. A* **2015**, *645*, 318–322. [CrossRef]
41. Kim, H.S. Fictive stress model based finite element analysis for bulk metallic glasses at an elevated temperature. *Met. Mater. Int.* **2004**, *10*, 461–466. [CrossRef]
42. Jun, H.J.; Lee, K.S.; Chang, Y.W.; Kim, H.S. Finite Element Analysis for Application to the Forming Process of Ti-based Bulk Metallic Glasses. *Mater. Manuf. Process.* **2014**, *29*, 801–807. [CrossRef]
43. Kim, H.S.; Kato, H.; Inoue, A.; Chen, H.S. Finite element analysis of compressive deformation of bulk metallic glasses. *Acta Mater.* **2004**, *52*, 3813–3823.
44. Cheng, M.; Zhang, S.H.; Wert, J.A. Finite element analysis of micro imprinting of bulk metallic glasses in supercooled liquid regime. *J. Mater. Sci.* **2007**, *42*, 5999–6003. [CrossRef]
45. Zhu, Z.; Zhang, H.; Hu, Z.; Zhang, W.; Inoue, A. Ta-particulate reinforced Zr-based bulk metallic glass matrix composite with tensile plasticity. *Scripta Mater.* **2010**, *62*, 278–281. [CrossRef]
46. Ding, H.; Bao, X.; Jamili-Shirvan, Z.; Jin, J.; Deng, L.; Yao, K.; Gong, P.; Wang, X. Enhancing strength-ductility synergy in an ex situ Zr-based metallic glass composite via nanocrystal formation within high-entropy alloy particles. *Mater. Des.* **2021**, *210*, 110108. [CrossRef]
47. Ott, R.T.; Sansoz, F.; Molinari, J.F.; Almer, J.; Ramesh, K.T.; Hufnagel, T.C. Micromechanics of deformation of metallic-glass-matrix composites from in situ synchrotron strain measurements and finite element modeling. *Acta Mater.* **2005**, *53*, 1883–1893. [CrossRef]
48. Ott, R.T.; Sansoz, F.; Jiao, T.; Warner, D.; Fan, C.; Molinari, J.F.; Ramesh, K.T.; Hufnagel, T.C. Yield criteria and strain-rate behavior of $Zr_{57.4}Cu_{16.4}Ni_{8.2}Ta_8Al_{10}$ metallic-glass-matrix composites. *Metall. Mater. Trans. A* **2006**, *37A*, 3251–3258. [CrossRef]
49. Zhang, X.Q.; Wang, L.; Fan, Q.B.; Xue, Y.F.; Wang, Y.D.; Nie, Z.H.; Zhang, H.F.; Fu, H.M. Micro-deformation mechanism of Zr-based metallic glass/porous tungsten composite by in-situ high-energy X-ray diffraction and finite element modeling. *Mater. Sci. Eng. A* **2014**, *598*, 407–412. [CrossRef]
50. Jiang, Y.; Qiu, K. Computational micromechanics analysis of toughening mechanisms of particle-reinforced bulk metallic glass composites. *Mater. Des.* **2015**, *65*, 410–416. [CrossRef]

51. Jiang, Y.; Shi, X.; Qiu, K. Numerical study of shear banding evolution in bulk metallic glass composites. *Mater. Des.* **2015**, *77*, 32–40. [CrossRef]
52. Jiang, Y.; Sun, L.; Wu, Q.; Qiu, K. Enhanced tensile ductility of metallic glass matrix composites with novel microstructure. *J. Non-Cryst. Solids* **2017**, *459*, 26–31. [CrossRef]
53. Jia, H.L.; Zheng, L.L.; Li, W.D.; Li, N.; Qiao, J.W.; Wang, G.Y.; Ren, Y.; Liaw, P.K.; Gao, Y. Insights from the Lattice-Strain Evolution on Deformation Mechanisms in Metallic-Glass-Matrix Composites. *Metall. Mater. Trans. A* **2015**, *46A*, 2431–2442. [CrossRef]
54. Bian, P.-L.; Liu, T.-L.; Qing, H.; Gao, C.-F. 2D Micromechanical Modeling and Simulation of Ta-Particles Reinforced Bulk Metallic Glass Matrix Composite. *Appl. Sci.* **2018**, *8*, 2192. [CrossRef]
55. Gong, P.; Wang, X.; Shao, Y.; Chen, N.; Liu, X.; Yao, K.F. A Ti-Zr-Be-Fe-Cu bulk metallic glass with superior glass-forming ability and high specific strength. *Intermetallics* **2013**, *43*, 177–181. [CrossRef]
56. Park, J.M.; Wang, G.; Pauly, S.; Mattern, N.; Kim, D.H.; Eckert, J. Ductile Ti-Based Bulk Metallic Glasses with High Specific Strength. *Metall. Mater. Trans. A* **2011**, *42*, 1456–1462. [CrossRef]
57. Perisanu, S.; Gouttenoire, V.; Vincent, P.; Ayari, A.; Choueib, M.; Bechelany, M.; Cornu, D.; Purcell, S.T. Mechanical properties of SiC nanowires determined by scanning electron and field emission microscopies. *Phys. Rev. B* **2008**, *77*, 165434. [CrossRef]
58. Chowdhury, E.H.; Rahman, M.H.; Hong, S. Tensile strength and fracture mechanics of two-dimensional nanocrystalline silicon carbide. *Comp. Mater. Sci.* **2021**, *197*, 110580. [CrossRef]
59. Maruthoor, S.; Ajayakumar, A.; Fuchs, T.; Jakovlev, O.; Reinecke, H.; Wilde, J. Mechanical Characterization of Polycrystalline and Amorphous Silicon Carbide Thin Films Using Bulge Test. *J. Microelectromech. S.* **2013**, *22*, 140–146. [CrossRef]
60. Wang, X. Surface Crystallization in Mg-Based Bulk Metallic Glass during Copper Mold Casting. *Adv. Mater. Sci. Eng.* **2014**, *2014*, 798479. [CrossRef]
61. Pabst, O.; Schiffer, M.; Obermeier, E.; Tekin, T.; Lang, K.D.; Ha-Duong, N. Measurement of Young's modulus and residual stress of thin SiC layers for MEMS high temperature applications. *Microsyst. Technol.* **2012**, *18*, 945–953. [CrossRef]
62. Yibibulla, T.; Jiang, Y.; Wang, S.; Huang, H. Size- and temperature-dependent Young's modulus of SiC nanowires determined by a laser-Doppler vibration measurement. *Appl. Phys. Lett.* **2021**, *118*, 043103. [CrossRef]
63. Liu, Y.; Wang, G.; Li, H.; Pang, S.; Chen, K.; Zhang, T. Ti-Cu-Zr-Fe-Sn-Si-Sc bulk metallic glasses with good mechanical properties for biomedical applications. *J. Alloys Compd.* **2016**, *679*, 341–349. [CrossRef]
64. Ma, C.; Soejima, H.; Ishihara, S.; Amiya, K.; Nishiyama, N.; Inoue, A. New Ti-based bulk glassy alloys with high glass-forming ability and superior mechanical properties. *Mater. Trans.* **2004**, *45*, 3223–3227. [CrossRef]
65. Rajagopal, G.; Thiyyakkandi, S. Numerical evaluation of the performance of back-to-back MSE walls with hybrid select-marginal fill zones. *Transp. Geotech.* **2021**, *26*, 100445. [CrossRef]
66. Hays, C.C.; Kim, C.P.; Johnson, W.L. Microstructure controlled shear band pattern formation and enhanced plasticity of bulk metallic glasses containing in situ formed ductile phase dendrite dispersions. *Phys. Rev. Lett.* **2000**, *84*, 2901–2904. [CrossRef] [PubMed]
67. Guo, S.F.; Liu, L.; Li, N.; Li, Y. Fe-based bulk metallic glass matrix composite with large plasticity. *Scr. Mater.* **2010**, *62*, 329–332. [CrossRef]
68. Wang, H.; Li, R.; Wu, Y.; Chu, X.M.; Liu, X.J.; Nieh, T.G.; Lu, Z.P. Plasticity improvement in a bulk metallic glass composed of an open-cell Cu foam as the skeleton. *Compos. Sci. Technol.* **2013**, *75*, 49–54. [CrossRef]
69. Shao, Y.; Zheng, W.; Guo, W.; Lu, S.; Wu, S. In situ Fe-rich particle reinforced Mg -based metallic glass matrix composites via dealloying in metallic melt. *Mater. Lett.* **2021**, *285*, 129165. [CrossRef]
70. Yang, S.; Li, M.; Cao, P.; Zhang, Q.; He, L. Enhanced mechanical properties of dendrite-reinforced Ti-based bulk metallic glass composites by tuning the microstructure. *Intermetallics* **2022**, *142*, 107458. [CrossRef]
71. Dlouhý, I.; Chlup, Z.; Boccaccini, A.R. Fracture Behaviour of Brittle (Glass) Matrix Composites. *Mater. Sci. Forum* **2005**, *482*, 115–122. [CrossRef]

materials

MDPI

Review

Development of Inoculants for Aluminum Alloy: A Review

Shuiqing Liu [1,2,*], Tong Zhao [3], Jinyuan Fu [2] and Qun Zu [2,*]

1 State Key Laboratory for Reliability and Intelligence of Electrical Equipment, Hebei University of Technology, Tianjin 300401, China

2 School of Mechanical Engineering, Hebei University of Technology, Tianjin 300401, China; fjy18238941292@163.com

3 Arizona College of Technology, Hebei University of Technology, Tianjin 300401, China; zhaotong20022021@163.com

* Correspondence: liushuiqing0824@126.com (S.L.); qzu@hebut.edu.cn (Q.Z.)

Abstract: Aluminum and its alloys are widely used in packaging, transportation, electrical materials, and many other fields because of their abundance, light weight, good mechanical properties, suitable corrosion resistance, excellent electrical conductivity, and other advantages. Grain refinement achieved by adding inoculant is important not only to reduce the segregation and thermal cracking of alloy castings but also to improve the mechanical properties of alloy castings. Therefore, fine equiaxed grain structure has always been one of the goals pursued by the aluminum alloy casting industry. For this reason, the selection and development of effective inoculants for aluminum alloy is a key technology in the aluminum processing industry. This paper summarizes the development history of inoculants for aluminum alloy, including Al-Ti-C, Al-Ti-B, Al-Ti, Al-Ti-B-(C)-Ce, Al-Sc, and the Fe-rich phase of Al-Si alloy. At the same time, the advantages and disadvantages of common inoculants are introduced and prospective future applications are reviewed.

Keywords: inoculate; grain refinement; Al-Ti-B; Al-Sc alloy; Fe-rich phase in Al-Si alloy

1. Introduction

In the aluminum alloy industry, grain refinement is often used to improve the strength and plasticity of the material, and for grain sizes of the product above the nano level, refining the grain can improve the dislocation density and increase the strength of the material. Compared with large grains, small, coordinated deformation per unit volume can improve the extension of the overall material and increase the plasticity of the material [1]. Grain refinement can also reduce the twinning caused by dendrite growth, reduce the tendency for segregation, cold isolation, and hot cracking in the process, improve the feeding during solidification, eliminate or better disperse loose grains, improve the air tightness and surface quality of the casting, and improve the casting success rate [2–4].

Grain refinement refers to the grain transformation process that inhibits the growth of columnar grains and promotes the formation of fine equiaxed grains in castings. Therefore, in order to expand the equiaxed grains region and reduce the columnar grains region, inoculants can be added to the aluminum alloy to promote the transformation of grain from columnar to equiaxed. According to the basic theory of material science, the methods of grain refinement can be classified into two major categories: physical methods and chemical methods. Physical methods include rapid cooling, physical field refinement, and mechanical-physical refinement. On the other hand, chemical methods involve the addition of grain refiners, which promote grain nucleation or hinder the growth of crystals to achieve fine-grained structures. Currently, adding inoculants to aluminum melts is considered the most effective and practical grain refinement method due to its fast action, excellent refining results, easy operation, and adaptability. The inoculants are mainly added in the form of aluminum base intermediate alloys. At present, grain refiners are mainly added in the form of aluminum base intermediate alloys. The existing common refiners mainly include Al-Ti-B, Al-Ti-C, Al-Ti-B-C, Al-B-Ce, Al-Sc, and Al-Sc-Zr; among which, Al-Ti-B wire is the most widely used in industry. The grain refiner used by more than 80% of aluminum processing enterprises in the world is Al-Ti-B wire intermediate alloy.

In 1930, pure Ti was added to aluminum ingots and then aluminum grains were refined [5]. With the development of grain-refining technology, mixed salts such as K_2TiF_6 and KBF_4 have been gradually applied to grain refining of aluminum alloys, but they produce a large amount of toxic fluoride gas and slag inclusion and have unstable refining efficiency [6–8]. In the 1960s, the era of intermediate alloys was entered. With the emergence and promotion of coreless induction furnaces, the Al-Ti binary alloy was first developed, and then a more effective Al-Ti-B intermediate alloy was successfully developed. The stability of its refinement effect was greatly improved by the addition of B element. At the beginning, Al-Ti-B is added in bulk. Although it can play a good thinning role, the second-phase particle density is large, which is easy to gather and precipitate during static so that the thinning effect declines, and it is "poisoned" when it is subjected to elements such as Zr. With the development of semicontinuous and continuous casting, by the mid-1970s, the United States developed a new Al-Ti-B refining technology, that is, in the form of a wirefeeder continuously added to the flow tank melt, not only to achieve the automation of refining processing, but also to effectively reduce the agglomeration of the second-phase particles. After the 1980s, various Al-Ti-B intermediate alloys were developed, of which Al-5Ti-1B intermediate alloy was the most widely used. In 1985, German researcher Banerji [9–11] and others developed Al-Ti-C intermediate alloy, which opened up a new field for grain refiners. Since the 1990s, the development of a new series of refiners, including Al-Ti-C-B, Al-Ti-B-Re, and Al-Ti-C-Re, has achieved certain results.

2. Specific Grain Refiner

2.1. Al-Ti-B

Al-Ti-B grain inoculant has a good refining effect and has a wide range of applications. It is the most important refiner, with remarkable refining effect in the aluminum alloy casting industry. The advantages of Al-Ti-B are its clear thinning effect, good stability, and lower price. The disadvantage is that the refining behavior of Al-Ti-B refiner is significantly affected by many factors such as particle distribution, refining process, alloy composition, and cooling rate.

At present, there are various theories on the refinement mechanism of Al-Ti-B, which complement each other. The main refinement theories include phase diagram theory, particle theory, dual nucleation theory, etc. The dual nucleation theory can explain the grain-refining process of Al-Ti-B grain refiner well. The phase diagram theory was first proposed by Crossley based on the Al-Ti binary phase diagram [12]. Peritectic reaction occurs at 0.15%Ti and 665 °C, which promotes α-Al nucleation and achieves grain refinement L + $TiAl_3 \rightarrow \alpha$-Al. Maxwell found that $TiAl_3$ particles existed in the grain center and considered that peritectic reaction was the main refinement mechanism [13]. Some scholars have proposed new explanations and opposing views. Based on the controversy, the phase diagram theory is not adequate to correctly explain the mechanism of Al-Ti-B grain refinement [14–16]. The boride particle theory was first proposed by Cibula [6–8]. According to the theory, TiB_2 has a high melting point and can become an effective nucleating particle in the melt. However, it is found that in the absence of abundant solutes, Ti in the melt, single TiB_2 particles, and clusters of TiB_2 particles in the Al matrix will be pushed to the grain boundary by α-Al dendrites, which makes it difficult for the particle to become an effective hetero-nucleation core of α-Al, and it cannot produce a grain-refinement effect [17]. Therefore, it is difficult for particle theory to perfectly explain the mechanism of Al-Ti-B grain refining. Combining the phase diagram theory with the particle theory, Mohanty proposed the dual nucleation theory to explain the grain refinement mechanism of Al-Ti-B [17]. Mohanty holds the view that TiB_2 with a high melting point is relatively stable after entering the melt, while $TiAl_3$ will dissolve quickly, and the dissolution of $TiAl_3$ will bring excess Ti to local areas of the melt. There is a concentration gradient of Ti on the surface of TiB_2, so Ti is easy to segregate on its surface, form a new layer of $TiAl_3$ after supercooling, and react with Al peritectic to form crystal nuclei after the temperature drops. Backerud put forward the theory of a peritectic shell, but the theory is inconsistent with practice [18]. In recent years, many scholars have conducted research under the framework of the dual nucleation theory [19–22]. In summary, dual nucleation can explain the grain refinement mechanism of Al-Ti-B more correctly, which is of great research significance.

The preparation methods of Al-Ti-B refining agent include the high-temperature self-spreading method (high cost, not suitable for industrial production), thermite reduction method (not easy to operate, slow reaction, high energy consumption, and B recovery rate is only 25%), electrolysis method (difficult to produce intermediate alloys with high Ti and B content), and fluorine thermite reaction method (simple process, low cost, and suitable for industrial promotion). However, the size of the second phase, the recovery rate of B, and the uniformity of the organization are difficult to control, and the impurities are difficult to completely remove, requiring strict air process conditions [23,24]. The powder metallurgy rule can effectively control the size, distribution, and element ratio of the second phase in the refiner by changing the raw material and preparation process and is expected to achieve a better thinning effect [25–29]. In recent years, nano-ceramic particles, with their good thermal and chemical stability, have been increasingly used for aluminum and aluminum alloy grain refinement. Nanoparticles can provide rich nucleation particles for α-A, and some particles in the solidification process will be adsorbed to the solid–liquid interface, which will hinder the liquid phase of the molten atom to the growth interface diffusion and inhibit the grain growth [30–33]. However, the difficulty of such refiners in refining grains lies in the agglomeration phenomenon caused by the high surface energy of nanoparticles, and their poor wettability with aluminum melt leads to the rapid settlement of agglomeration in aluminum melt, which seriously weakens the refining effect. Improving the dispersion of nanoparticles in refiners is an important problem to be solved in grain refinement of this type of refiners, and various attempts, such as electrical pulse and ultrasonic vibration, have not obtained ideal results [34–36]. Recently, high-energy ball milling has been increasingly used to prepare micron metal/nano second-phase composite powder refiners [22,37,38]. The particle agglomeration phenomenon is effectively reduced, the dispersion of nanoparticles in Al melt is more uniform, and a better refinement effect is obtained. However, current studies mainly focus on the refinement effect and refinement mechanism, among which the key dispersion of nano second-phase particles and its influence on the refinement effect and mechanism are still unclear. Figure 1 shows the effect diagram of aluminum titanium-boron grain refiner prepared by the SLM (selective laser melting) method when it is actually applied to alloy refining.

The forming methods of Al-Ti-B wire mainly include the semicontinuous casting extrusion method (increased heat treatment process, low efficiency, high energy consumption, uneven composition and organization, and unstable product quality), continuous casting extrusion method (low energy consumption and stable product quality but slow casting speed and low production efficiency), continuous casting method, and continuous casting and rolling method (low energy consumption, high production efficiency, and product quality is stable). Extrusion casting, also known as liquid metal forging, is one of the effective methods of grain refinement. Extrusion casting is a casting and forging combined production process; that is, liquid metal is formed under high pressure and accompanied by plastic deformation so that the grain inside the material is broken and refined, and the defect density is reduced, thereby improving the performance of the material. Extrusion casting has a wide range of applications, including in automotive, high-speed rail, communications, and other fields. Compared with traditional die casting, extrusion casting has fewer defects such as pores and shrinkage holes, and the internal organization of the product is tighter and more uniform. At the same time, the extrusion casting process is suitable for large-scale operation, mechanized operation, and short production cycles.

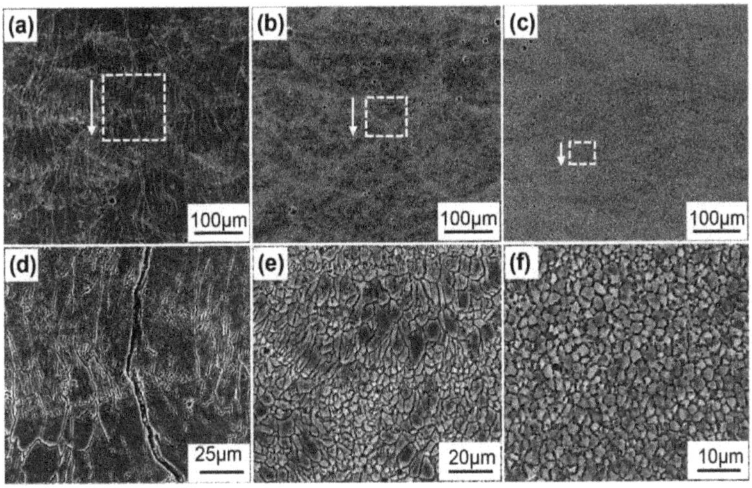

Figure 1. SEM micrograph of AA7075 Al alloy sample prepared by SLM using Al-Ti-B as grain refiner. (**a**) AA7075 sample without inoculant; (**b**) TiH2/AA7075 sample; (**c**) (TiH2+B)/AA7075 sample; (**d**–**f**) are the enlarged view of the rectangle region corresponding to (**a**–**c**) [39].

It was found that the second phase ($TiAl_3$ and TiB_2) of the refiner is a good nucleating matrix of α-Al, which plays a role in refining the grain [40–42]. The grain-refining effect of Al-Ti-B refiner is mainly determined by the microstructure of the alloy, including the shape, size, and distribution of $TiAl_3$ phase and TiB_2 particles. As can be seen in Figure 2, the results show that the Al-5Ti-1B intermediate alloy is composed of Al matrix, TiB_2, and a small amount of the $TiAl_3$ phase. When the addition of TiB_2 is constant, the grain size decreases with the increase in Al alloy solute content (i.e., the larger growth limiting factor Q). When the alloy composition is constant, the grain density increases, and the grain size decreases with the addition of TiB_2. When there are a large number of dispersed and evenly distributed TiB_2 and irregular lumps of $TiAl_3$ in Al-Ti-B refiner, the refining effect is the best.

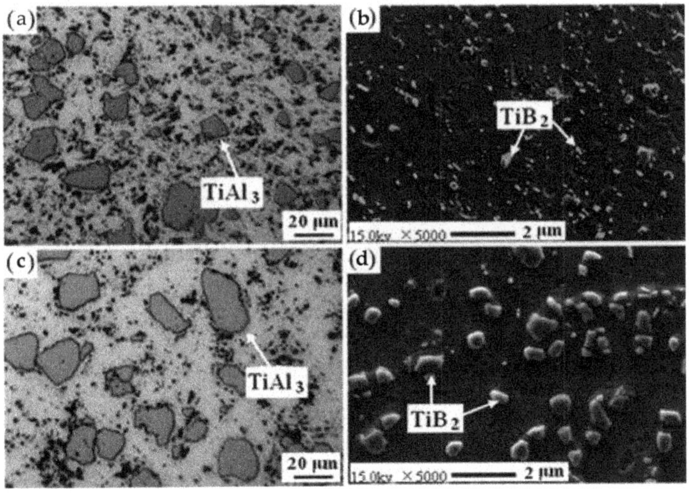

Figure 2. The Al-5Ti-1B ribbon and SEM image of its microstructure: (**a**) $TiAl_3$ phase; (**b**,**d**) TiB_2 phase; (**c**) $TiAl_3$ phase [42].

The more dispersed and smaller the nucleating particles in Al-Ti-B grain refiner, the stronger the grain-refining ability is. The microstructure of the refiner can be controlled in various ways. For example, a rapid solidification process can provide a large degree of supercooling, promote the nucleation of $TiAl_3$ and TiB_2, and reduce gravity segregation. The addition of ultrasonic and electromagnetic energy fields can stir the melt, disperse the nucleating particles, and break the dendrites, thus reducing the size of the nucleating particles. Therefore, certain methods can be used to homogenize the refiner structure and reduce the particle size of $TiAl_3$ and TiB_2, which is very beneficial to improve the grain-refining effect of Al-Ti-B.

The TiB_2 particles are easy to aggregate and precipitate, which degrades the thinning effect. In the case of Zr, Cr, or Mn, B in TiB_2 particles will be taken away, forming the corresponding borides, making TiB_2 particles "poisoned", and weakening or eliminating the thinning effect. On the contrary, Mg, Nb, and other elements can promote the grain-refining effect, and Mg can effectively inhibit the poisoning phenomenon of Al-5Ti-1B refiner caused by Si element. Therefore, revealing the influence of alloy composition on the refining ability of Al-Ti-B can clarify the applicability of Al-Ti-B refiner and avoid the increase in production cost. By adjusting the addition method, addition temperature, and retention time of Al-Ti-B refiner, the agglomerative precipitation phenomenon of TiB_2 and the "poisoning" phenomenon of related elements can be effectively weakened. Some studies consider that the reaction of Zr with Si or impurity elements is the main factor causing toxicity [43].

However, Zr has a negative effect on the grain-refining ability of Al-Ti-B refiner [44], resulting in the grain-refining effect becoming worse and irreversible and the grains gradually coarsening with the extension of holding time, which is referred to as the Zr poisoning effect. Although there is no precise theory to explain the mechanism of Zr toxicity [44], it can be explained from two perspectives: (1) the influence of the reaction between Zr and TiB_2 and $TiAl_3$ on the nucleation ability of Al-Ti-B is studied. According to the nucleation theory, TiB_2 and $TiAl_3$ are important nucleating particles in Al-Ti-B refiner, and their existence states directly affect the grain-refining ability of the refiner. Therefore, the reaction between Zr and TiB_2 and $TiAl_3$ may reduce the refining ability of the refiner. According to Jones et al., this is because Zr replaces Ti in TiB_2 and generates ZrB_2 covering the surface of TiB_2, which greatly weakens the nucleation ability of TiB_2 and poisons the refiner. Abdel-Hamid considers that Zr may diffuse into TiB_2, gradually forming a stable solid solution phase, thereby reducing the hetero-nucleation ability of TiB_2. Murty holds the view that the reaction generates Al-Ti-Zr ternary compounds with weaker nucleation ability than TiB_2. Bunn thinks that this toxic effect is due to Zr replacing Ti atoms in the $TiAl_3$ layer on the surface of TiB_2 [45]. Recent studies based on the dual nucleation theory suggest that the toxicity is caused by the replacement of Ti by Zr in the $TiAl_3$ layer of TiB_2 [46,47]. In addition, molecular dynamics simulation results show that the Ti_2Zr_2 compound roughens the surface of TiB_2 and reduces the atomic order of the surface [48]. (2) The effect of the reaction between Zr and Ti or impurity elements on grain growth was studied. Spittle showed that adding trace elements such as Zr, Fe, Si, or Cr to pure Al melt alone could refine the grains [49,50]. However, when Zr and Fe or Zr and Si solute elements are present in the pure Al melt added to Al-Ti alloy, Zr will interact with other elements to form intermetallic phase, and the reduction of these solute elements will reduce the growth restriction effect of solute, resulting in grain coarsening. Johnsson considered that Zr, Ti, and Al formed ternary compounds, which reduced the Ti solute in the melt and the inhibitory effect of Ti solute on the growth of Al grains [51]. Qiu calculated the matching relationship between Al_8Fe_4Zr and Al substrate using the edge–edge matching model, indicating that Al_8Fe_4Zr reduces the nucleation ability of $TiAl_3$ [52,53].

In forged aluminum alloys, the grain-refinement effect of Al-Ti-B is more significant due to the absent or low Si [54], but for the most widely used cast aluminum alloys, the high content of solute Si is prone to greatly attenuating the refining ability of Al-Ti-B and resulting in a toxic effect [55,56]. At present, it is widely recognized that TiB_2 and $TiAl_3$

particles in grain refiners interact with Si solves to form intermetallic compounds, which causes TiB_2 to lose its nucleation ability and blocks the epitaxial nucleation of α-Al. The latest research shows that the cause of Si toxicity may be the segregation of Si solute at the TiB_2/Al interface [57].

To sum up, Zr toxicity and Si toxicity are the major diseases restricting the thinning effect of Al-Ti-B. Although some researches have explained them, there is still a lack of a perfect unified theoretical mechanism. Recent studies on the toxic effect show that the reduction of nucleation efficiency of TiB_2 is mainly related to the newly formed compounds with low nucleation ability on the surface of the upper Al_3Ti layer (such as Ti_2Zr, Al_8Fe_4Zr, TiSi, $TiSi_2$, Ti_5Si_3, etc.). At the same time, it may also be caused by the expansion of the invisible nuclear zone during solidification. Based on the importance of Al-Ti-B refiner and its toxic effect, ultrasonic treatment with melt is expected to inhibit or eliminate the toxic effect so as to significantly improve the grain-refining technology level of aluminum alloy.

In order to improve the production conditions of aluminum alloy grain refiners and solve the problems of thinning decline and poisoning, domestic and foreign efforts are devoted to the research of new grain refiners including Al-Ti-C, Al-Ti-C-B, Al-Ti-B-Re, and Al-Ti-C-Re. The addition of C and Re can greatly improve the precipitation and aggregation of $TiAl_3$ and TiB_2 and increase the viscosity of aluminum melt. The addition of C can not only avoid the pollution of the environment by harmful gases in the production process but also eliminate the inclusion generated by the reaction with salt. The addition of Re can improve the decay delay of Al-Ti-B and greatly improve the thinning efficiency. However, due to the poor wettability of C and the alloy matrix, the yield of C is low, which reduces the production efficiency. Furthermore, Re is expensive, and the preparation process is complex. So, it is still necessary to develop more efficient process technology to achieve large-scale industrial production.

2.2. Al-Ti-C

The refining capacity of Al-Ti-carbon grain refiner is lower than that of Al-Ti-boron grain refiner, and it has obvious attenuation. Coupled with the complex preparation process, Al-Ti-C is a terpolymer intermediate alloy refiner that was proposed before Al-Ti-B. In the 1950s, the carbide and boride nucleation theory proposed by Cibula [58] suggested that TiB_2 and TiC were good heterogeneous nucleation bases for α-Al. TiB_2 and TiC particles with a high melting point can still exist stably in aluminum melt after being melted by Al-Ti-B and Al-Ti-C grain refiners and become heterogeneous nucleation substrates during α-Al solidification. Subsequently, Cibula tried to prepare Al-Ti-C intermediate alloy by various methods but was not successful due to the poor wettability of carbon in aluminum liquid and the small density of carbon that floated easily in aluminum liquid. In 1985, Banerji obtained Al-Ti-C intermediate alloy using the strong stirring method in the laboratory. Since then, Al-Ti-C intermediate alloys have once again attracted wide interest [9].

Al-Ti-B is widely used as a grain refiner for aluminum alloy, but TiB_2 particles interact easily with Zr, Mn, Cr, and other elements in aluminum alloy, resulting in the "refiner poisoning" phenomenon, which leads to the decline of its refining effect, causing uneven grain structure. In addition, in Al-Ti-B refiner, TiB_2 particles, which can be used as an effective nucleating core, agglomerate and precipitate easily, which will damage the surface quality of rolled sheet and foil [59–61]. TiB_2 particles are usually less than 1% of the total, which means the refiner's nucleation ability is limited, so it is necessary to add a lot of Al-Ti-B refiner to reduce the grain size of aluminum alloy to 100 μm. Al-Ti-C grain refiner has an excellent grain-refining effect [50,62]. TiC is the main heterogeneous nucleation particle and plays the role of refining α-Al grains, and $TiAl_3$, as a good flux, can promote the wetting of carbide and aluminum melt, thus promoting the formation of TiC, providing conditions for refining. TiC particles are much smaller than TiB_2 particles, with an average size of about one-third that of TiB_2 particles. The small size of TiC particles makes TiC in aluminum melt difficult to precipitate to the bottom of the melt, which makes Al-Ti-C grain refiner have good resistance to refining decline [63]. In addition to the advantages of TiC particles, Al-Ti-C grain refiner has a certain degree of resistance to Si and Zr poisoning, unlike Al-Ti-B grain refiner, which loses its refining ability in the case of Si and Zr (only 0.2 wt.%). Zr can eliminate the refining effect of Al-Ti-B in a short time [64,65]. TiC particles are too stable in liquid aluminum to react with alloying elements such as Mn, Cr, and Zr, so TiC particles are immune to the phenomenon of "poisoning" of elements such as Zr, Cr, V, Si, and Mn. The aggregation tendency of TiC particles is smaller than that of TiB_2 particles. Al-Ti-C intermediate alloys contain a large number of $TiAl_3$ and TiC phases and are currently the best choice to replace Al-Ti-B refiners [66].

Al-Ti-C series grain refiners avoid Zr poisoning. It is generally proposed that TiC particles in Al-Ti-C grain refiners have a smaller aggregation tendency than TiB_2 particles in Al-Ti-B and have toxic "immunity" to Zr, Cr, Si, V, and Mn [63]. However, Ding showed that Zr could also poison Al-Ti-C grain refiner [48]. Studies have shown that the Zr poisoning of Al-Ti-B and Al-Ti-C grain refiners or the degree of weakening or disappearance of their refining effect is related to many factors such as melting holding time and temperature [67]. The effect of Zr on the grain refinement of the commercial purity Al is shown in Figure 3. It can be observed that at 730 °C, the grain coarsens gradually with increasing Zr content. The poisoning is mainly due to the reaction between the Zr-containing phase in aluminum alloy and the Al_3Ti phase or related phase (TiB_2/TiC) in the refiner [60,68–70]. And the substitution of Zr elements on the surface of TiB_2/TiC particles forms ZrB_2/ZrC [49]. These two theories are affected by the low content, wide distribution, or unequal distribution of Zr elements or grain refiners in the alloy, and it is difficult to observe the phenomenon whereby the Zr-containing phase reacts with the related phase in the refiner and forms a composite phase. There is no direct evidence to support either theory.

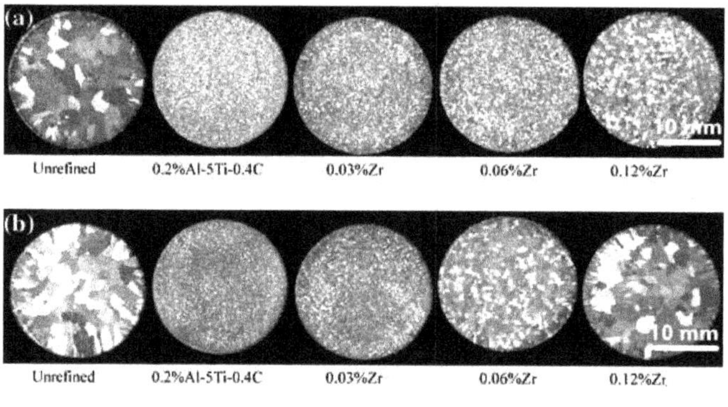

Figure 3. Macrostructures of samples refined with 0.2%Al–5Ti–0.4C for different content of Zr. (**a**) The commercial purity Al melted at 730 °C; (**b**) The commercial purity Al melted at 800 °C [63].

At present, the main preparation methods are the solid–liquid reaction method, reaction sintering method, SHS (self-propagating high-temperature synthesis) casting method, XD (exothermic reaction) method, VLS (gas–liquid–solid contact reaction) method, etc. In the 1980s, Banerji and Reif prepared Al-Ti-C grain refiner by adding preheated graphite powder to superheated Al-Ti melt using the solid–liquid reaction method [9,11]. The SHS method mixes the matrix metal powder with the solid–liquid powder containing the internal phase components and ignites the sample in the test device with the ignition tungsten wire so that the chemical reaction between Al, Ti, and C is generated and the combustion wave is spread, and when the combustion wave is pushed forward, the reactant becomes the Al-Ti-C refiner. The XD method mixes both the solid reactive element powder and metal substrate powder evenly after compacting and degassing, pressing the block under vacuum conditions, and rapidly heating to the temperature above the melting point of the metal substrate. In the metal melt medium, the two solid reaction elements diffuse with each other, contact, and constantly react to precipitate a stable reinforcement phase, and then the melt is cast and extruded. The VLS method was invented by Koczak in 1989.

The VLS process mainly includes two reaction processes: first, under the action of the high-temperature alloy liquid, the gas in the alloy liquid is dispersed. The second is the chemical reaction of a gas or element obtained after decomposition in the alloy liquid.

The preparation technology of Al-Ti-C refiner is seriously lagging and still has not been applied on a large scale, mainly because of the poor wettability of liquid aluminum to carbon during the preparation process, which makes alloying difficult and makes it difficult to realize industrial production and practical application. At 1000 °C, the wetting angle between graphite and liquid aluminum is still greater than 90°, which is a nonwetting state. Improving wettability has become a key technology for preparing Al-Ti-C grain refiners.

2.3. *Al-Ti-B-C*

Al-Ti-B and Al-Ti-C series intermediate alloys have been widely used in the grain-refining process of aluminum alloy, and the refining effect is obvious. The main reason is that there is excellent nucleation base TiB_2 or TiC in the refiner, and the solute Ti can significantly inhibit the grain growth of Al. However, the agglomeration effect of TiB_2 particles in Al-Ti-B intermediate alloy can cause pinhole defects in aluminum foil and strip defects in anodized sheet. In addition, Zr, Cr, Si, and other elements will also lead to TiB_2 surface "poisoning", so that the refining performance rapidly declines. These disadvantages limit the application of Al-Ti-B intermediate alloys. Al-Ti-C series intermediate alloy is considered to be an ideal substitute for it. However, it was found that the long-term thermal insulation treatment in the refining process may cause the decomposition of TiC particles, thus reducing the refining performance of the intermediate alloy. In addition, ternary intermediate alloys are often produced by the fluorine salt method in the process of preparation, but the production process has environmental problems. Moreover, the refining effect of the traditional Al-Ti-B series intermediate alloy on Al alloy has reached the limit, and the question of how to break this refining bottleneck has become an important topic. A series of Al-Ti-B-C intermediate alloys were prepared by reacting liquid aluminum with pure Ti, graphite powder, and Al-3B intermediate alloys. The results show that Al-Ti-B-C quaternary intermediate alloys containing $TiAl_3$, TiB_2, and TiC particles can synthesize the advantages of Al-Ti-B and Al-Ti-C intermediate alloys. Nie avoids the poisoning phenomenon of TiB_2 in aluminum alloys containing Cr, Zr, and Mn and the decline of the anti-decay phenomenon caused by the reaction of TiC particles with aluminum melts, obtaining better refining properties [71].

Compared with Al-Ti-B and Al-Ti-C intermediate alloys, Al-Ti-B-C has stronger resistance to Zr "poisoning". Both Al-Ti alloy and Al-Zr alloy have a good refining effect when refining aluminum alloy alone, but when they are added to alloy melt at the same time, the grain-refining performance will be significantly weakened; that is, the two appear to produce a mutual "poisoning" phenomenon [46]. Al-Ti-B-C quaternary intermediate alloys containing $TiAl_3$, TiB_2, and TiC phases have excellent refining properties.

The wettability of Al and B_4C is the key to the preparation of Al-Ti-B-C. Some studies consider that in a vacuum state of 1200 °C and 1300 °C for 15 min, the contact angle of Al/B_4C is 45° and 20°, respectively, but with the addition of 10% Ti element, the contact angle under the same state is rapidly reduced to 0°. High-energy ultrasonic treatment and induction melting can change the local or whole reaction temperature of the melt, which has potential positive effects on improving the wettability of Al/B_4C and promoting the reaction process. While meeting the requirements of grain refinement, it can effectively avoid the pollution and disposal problems of harmful waste residue in the traditional process.

2.4. Al-Ti-B-(C)-Ce

With special electron layer structure and active chemical properties, rare-earth elements have the reputation of being "industrial vitamins". They can form rare-earth compounds with almost all kinds of elements. Rare earth can not only play a refining and metamorphic role in aluminum alloy but can also remove gas and impurities, reduce the pinhole rate, and play a refining role. Rare earth is the only alloying element that can improve the corrosion resistance of aluminum and its alloys. Among the alloys, Al-Ti-C, Al-Ti-B, and other intermediate alloys are refiners used for aluminum and aluminum alloys, and remarkable results have been achieved. However, the nucleation particles of these refiners are often obtained by in situ reaction, and their morphology, size, and aggregation state are difficult to control, which increases the difficulty of regulation. Rare earth can improve the morphology and distribution of the second-phase particles in Al-Ti-B or Al-Ti-C refiners, thereby increasing their refining effect. Therefore, rare-earth composite refinement is regarded as an important direction for grain refinement of aluminum alloys in the future.

The advantages of rare earths as refining agents are currently divided into two theories. One is that in the high melting point compounds formed by rare-earth elements and metallic aluminum, certain alloying elements or impurities (Fe, Si, S, etc.) act as heterogeneous nuclei and precipitate at the grain boundaries. Due to the pinning effect and organization of grain growth, the high surface activity of rare earths can reduce the surface energy of the nucleus and increase the rate of crystalline nucleation. The other is that due to the low solid solubility of rare-earth elements in aluminum melt and of surface active substances, it is easy to adsorb the polarization of grain boundaries and phase surfaces and fill the defects on the interface, thus hindering the growth of nucleation cores. The lattice mismatch between Ce and Al is 27.3%, which means that they are insoluble in Al and do not form a replacement solid solution with Al. Owing to its high surface energy, the aluminum melt is wrapped on the surface of the aluminum grain during crystallization, which hinders the growth of the grain. Rare earth and aluminum form a compound with a high melting point, which stably exists in aluminum melt as the core of heterogeneous nucleation, hindering the growth of grains and refining the grains of the matrix. In addition, the existence of rare-earth phase distributed at the grain boundary forms a compositional supercooling mechanism to metamorphize the aluminum melt and refine the grain size. Adding trace rare earth to high purity aluminum has the effect of refining pure aluminum grains. Adding about 0.3% aluminum–cerium alloy refiner can achieve the best refining effect. $Al_{11}Ce_3$ and α-Al produced by eutectic reaction during the preparation of Al-Ce intermediate alloy have similar crystal structure, and the lattice constant can be corresponding, so $Al_{11}Ce_3$ can be used as the heterogeneous nucleation point during the solidification of α-Al, thus promoting thinning. By refining the pure aluminum grains, the pure aluminum materials can be strengthened and toughened. The mechanism of rare-earth refining of pure aluminum affects the solute distribution within the grain and at the liquid–solid interface.

An Al-B-Ce intermediate alloy containing Al, CeB_6, and $Al_{11}Ce_3$ phases was prepared by vacuum arc furnace. The particle size of CeB_6 was detected to be about 2–4 μm and was found in clusters in the aluminum matrix. Al-B-Ce intermediate alloy was rapidly solidified to prepare a thin strip of CeB_6/Al inoculant. As shown in Figure 4, CeB_6/Al with different cooling rates has different inoculation effects. The results showed that compared with Al-B-Ce intermediate alloy, CeB_6 particles became smaller and dispersed in the aluminum matrix. The enhancement effect of CeB_6/Al inoculant depends on the size of CeB_6 phase. The thin strip CeB_6/Al inoculant, with an average particle size of about 500 nm, has a better refining and strengthening effect. According to the calculation of mismatch theory, the mismatch between CeB_6 particles and Al is small and the lattice match is good. Therefore, CeB_6 particles can be used as an effective and stable nucleation core of Al.

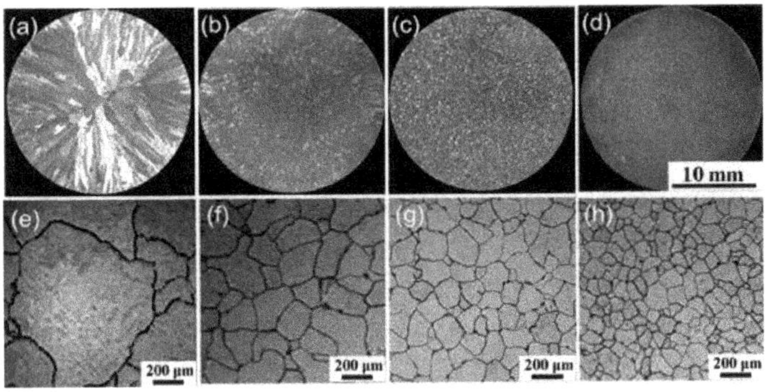

Figure 4. Optical images of as-cast aluminum samples: (**a**,**e**) pure Al without inoculation; (**b**–**d**,**f**–**h**) are inoculation with in situ CeB_6/Al composite inoculant with different cooling rates, respectively [72].

The addition of rare earth has a significant effect on the size, morphology, and distribution of $TiAl_3$ phase of Al-Ti-B intermediate alloy, and the addition of an appropriate amount of rare earth can reduce the wetting angle between Al melt and TiB_2, make TiB_2 difficult to aggregate and precipitate, and enhance the refining property of the intermediate alloy. Therefore, the combination of rare earth and Al-Ti-B is regarded as an effective method to improve the refining effect of aluminum, titanium, and boron.

The rare-earth element Ce can improve the wettability between graphite and liquid aluminum and solve the key problem of the wettability of Al-Ti-C series grain refiners. The surface activity and catalytic action of rare-earth element Ce effectively improves the wettability between graphite and liquid aluminum, promotes the reaction of graphite, and increases the amount of TiC production. Rare-earth and Al-Ti-C composite thinning $TiAl_3$ phase not only maintain blocky and uniform distribution but also reduce the agglomeration of TiB_2 particles, resulting in an enhanced thinning effect. For example, it has an "immune" effect on Mn, Cr, Zr, and other elements in aluminum alloy, is difficult to decompose, and is difficult to aggregate with TiC particles. However, Al-Ti-C intermediate alloy research started late, so the refining effect is unstable, and the wettability between aluminum liquid and carbon is poor, which seriously limits the application of Al-Ti-C intermediate alloy in actual production. Ding studied the effect of rare-earth element Ce on the microstructure and refining effect of Al-5Ti-0.6C-1Ce grain refiner [73]. Rare-earth Ce improves the refining effect of Al-5Ti-0.6C-1Ce grain refiner because it promotes the formation of TiC particles, but it has a great influence on the synthesis of TiC particles; Xu prepared Al-Ti-C-La grain refiner containing La using self-propagating high-temperature technology, and found that La can improve the wettability of liquid aluminum and graphite, and the addition of La leads to the dispersion of $TiAl_3$ and TiC particles in the aluminum matrix. The rare-earth element Ce was introduced into Al-5Ti-0.25C grain refiner, the wettability between graphite and liquid aluminum was improved by the surface activity and catalysis of rare-earth element, and the refining effect was improved. The microstructure of Al-5Ti-0.25C-1Ce grain refiner not only contains more $TiAl_3$ and $Ti_2Al_{20}Ce$ phases, but also ensures that the phase size of $TiAl_3$ and $Ti_2Al_{20}Ce$ is not too large. The contents of $TiAl_3$ and TiC are 44.52% and 60.87%, respectively, higher than those of Al-5Ti-0.25C grain refiner.

Two different forms of Al-Ti-B-C-Ce inoculant were prepared by vacuum arc furnace and vacuum rapid quenching furnace. The reinforced particle size of Al-Ti-B-C-Ce intermediate alloy can reach several microns. It is then quickly melted and cooled to obtain Al-Ti-B-C-Ce inoculant ribbons in which the size of the reinforcing particles is about a few hundred nanometers. Compared with the intermediate alloy, the phase composition of the inoculant ribbon did not change, but the reinforcement particles became smaller, with a size of about 600 nm, and dispersed in the Al matrix. Due to the similarity of crystal structure, Al_3Ti and Al have a good lattice matching relationship. The addition of rare-earth elements promoted the formation of $Ti_2Al_{20}Ce$ phase. The presence of rare earth not only plays the role of dehydrogenation, deoxygenation, and purification of melt, but also changes the morphology of Al_3Ti from acicular to massive and becomes a reinforcement that does not split the matrix and is dispersed in the Al matrix, which not only increases the dispersion strengthening effect, but also improves the grain-refinement effect.

Song conducted a study on adding (Pr + Ce) rare-earth mixture to Al-7Si-0.7Mg alloy and added 0.6 wt% (Pr + Ce) rare-earth mixture to obtain the best refining effect. The best grain-refining effect can be obtained by adding different contents of La and Ce composite rare earths to La and Ce alloys to form a variety of compounds with other alloying elements. These compounds can be used as heterogeneous nucleation substrate to improve the nucleation rate of alloys. At the same time, La and Ce are enriched at the front of the solid–liquid interface, which hinders the growth of α-Al.

2.5. Al-Sc Alloy

Scandium (Sc) is the most effective alloying element for aluminum alloy found so far. It can significantly improve the properties of aluminum alloy. Sc is not only a rare-earth element but also a transition metal element, which has the advantages of both. Sc can not only refine the aluminum alloy structure but also increase the recrystallization temperature and hinder grain growth, thereby improving the strength of the material, corrosion resistance, and heat resistance. The grain-refining effect of scandium is due to the fact that the intermetallic compound Al_3Sc and α-Al have the same FCC (face-centered cubic) structure. Its lattice constant is about 0.414 nm, while that of α-Al is 0.412 nm at 659 °C [74]. At this time, the mismatch is only about 0.5%, and the interface energy and distortion energy needed to be overcome by the nucleation of aluminum atoms at the Al_3Sc interface are small, which plays a very effective role in the nucleation core. The Al_3Sc precipitated from the solid solution is dispersed in fine particles, which can significantly improve the strength and recrystallization temperature of aluminum alloy semifinished products.

The grain-refinement effect of Sc and Ti combined in aluminum alloy is better than that of a single addition. Sc has an excellent grain-refining effect on Al alloy and can improve the as-cast microstructure of Al alloy. Figure 5 shows the optical image, EBSD image, and statistical analysis results of the as-cast Al-Sc alloy. It can be seen that the grain size decreases with the increase in Sc content. Without changing the grain size, Sc content has a significant effect on the number of grains. Li found that with an increase in Sc content, not only the as-cast grains of an Al-Zn-Mg-Mn alloy were refined, but also the microstructure gradually changed from dendrite to equiaxed. It is worth mentioning that a semicontinuous ingot of Al-Mg-Mn-Zr aluminum alloy containing Sc has a uniform equiaxed crystal structure, and the average grain size is only about 22 μm [75].

Sc and Zr added to aluminum at the same time will also produce an excellent grain-refinement effect [76]. The mechanism of Sc and Zr composite grain refinement is that the generated Al_3Zr and Al_3Sc phases have a lattice structure similar to α-Al and can be used as nucleation particles, which increases the nucleation rate of the alloy. The combination of rare-earth and non-rare-earth elements can reduce the amount of rare-earth metals and thus reduce costs. Yin studied the effect of the composite effect of rare-earth Sc and Zr on the microstructure and mechanical properties of Al-Mg alloy. The results show that the grain size of Al-5Mg alloy is reduced from 370 μm to 42 μm, and the tensile strength is increased from 260 MPa to 398 MPa. At the same time, the elongation of 18% is maintained. It is found that the as-cast grain of Al-Zn-Mg-Cu alloy can be significantly refined by adding trace Sc and Zr, and the dendrite structure can be reduced. The average grain of Al-Zn-Mg-Cu alloy can be refined to 35 μm. In summary, the advantages and disadvantages of different types of inoculants are shown in Table 1.

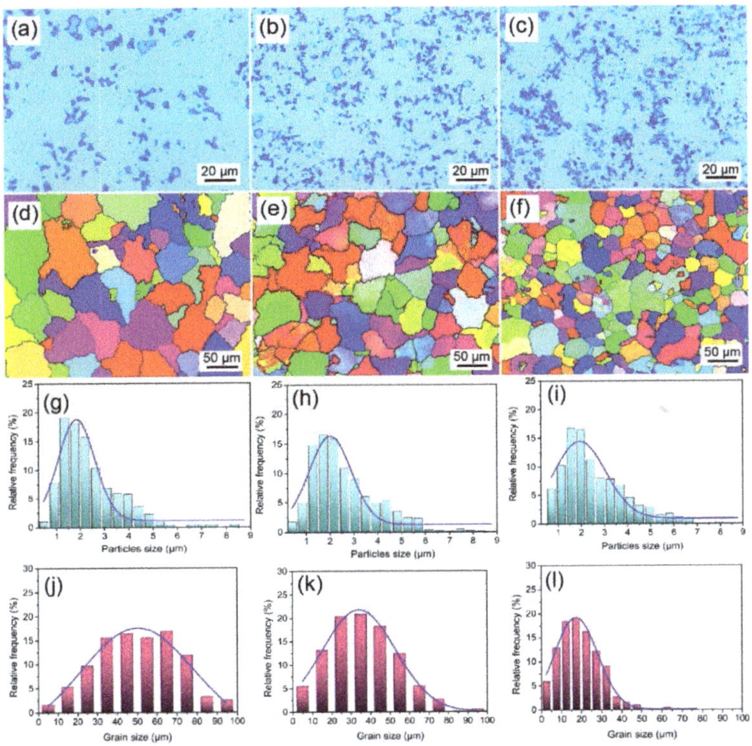

Figure 5. (**a–c**) Representative OM images of as-cast Al-xSc (x = 2, 4, 8) alloys, with corresponding EBSD maps of (**d–f**). (**g–i**) are the particle size distribution maps of as-cast Al-xSc alloys. (**j–l**) are the statistical distribution of grain size of as-cast Al-xSc alloys [74].

Table 1. Disadvantages and advantages of different kinds of inoculants.

Refiner Type	Obvious Advantage	Analysis of Disadvantages
Al-Ti-B	• High refining efficiency: 12 times higher than Al-Ti • Al3Ti and TiB2 are excellent nucleation cores of Al [40,41] • TiB2 particle size is 4~5 µm [42] • The addition of 1.5 wt.%~2.5 wt.% has an effective grain-refinement effect	• TiB_2 is easy to accumulate and deposit, causing quality problems [43] • In the case of elements such as Zr, Cr and Mn, it will be "poisoned" and lead to failure [43]
Al-Ti-C	• Pure and pollution-free • Immune to elements such as Zr, Cr, and Mn [63] • TiC particle size is 2~3 µm. TiC particle size is smaller [62,65]	• Poor anti-attenuation performance [58] • The refinement effect is slightly poor [58] • C has poor wettability with matrix [9]
Al-Ti-B-C	• Refining efficiency has been further improved [46] • Strong wettability with matrix • Immune to elements such as Zr, Cr, and Mn [71]	• Borides tend to segregate [43]. • Carbide stability is poor
Al-Ti-B(C)-Ce	• Includes the advantages of the above three refiners [73] • Effective control of particle size, morphology, and aggregation state [72] • The grain size of industrial pure aluminum can be refined from 1430 µm to 227 µm by adding 0.3 wt.%	• Moderate dosage control • High temperature preparation can lead to oxidation caking and other problems

Table 1. *Cont.*

Refiner Type	Obvious Advantage	Analysis of Disadvantages
Al-Sc alloy	• Has the advantages of both rare-earth elements and transition metals [74] • Adding 0.1~0.2% Sc can increase the recrystallization temperature of the alloy by 150~220 °C • The addition of 0.05~0.1 wt.% has an effective grain-refinement effect [75]	• Resources are scarce, dispersed and expensive • Dosage control. The performance deteriorates if the Sc content exceeds 4%

2.6. Fe-Rich Phase in Al-Si Alloy

In conventional cast Al-Si alloys, the primary α(Al) phase in the room temperature solidification structure is usually dendritic with good strength and toughness, while the Si phase is hard and brittle. The primary Si is generally a polygonal block, and the eutectic Si is mostly of a long acicular or lamellar structure. Al-Si alloys with lower Si content usually have excellent casting properties, fatigue resistance, and corrosion resistance [45]. With the increase in Si content, when eutectic silicon and primary silicon are precipitated in large quantities, the Al matrix will be cut, and the ductile strength of the alloy will be seriously reduced. In industrial production, Al-Si alloys with a Si mass fraction greater than 6% are usually modified. After modification treatment, the primary Si will become fine and rounded from a large polygonal block, and the eutectic silicon will become fine lamellar, or even fibrous, from sharp, long needles. The transformation of the Si phase microstructure greatly promotes the improvement of the mechanical properties of the alloy. Modification treatment has become the most extensive and effective means to improve the properties of Al-Si alloys. Subeutectic and eutectic Al-Si alloys are usually modified by chemical modification, rapid solidification, ultrasonic treatment, superheating treatment, melt stirring (including mechanical stirring and electromagnetic stirring), melt pretreatment, current, or electric field treatment. The metamorphic mechanism lies in controlling the nucleation and growth of eutectic Si [45]. At present, the chemical metamorphism method is mainly used in industry, and the addition of the metamorphic agent method has the advantages of stable metamorphic effect, strong operability, mature process, and no additional equipment.

The earliest chemical modifier used for hypoeutectic and eutectic Al-Si alloys was metal Na, but the low boiling point and overactive chemical properties of metal Na caused a lot of inconvenience to the production. Because of its small density, the modified alloy structure showed obvious specific gravity segregation and, generally, has been replaced by sodium salt metamorphism. At present, binary (67 wt.% NaF + 33 wt.% NaCl) or ternary (25 wt.% NaF + 62 wt.% NaCl + 13 wt.% KCl) sodium salt mixtures containing NaF are mainly used in industry as eutectic silicon metamorphisms. The advantages of this method are low application cost, good metamorphic effect, no latency, and insensitivity to cooling rate. Its disadvantages are short effective metamorphic time, remelting failure, and corrosion of crucible equipment, which will cause certain harm to the environment and human body. The addition of Sb could change the morphology of eutectic Si from needle-like to rod-like, but it was eliminated due to its high cost and toxic properties. With the deepening of the study of metamorphic agents, many countries began to use Sr metamorphic instead of Na. The metamorphic effect of Sr is similar to that of Na, and eutectic Si can also be refined into fiber [77]. Sr modification is usually added in the form of Al-10Sr intermediate alloy, and the addition of Sr near 0.02% can obtain a good metamorphic effect. Sr deterioration has many advantages such as its 5~7 h long-term performance, good remelting, small dosage, and production that is easy to control. Furthermore, it does not cause excessive deterioration, does not pollute the environment, does not corrode equipment, etc. Its disadvantages are that Sr is easy to oxidize and burn, it increases the alloy suction tendency, and its deterioration has a latency of about 40 min.

Sr has a good metamorphic effect on eutectic Si and promotes the nucleation growth of coarse columnar or dendritic α-Al grains, which indicates that it is not enough to only degrade eutectic Si for casting Al-Si alloys, and it is necessary to equiaxial and fine the α-Al dendrites to eliminate the adverse effects of this microstructure on the mechanical properties of the alloys.

Al-Ti-C and Al-Sr are added at the same time. The two do not interfere with each other, and when Al-Ti-C refines α-Al, it can promote the metamorphism of eutectic Si to a certain extent. This is because during the solidification and growth of Al-Si alloy, α-Al dendrites are precipitated, first, and then grown during the subsequent eutectic transformation. Eutectic Si is precipitated and grown in the gap between α-Al dendrites. Therefore, the size of the eutectic Si must depend on the size of the α-Al dendrites. Thus, while refining α-Al dendrites, it is shown that refining treatment has a certain metamorphic effect on eutectic Si. This mutual promotion of refinement and metamorphism can reduce the amount of modifier or refiner added, to a certain extent, and reduce production costs when the same refinement and metamorphism are obtained.

Iron is an unavoidable impurity in Al-Si alloys and has long been regarded as a harmful element. These iron impurities are originally derived from electrolytic brocade and are continuously increased in the later continuous melting process through the corrosion of the cauldron and various tools. Additional iron impurities are caused by incomplete sorting in the recycling process. Since the solid solubility of Fe in aluminum and aluminum alloys is only 0.05%, Fe basically exists in the form of iron-rich phase in aluminum, such as $AlFe_3$, Al_5FeSi, Al_8Fe_2Si $(MnFe)Al_6$, $(CuFe)Al_6$, etc. [78]. The iron impurities in the alloy can form a variety of complex lamellar intermetallic chemicals with other elements, resulting in serious separation of the matrix [45]. In addition, the large size of the iron-rich phase has an adverse effect on the feeding of the alloy after solidification, thereby increasing the formation tendency of porosity, so the iron content is often required to be low in the cast alloy to avoid the deterioration of its mechanical properties [79–81]. When the iron content is high, many methods can be adopted to change the shape of the iron-rich phase or limit its size to reduce its harm. Metal elements can usually be added to transform the iron-rich phase from a sheet-like phase to a kanji-like quaternary or multicomponent iron-rich phase [82,83]. However, the addition of these elements leads to a further increase in the content of the iron-rich phase, which makes them impossible to use without restriction [84,85]. In addition, the use of high cooling speed, ultrasonic treatment, friction stir processing, and other methods can limit the size of the iron-rich phase and can transform the iron-rich phase of the crystal type and improve the mechanical properties of the alloy [86–88]. As shown in Figure 6, the optical micrographs of Al-Si-Fe at different cooling rates show different stages of iron-rich phase [89].

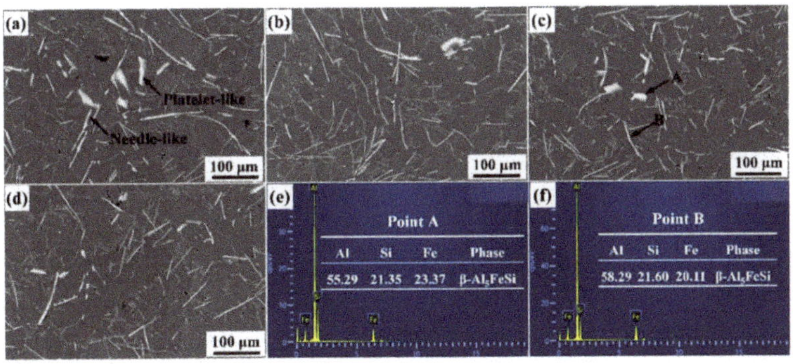

Figure 6. The morphology of the Fe-rich phases with different B contents ((**a**) B00 alloy; (**b**) B15 alloy; (**c**) B30 alloy; and (**d**) B60 alloy), and chemical compositions at Point A (**e**) and B (**f**) [89].

The common method of removing iron is to form the first precipitated compound with the iron element and other additives, and then use filtration, centrifugation, electromagnetic separation, precipitation, and other methods to separate these compounds. Electrolysis methods were invented by Hoopes, but these methods have some problems; e.g., they are time-consuming and have a low impurity removal efficiency. Recently, the newly developed electroslag remelting method has obtained good iron removal efficiency. This method uses a slagging agent to react with iron to form an iron–boron intermediate compound during the electroslag remelting process, which is then captured by the molten slag agent [90,91].

3. Conclusions

(1) In recent years, efforts have been made to develop new efficient and environmentally friendly grain refiners for aluminum alloys, including Al-Ti-C, Al-Ti-C-B, and Al-Ti-B-C-Ce, and adding rare-earth elements. The addition of Re element can improve the decay delay of Al-Ti-B and greatly improve the refining efficiency. However, due to the poor wettability of C and the Al matrix, the yield of C is low and the production efficiency is reduced, while Re is expensive and easy to oxidize, and more efficient process technology still needs to be developed to achieve large-scale industrial production.

(2) Sc is the most effective element for grain refinement of aluminum alloy, and its mismatch with aluminum matrix is only 0.5%. The interface energy and distortion energy needed to be overcome by the nucleation of aluminum atoms at the Al3Sc interface are very small, and it plays a very effective role in nucleation. However, its application as nucleation substrata in production requires further consideration of cost and environmental issues. Therefore, Sc acts together with other elements, such as Ti and Zr, which is a very promising research direction for the future.

(3) The excessive iron element in Al-Si alloy is the key problem restricting its mechanical properties and recycling. Whether it is chemical modification, rapid solidification, ultrasonic treatment, or overheating treatment, there are complex processes and environmentally unfriendly problems; therefore, Al-Si alloy will develop in the direction of high efficiency, low addition, high purity, and high stability, and the research work of the grain refinement mechanism will be gradually advanced.

Author Contributions: Conceptualization, S.L.; methodology, S.L.; validation, S.L.; writing—original draft preparation, T.Z.; writing—review and editing, J.F. and Q.Z.; supervision, S.L. and Q.Z. All authors have read and agreed to the published version of the manuscript.

Funding: This research was funded by the Natural Science Foundation of Hebei Province (No. E2021202140), Ministry of Education "Chunhui plan" Cooperative Scientific Research Project (HZKY20220253), Outstanding Youth Innovation Fund of State Key Laboratory for Reliability and Intelligence of Electrical Equipment (EERI_OY2022009), and S&T Program of Hebei (215676146H and 225676163GH).

Institutional Review Board Statement: Not applicable.

Informed Consent Statement: Not applicable.

Data Availability Statement: Not applicable.

Conflicts of Interest: The authors declare no conflict of interest. We declare that we have no financial and personal relationships with other people or organizations that can inappropriately influence our work, and that there is no professional or other personal interest of any nature or kind in any product, service, and/or company that could be construed as influencing the position presented in, or the review of the manuscript submitted.

References

1. Quested, T.E. Understanding mechanisms of grain refinement of aluminium alloys by inoculation. *Mater. Sci. Technol.* **2004**, *20*, 1357–1369. [CrossRef]
2. Wang, F.; Liu, Z.; Qiu, D. Revisiting the role of peritectics in grain refinement of Al alloys. *Acta Mater.* **2013**, *61*, 360–370. [CrossRef]
3. Minagawa, A.; Piper, M. *Mechanism of High Grain Refinement Effectiveness on New Grain Refiner "TiBAl Advance"*; Springer International Publishing: Cham, Switzerland, 2021; pp. 844–849.
4. Wang, T.; Chen, Z.; Fu, H. Grain refining potency of Al–B master alloy on pure aluminum. *Scr. Mater.* **2011**, *64*, 1121–1124. [CrossRef]

5. Chen, Z.; Yan, K. Grain refinement of commercially pure aluminum with addition of Ti and Zr elements based on crystallography orientation. *Sci. Rep.* **2020**, *10*, 16591. [CrossRef]
6. Zhang, L.; Zheng, Q.; Jiang, H. Interfacial energy between Al melt and TiB_2 particles and efficiency of TiB_2 particles to nucleate α-Al. *Sci. Mater.* **2019**, *160*, 25–28. [CrossRef]
7. Xie, H.; Lv, J. Precipation of $TiAl_3$ in remelting Al-5Ti- 1B and the grain refinement of 7050 alloy. *Mater. Res. Express.* **2021**, *8*, 066513. [CrossRef]
8. Qian, P.; Tang, Z.; Yuan, M. Microstructure and refinement mechanism of $TiB_2/TiAl_3$ in remelted Al-5Ti- 1B system. *Mater. Sci. Technol.* **2019**, *35*, 1563–1571. [CrossRef]
9. Banerji, A.; Reif, W. Grain refinement of aluminum by TiC. *Metall. Trans. A* **1985**, *16*, 2065–2068. [CrossRef]
10. Banerji, A.; Reif, W. Metallographic investigation of TiC nucleants in the newly developed Al-Ti-C grain refiner. *J. Mater. Sci.* **1994**, *29*, 1958–1965. [CrossRef]
11. Banerji, A.; Reif, W. Development of Al-Ti-C grain refiners containing TiC. *Metall. Trans. A* **1986**, *17*, 2127–2137. [CrossRef]
12. Jiang, R.P.; Li, X.Q.; Zhang, M. Investigation on the mechanism of grain refinement in aluminum alloy solidified under ultrasonic vibration. *Met. Mater. Int.* **2015**, *21*, 104–108. [CrossRef]
13. Maxwell, I.; Hellawell, A. An analysis of the peritectic reaction with particular reference to Al-Ti alloys. *Acta Metall.* **1975**, *23*, 901–909. [CrossRef]
14. Tan, Q.Y.; Zhang, J.Q.; Sun, Q. Inoculation treatment of an additively manufactured 2024 aluminium alloy with titanium nanoparticles. *Acta Mater.* **2020**, *196*, 1–16. [CrossRef]
15. Hu, M.L.; Wang, F.; Jiang, B. Fabrication of Al–Ti–B Grain Refiner Using Machining Ti Chips, Reaction Mechanisms and Grain Refinement Performance in Pure Al. *Met. Mater. Int.* **2021**, *28*, 1471–1479. [CrossRef]
16. Zaid, A.I.O.; Al-Qawabah, S.M.A. Effect of Zirconium Addition on the Grain Size and Mechanical Behavior of Aluminum Grain Refined by Titanium Plus Boron (Ti plus B) in the as Cast and Cold Extruded Conditions. *Adv. Mater. XII* **2012**, *510–511*, 241. [CrossRef]
17. Wang, F.; Chiu, Y.L.; Eskin, D. A grain refinement mechanism of cast commercial purity aluminium by vanadium. *Mater. Charact.* **2021**, *181*, 111468. [CrossRef]
18. Baeckerud, L.; Johnsson, L.; Gustafson, P. Grain refining mechanisms in aluminium as a result of additions of titanium and boron. *Pt. Alum.* **1991**, *67*, 780–785.
19. Mao, G.; Liu, S.; Wu, Z. The effects of Y on primary α-Al and precipitation of hypoeutectic Al-Si alloy. *Mater. Lett.* **2020**, *271*, 127795. [CrossRef]
20. Men, H.; Fan, Z. Effects of lattice mismatch on interfacial structures of liquid and solidified Al in contact with hetero-phase substrates: MD simulations. *IOP Conf. Ser. Mater. Sci. Eng.* **2012**, *27*, 012007.
21. Han, Y.F.; Dai, Y.B.; Shu, D. First-principles calculations on the stability of Al/TiB_2 interface. *Appl. Phys. Lett.* **2006**, *89*, 144107. [CrossRef]
22. Fan, Z.; Wang, Y.; Zhang, Y. Grain refining mechanism in the Al/Al-Ti-B system. *Acta Mater.* **2015**, *84*, 292–304. [CrossRef]
23. Cao, Y.D.; Chen, X.H.; Wang, Z.D. Synergistic influence of La and Zr on microstructure and mechanical performance of an Al-Si-Mg alloy at casting state. *J. Alloys Compd.* **2022**, *902*, 163829. [CrossRef]
24. Xue, J.; Wang, J.; Han, Y.F. Behavior of CeO_2 additive in in-situ TiB_2 particles reinforced 2014 Al alloy composite. *Trans. Nonferrous Met. Soc. China* **2012**, *22*, 1012–1017. [CrossRef]
25. Ding, W.W.; Zhao, X.Y.; Zhao, W.J. Effects of Al-Ti-C-Ce Master Alloy on Microstructure and Mechanical Properties of Hypoeutectic Al–7%Si Alloy. *Int. J. Metalcast.* **2019**, *13*, 426–437. [CrossRef]
26. Davis, T.A.; Bochler, L.; Delia, F. Effect of TiB or on the grain refinement and hot tearing susceptibility of AZ91D magnesium alloy. *J. Alloys Compd.* **2018**, *759*, 70–79. [CrossRef]
27. Xu, Z.Y.; Fang, C.F.; Wang, R. Microstructural evolution, strengthening and toughening mechanisms of AZ80 composite sheet reinforced by TiB2 with fiber-like distribution. *J. Alloys Compd.* **2021**, *877*, 160278. [CrossRef]
28. Bae, J.H.; You, B.S. Feasibility of grain refinement method for AZ91 alloy using commercial Al-SiC composite. *Mater. Sci. Technol.* **2020**, *36*, 194–201. [CrossRef]
29. Zhao, Y.L.; He, W.X.; Song, D.F. Effect of ultrasonic melt processing and Al-Ti-B on the microstructural refinement of recycled Al alloys. *Ultrason. Sonochem.* **2022**, *89*, 106139. [CrossRef]
30. Wang, K.; Jiang, H.; Wang, Q. A novel method to achieve grain refinement in aluminum. *Metall. Mater. Trans. A* **2016**, *47*, 4788–4794. [CrossRef]
31. Zhao, K.; Gao, T.; Yang, H. Enhanced grain refinement and mechanical properties of a high-strength Al-Zn-Mg-Cu-Zr alloy induced by TiC nano-particles. *Mater. Sci. Eng. A* **2021**, *806*, 140852–140864. [CrossRef]
32. Wang, K.; Jiang, H.Y.; Wang, Q.D. Nanoparticle- induced nucleation of eutectic silicon in hypoeutectic Al-Si alloy. *Mater. Charact.* **2016**, *117*, 41–45. [CrossRef]

33. Wang, K.; Jiang, H.Y.; Jia, Y.W. Nanoparticle- inhibited growth of primary aluminum in Al–10Si alloys. *Acta Mater.* **2016**, *103*, 252–263. [CrossRef]
34. Lyu, Y.; Ding, R.; Xu, G. Effect of electric current pulses on TiB_2 particle dispersion in aluminum alloy melt. *Mater. Lett.* **2020**, *274*, 127891–127894. [CrossRef]
35. Gao, Q.; Wu, S.; Lü, S. Improvement of particles distribution of in-situ 5 vol% TiB_2 particulates reinforced Al-4.5Cu alloy matrix composites with ultrasonic vibration treatment. *J. Alloys Compd.* **2017**, *692*, 1–9. [CrossRef]
36. Zhao, B.; Cai, Q.; Li, X. Effect of TiC nanoparticles supported by Ti powders on the solidification behavior and microstructure of Pure Aluminum. *Met. Mater. Int.* **2018**, *24*, 945–954. [CrossRef]
37. Li, X.; Cai, Q.; Zhao, B.; Li, B.; Liu, B.; Ma, W. Grain refining mechanism in pure aluminum with nanosized TiN/Ti composite refiner addition. *J. Alloys Compd.* **2017**, *699*, 283–290. [CrossRef]
38. Karbalaei Akbari, M.; Baharvandi, H.R.; Mirzaee, O. Fabrication of nano-sized Al_2O_3 reinforced casting aluminum composite focusing on preparation process of reinforcement powders and evaluation of its properties. *Compos. Part B* **2013**, *55*, 426–432. [CrossRef]
39. Bin, H.; Liu, Y.Z.; Zhou, Z.G.; Cheng, W.; Liu, X.H. Selective laser melting of 7075 aluminum alloy inoculated by Al–Ti–B: Grain refinement and superior mechanical properties. *Vacuum* **2022**, *200*, 111030.
40. Men, H.; Fan, Z. Effects of solute content on grain refinement in an isothermal melt. *Acta Mater.* **2011**, *59*, 2704–2712. [CrossRef]
41. Li, J.; Hage, F.S.; Ramasse, Q.M. The nucleation sequence of α-Al on TiB_2 particles in Al-Cu alloys. *Acta Mater. A* **2021**, *206*, 116652–116661. [CrossRef]
42. Zhao, Z.Y.; Guan, R.G.; Guan, X.H. Effects of Electromagnetic Stirring, Shearing, and Extrusion on TiB_2 and $TiAl_3$ Particles in Al-5Ti-1B(wt.%) Alloy. *Mater. Manuf. Process.* **2015**, *30*, 1223–1228. [CrossRef]
43. Wang, S.C.; Chou, C.P. Effect of adding Sc and Zr on grain refinement and ductility of AZ31 magnesium alloy. *J. Mater. Process. Technol.* **2008**, *197*, 116–121. [CrossRef]
44. Rao, A.A.; Murty, B.S.; Chakraborty, M. Role of zirconium and impurities in grain refinement of aluminium lNith Ai-Ti-B. *Mater. Sci. Technol.* **1997**, *13*, 769–777. [CrossRef]
45. Jin, Y.; Li, Y.; Ke, M. In-situ observation of grain refinement dynamics of hypoeutectic Al-Si alloy inoculated by Al-Ti-Nb-B alloy. *Scr. Mater.* **2020**, *187*, 142–147.
46. Tamirisakandala, S.; Bhat, R.B.; Tiley, T.S.; Miracle, D.B. Grain refinement of cast titanium alloys via trace boron addition. *Scr. Mater.* **2005**, *53*, 1421–1426. [CrossRef]
47. Abdel-Hamid, A.A. Effect of other elements on the grain refinement of Al by Ti or Ti and B. *Int. J. Mater. Res.* **1989**, *80*, 566–569. [CrossRef]
48. Murty, B.S.; Kori, S.A.; Chakraborty, M. Grain refinement of aluminium and its alloys by heterogeneous nucleation and alloying. *Int. Mater. Rev.* **2002**, *47*, 3–29. [CrossRef]
49. Bunn, A.M.; Schumacher, P.; Kearns, M.A. Grain refinement by Al-Ti-B alloys in aluminium melts: A study of the mechanisms of poisoning by zirconium. *Mater. Sci. Technol.* **1999**, *15*, 1115–1123. [CrossRef]
50. Wang, Y.; Fang, C.M.; Zhou, L. Mechanism for Zr poisoning of Al-Ti-B based grain refiners. *Acta Mater.* **2019**, *164*, 428–439. [CrossRef]
51. ZHANG, L.L.; JIANG, H.X.; Jie, H. Kinetic behaviour of TiB_2 particles in Al melt and their effect on grain re-finement of aluminium Alloys. *Trans. Nonferrous Met. Soc. China* **2020**, *30*, 2035–2044. [CrossRef]
52. Johnsson, M. Influence of Zr on the grain refinementof aluminium. *Int. J. Mater. Res.* **1994**, *85*, 786–789. [CrossRef]
53. Qiu, D.; Taylor, J.A.; Zhang, M.X. Understanding the co-poisoning effect of Zr and Ti on the grain refinement of cast aluminum alloys. *Metall. Mater. Trans. A* **2010**, *41*, 3412–3421. [CrossRef]
54. Ohnsson, M. Influence of Si and Fe on the grain refinement of aluminium. *Int. J. Mater. Res.* **1994**, *85*, 781–785. [CrossRef]
55. Lee, Y.C.; Dahle, A.K.; Stjohn, D.H. The effect of grain refinement and silicon content on grain formation in hypoeutectic Al-Si alloys. *Mater. Sci. Eng. A* **1999**, *259*, 43–52. [CrossRef]
56. Hernandez, F.C.R.; Sokolowski, J.H. Comparison among chemical and electromagnetic stirring and vibration melt treatments for Al-Si hypereutectic alloys. *J. Alloys Compd.* **2006**, *426*, 205–212. [CrossRef]
57. Wang, Y.; Que, Z.P.; Hashimoto, T. Mechanism for Si poisoning of Al-Ti-B grain refiners in Al alloys. *Metall. Mater. Trans. A* **2020**, *51*, 5743–5757. [CrossRef]
58. Lim, Y.P.; Sham, S.; Abdel, M.H. Grain refinement of LM6 Al–Si alloy sand castings to enhance mechanical properties. *J. Mater. Process. Technol.* **2005**, *162–163*, 435–441.
59. Li, H.; Sritharan, T.; Seow, H.P. Grain refinement of DIN226 alloy at high titanium and boron inoculation levels. *Scr. Mater.* **1996**, *35*, 869–872. [CrossRef]
60. Taylor, J.A.; Qiu, D.; Zhang, M.X. A mechanism for the poisoning effect of silicon on the grain refinement of Al-Si alloys. *Acta Mater.* **2007**, *55*, 1447–1456.
61. Ding, W.; Chen, T.; Zhao, X.; Cheng, Y.; Liu, X.; Gou, L. Investigation of Microstructure of Al-5Ti-0.62C System and Synthesis Mechanism of TiC. *Materials* **2020**, *13*, 310. [CrossRef]
62. Jiang, H.; Sun, Q.; Zhang, L. Al-Ti-C Master Alloy with Nano-Sized TC Particles Dispersed in the Matrix Prepared by Using Carbon Nanotubes as C Source. *J. Alloys Compd.* **2018**, *748*, 774–782.

63. Ding, H.; Liu, X.; Yu, L. Influence of Zirconium on Grain Refining Efficiency of Al-Ti-C Master Alloys. *J. Mater. Sci.* **2007**, *42*, 9817–9821. [CrossRef]
64. Dong, Y.; Wang, M.; Zhang, G. Influence of Ti/C Mass Ratio on the Microstructure of Al-Ti-C Master Alloy and Refinement Effect on Pure Aluminum. *Results Phys.* **2021**, *23*, 104000. [CrossRef]
65. Vinod, G.S.; Murty, B.S.; Chakr, M. Development of Al–Ti–C grain refiners and study of their grain refining efficiency on Al and Al–7Si alloy. *J. Alloys Compd.* **2005**, *396*, 143–150. [CrossRef]
66. Mccartney, D.G. Grain refining of aluminium and its alloys using inoculants. *Int. Mater. Rev.* **1989**, *34*, 247–260. [CrossRef]
67. Jiang, H.T.; Hui, X.; Xu, Z.H. Effect of Zn content and Sc, Zr addition on microstructure and mechanical properties of Al-Zn-Mg-Cu alloys. *J. Alloys Compd.* **2023**, *947*, 169246. [CrossRef]
68. Guo, Y.W.; Wei, W.; Shi, W. Effect of Er and Zr additions and aging treatment on grain refinement of aluminum alloy fabricated by laser powder bed fusion. *J. Alloys Compd.* **2022**, *912*, 165237. [CrossRef]
69. Yang, H.B.; Gao, T.; Wang, H.C. Influence of C/Ti stoichiometry in TiCx on the grain refinement efficiency of Al–Ti–C master alloy. *J. Mater. Sci. Technol.* **2017**, *33*, 616–622. [CrossRef]
70. Sha, S.G.; Ghol, A.; Nag, S. Effect of Al-5Ti-1B and Zr grain refiners on the solidification characteristics and microstructure of Al-6Mg alloy studied through thermal analysis. *Thermochim. Acta* **2022**, *711*, 179200.
71. Ding, W.; Zhang, X.; Zhao, W. Microstructure of A1-5Ti-0.6C-1Ce Master Alloy and Its Grain-Refining Performance. *Int. J. Mater. Res.* **2015**, *106*, 1240–1243. [CrossRef]
72. Liu, S.Q.; Xin, W.; Cui, C.X. Enhanced grain refinement of in situ CeB_6/Al composite inoculant on pure aluminum by microstructure control. *J. Alloys Compd.* **2017**, *701*, 926–934. [CrossRef]
73. Zhang, L.; Song, Y.; Yang, L. Effect of La and TiB_2 Particles on Grain Refinement in Aluminum Alloy. *Materials* **2022**, *15*, 600. [CrossRef] [PubMed]
74. Liu, S.Q.; Xin, W.; Qun, Z.; Han, B.; Han, X. Significantly improved particle strengthening of Al-Sc alloy by high Sc composition design and rapid solidification. *Mater. Sci. Eng. A* **2021**, *800*, 140304. [CrossRef]
75. Fridlyander, I.N.; Tkachenko, E.A.; Bersteney, V.V.; Cherepok, G.V.; Latushkina, L.V.; Zhegina, I.P.; Grinevich, V.A. Effect of microstructure on the cracking resistance characteristics of Al-Zn-Mg-Cu-Zr wrought high-strength alloy. *Mater. Sci. Forum* **2002**, *396–402*, 396–402.
76. Vlach, M. Phase transformations in isochronally annealed mould cast and cold-rolled Al-Sc-Z-based alloy. *J. Alloys Compd.* **2010**, *492*, 143–148.
77. Samuel, A.; Doty, H.; Valtierra, S. A Metallographic Study of Grain Refining of Sr-Modified 356 Alloy. *Int. J. Metalcast.* **2017**, *11*, 305–320.
78. Ferreira, T.; Oliveira, I.L.D.; Zepon, G. Rotational outward solidification casting: An innovative single step process to produce a functionally graded aluminum reinforced with quasicrystal approximant phases. *Mater. Des.* **2020**, *189*, 108544.
79. Puncreobutr, C.; Phillion, A.; Fife, J.; Lee, P. Coupling in situ synchrotron X-ray tomographic microscopy and numerical simulation to quantify the influence of intermetallic formation on permeability in aluminium-silicon-copper alloys. *Acta Mater.* **2014**, *64*, 316–325. [CrossRef]
80. Seifeddine, S.; Svensson, I.L. Prediction of mechanical properties of cast aluminium components at various iron contents. *Mater. Des.* **2010**, *31*, S6–S12.
81. Sha, M.; Wu, S.; Wan, L. Combined effects of cobalt addition and ultrasonic vibration on microstructure and mechanical properties of hypereutectic Al-Si alloys with 0.7% Fe. *Mater. Sci. Eng. A* **2012**, *554*, 142–148.
82. Shabestari, S. The effect of iron and manganese on the formation of intermetallic compounds in aluminum-silicon alloys. *Mater. Sci. Eng. A* **2004**, *383*, 289–398.
83. Sreeja; Kumari, S.S.; Pillai, R.M.; Rajan, T.P.D.; Pai, B.C. Effects of individual and combined additions of Be, Mn, Ca and Sr on the solidification behaviour, structure and mechanical properties of Al-7Si-0.3Mg-0.8Fe alloy. *Mater. Sci. Eng. A* **2007**, *460–461*, 561–573.
84. Huang, H.J.; Cai, Y.H.; Cui, H.; Huang, J.F.; He, J.P.; Zhang, J.S. Influence of Mn addition on microstructure and phase formation of spray-deposited Al-25Si-xFe-yMn alloy. *Mater. Sci. Eng. A* **2009**, *502*, 118–325.
85. Zhong, G.; Wu, S.; Jiang, H.; An, P. Effects of ultrasonic vibration on the iron-containing intermetallic compounds of high silicon aluminum alloy with 2% Fe. *J. Alloys Compd.* **2010**, *492*, 482–487. [CrossRef]
86. Srivastava, A.K.; Srivastava, V.C.; Gloter, A.; Ojha, S.N. Microstructural features induced by spray processing and hot extrusion of an Al-18% Si-5% Fe-1.5% Cu alloy. *Acta Mater.* **2006**, *54*, 1741–1748.
87. Xiao, S.; Liu, S.Q.; Wang, X.; Cui, C.X.; Li, Z.R. Effect of cooling rate on the microstructure evolution and mechanical properties of iron-rich Al-Si alloy. *Materials* **2022**, *15*, 411.
88. Johannes, P.S.; Eric, S.; Paul, S. Investigation of the Formation of Iron-Rich Intermetallic Phases in Al–Si Alloys via Thermal Analysis Cooling Curves, Including a Real-Time Detection for Filtration Process. *Adv. Eng. Mater.* **2023**, *25*, 2201576.

89. Song, D.F.; Zhao, Y.L.; Jia, Y.W. Effect of B addition on the formation of Fe-rich phases in Al-Si-Fe alloys. *J. Alloys Compd.* **2023**, *930*, 167426.
90. Anna, D.; Lu, P.Y.; Tang, N.K. Iron Reduction in 356 Secondary Aluminum Alloy by Mn and Cr Addition for Sediment Separation. *Int. J. Metalcast.* **2021**, *15*, 182–192.
91. Zheng, D.L.; Li, J.; Shi, C.B. Crystallization Characteristics and In-Mold Performance of Electroslag Remelting-Type TiO_2-Bearing Slag. *Metall. Mater. Trans. B* **2019**, *50*, 1148–1160.

materials

MDPI

Article

Study on the Microstructure and Magnetic Properties of Nd-Fe-B/Fe-Co Composite Nanowires

Licong Kang [1], Wei Yang [1,*], Lichen Zhao [1], Chunxiang Cui [1] and Feng Cao [2]

[1] Hebei Key Laboratory of New Functional Materials, School of Material Science and Engineering, Hebei University of Technology, No. 5340, Xiping Road, Tianjin 300401, China; klc910808@126.com (L.K.); zhlch@hebut.edu.cn (L.Z.); hutcui@hebut.edu.cn (C.C.)

[2] State Nuclear Power Demonstration Plant Company Limited, No. 666, Fujia Road, Weihai 264200, China; caofeng@spic.com.cn

* Correspondence: yangwei@hebut.edu.cn; Tel.: +86-022-6020-4125

Abstract: To solve the problem of the low coercivity of Nd-Fe-B-based nanowires impeding their application in magnetic storage media, highly ordered Nd-Fe-B/Fe-Co composite nanowires were fabricated in an anodic alumina template by means of the alternating electrochemical deposition method. In this paper, the effect of soft and hard magnetic phase compositing on the magnetic properties of Nd-Fe-B-based nanowires was investigated, and the coercivity improvement mechanism was demonstrated. The results show that after annealing at 600 °C for 2 h, Nd-Fe-B/Fe-Co nanowires crystallize into a multiphase structure containing a hard $Nd_2(Fe, Co)_{14}B$ phase and soft NdB_4, NdB_6, Fe_7Nd, and Fe_7Co_3 phases. It is characterized that the $Nd_2(Fe, Co)_{14}B$ phase preferentially nucleates, followed by NdB_4 + NdB_6 + Fe_7Nd, while Fe_7Co_3 has been formed in as-deposited nanowires. The existence of a $Nd_2(Fe, Co)_{14}B$ phase with high anisotropy fields, the remanence enhancement effect produced by exchange coupling between hard–soft magnetic phases, and the pinning effect between different phases make the composite nanowires approximately exhibit single hard magnetic phase characteristics with coercivity and remanence ratio as high as 4203.25 Oe and 0.89. The results indicate that synthesizing Nd-Fe-B/Fe-Co exchange-coupled composite nanowires via alternating electrodeposition is an effective way to optimize the magnetic performance of Nd-Fe-B-based nanowires.

Keywords: Nd-Fe-B; Fe-Co; electrochemical deposition; magnetic properties; exchange coupling

1. Introduction

Magnetic nanowires with large aspect ratios and apparent shape anisotropy are one of the research focuses in the new magnetic functional materials field, which has good application prospects in high-density linear magnetic storage media. As the third-generation rare earth magnets, Nd-Fe-B has outstanding magnetic properties that other materials cannot surpass. However, the low coercivity of currently manufactured Nd-Fe-B nanowires hinders their application in magnetic storage [1–3]. Therefore, how to prepare Nd-Fe-B-based nanowires with high magnetism has certain theoretical significance for developing new-generation vertical magnetic storage media. There has been a great deal of research on how to improve the hard magnetism of Nd-Fe-B nanowires through doping alloy elements [4–11]. In order to be better suitable for high-density linear perpendicular magnetic recording media, in addition to improving the coercivity of Nd-Fe-B nanowires, higher saturation magnetization should be expected to enhance their storage density.

The fabrication of nanocrystalline composites is an effective method to obtain both high coercivity and high saturation magnetization, which could obtain high magnetic energy products through the exchange coupling of hard–soft magnetic phases [12–15]. Fe-Co alloy is an excellent soft magnetic material that has a high saturation magnetic moment and initial permeability [16–18]. At the same time, the low coercivity and remanence ratio of soft magnetic materials limits their application in magnetic storage media [19]. Mansouri et al. [20] fabricated FeCo and FeNi nanowires with a coercivity of 627 Oe and 279 Oe through electrochemical deposition. Han et al. [21] successfully prepared a series of Fe_xCo_{1-x} (x = 0.0, 0.5, 1.0) nanowires with coercivity in the range of 450 Oe~570 Oe and a squareness of 0.23~0.41. Kaniukov et al. [22] researched the structural and magnetic performances of FeCo-alloy nanotubes deposited under different potentials and found that the nanotubes exhibited more excellent properties with a coercivity of 500 Oe and squareness of 0.2 due to the high crystallinity degree under the higher potential of 2 V. It is observed that most Fe-Co magnetic nanowires exhibit typical soft magnetic properties with low coercivity and a low magnetic energy product.

To simultaneously possess high coercivity and high saturation magnetization, there have been many studies about the combination of Fe-Co and rare earth permanent magnetic alloys. Chuanto et al. [23] fabricated $SmCo_5$/FeCo core–shell composite materials, showing how the magnetic energy product of a composite material with 10 wt.% FeCo increased by 25% compared to a single hard magnetic phase $SmCo_5$. Cui et al. [24] prepared Sm_2Co_{17}/Fe_7Co_3 duplex-phase magnetic nanowires with excellent shape anisotropy via alternating electrodeposition. They found that the duplex-phase Sm_2Co_{17}/Fe_7Co_3 magnetic nanowires have better remanence and coercivity in both annealed and deposited states than Fe_7Co_3 and Sm_2Co_{17} nanowires. However, research on Nd-Fe-B-based nanocomposites mainly focuses on composites with soft magnetic α-Fe and Fe_3B phases, currently [25–27]. Therefore, Nd-Fe-B/Fe-Co composite nanowires are an exciting route to solve the low coercivity of Nd-Fe-B-based nanowires.

The present work demonstrates an alternating electrochemical deposition for fabricating aligned Nd-Fe-B/Fe-Co composite nanowires within AAO templates. It is expected that a Nd-Fe-B/Fe-Co hybrid magnet, which is more suitable for magnetic storage media, can be obtained through soft–hard magnetic exchange coupling. This article also discusses the effect of the Fe-Co alloy on the phase component, microstructure, and properties of the Nd-Fe-B/Fe-Co composite nanowires before and after annealing. Moreover, this article presents the evolution of microstructure phase transition and phase distribution. And finally, it investigates the effects of phase evolution, dislocation, and the coupling of the hard–soft phase exchange on the magnetic performance of Nd-Fe-B/Fe-Co nanowires.

2. Materials and Methods

The methodology flow chart of this work is illustrated in Figure 1. The detailed procedure is briefly described below.

2.1. Synthesizing Process of AAO Template

The highly ordered alumina (AAO) templates used in this experiment were prepared via secondary anodizing in 0.3 mol/L oxalic acid solution under 40 V DC. During the anodization process, the oxidation tanks were placed in an ice–water mixture, preventing the excessive temperature caused by the exothermic reaction from affecting the quality of the oxide film. Between the first (4 h) and second (6 h) anodization, we used chromic acid and phosphoric acid to remove the oxide layer with lower order generated during the first anodization process. Highly ordered AAO templates with a pore size of 70 nm and length of 30 μm were successfully prepared through removing the bottom and reaming (Figure 2). It can be seen that AAO templates contain self-assembled, highly uniform quasi-hexagonal ordered holes. As shown in Figure 2c,d, the pore channels are parallel to each other and perpendicular to the membrane surface.

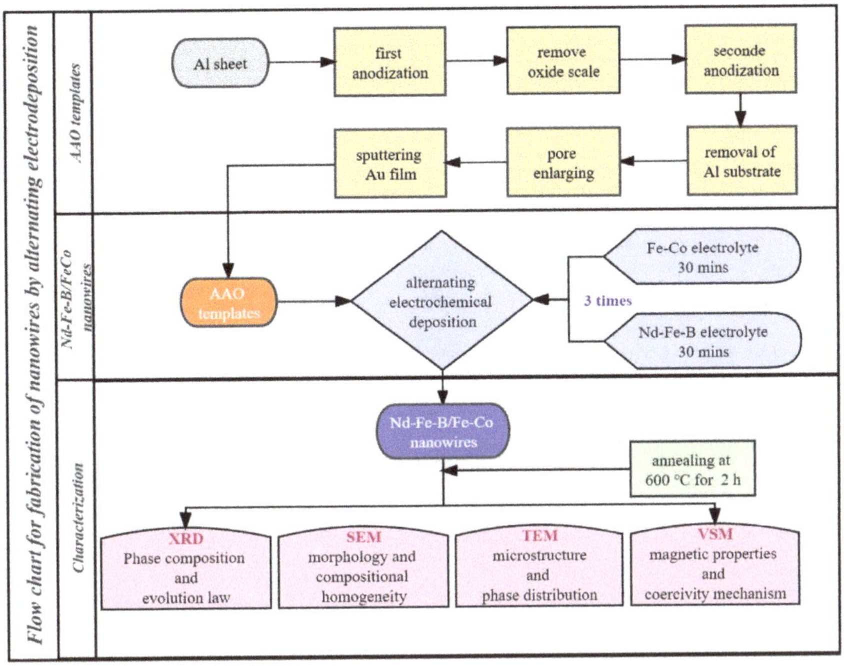

Figure 1. The flow chart describing the procedure for preparing nanowires based on electrochemical deposition in an AAO template.

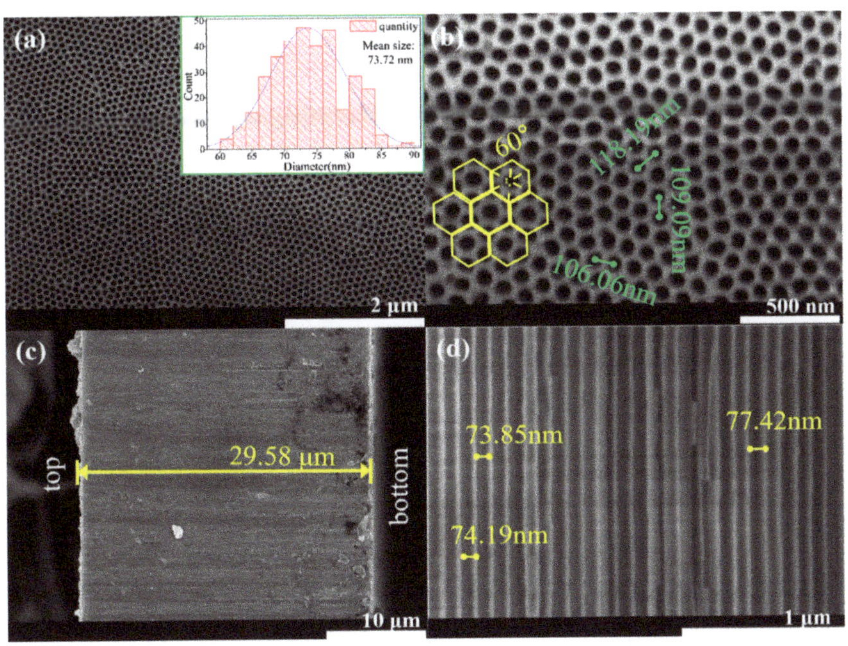

Figure 2. SEM images of AAO template prepared via two-step oxidation method: (**a**) top-view SEM image, the inset showing pore size distribution in AAO template; (**b**) top-view SEM image in high magnification; (**c**) cross-sectional SEM image; (**d**) cross-sectional SEM image in high magnification.

2.2. Electrochemical Deposition of Nd-Fe-B/Fe-Co Composite Nanowires

Before electrodeposition, a thin layer of gold film was plated on the back of the template to improve the conductivity of the alumina template. The deposition was based on graphite as the anode and the AAO template as the cathode, conducted under 2 V DC. The alternating deposition was completed through depositing in Fe-Co electrolyte for 1 min followed by Nd-Fe-B electrolyte for 30 min, alternating deposition three times. Table 1 lists the composition of the Fe-Co and Nd-Fe-B solutions. The deposition process was carried out on a magnetic stirrer, and the deposited solution was constantly stirred to accelerate ion diffusion and ensure a uniform concentration of ions in the electrolyte. A portion of the nanowires within the AAO template was then annealed at 600 °C for 2 h in a vacuum heat treatment furnace at a heating rate of 5 °C/min in an atmosphere of 99.99% argon.

Table 1. The electrolyte composition of Nd-Fe-B/Fe-Co composite nanowires prepared via alternate deposition.

Electrolyte	Concentration/(mol/L)							
	$NdCl_3$	$FeCl_2$	$CoCl_2$	$C_6H_8O_7$	$C_6H_8O_6$	H_3BO_3	$C_2H_5NO_2$	NH_4Cl
Fe-Co	0	0.35	0.15	0.073	0.007	0.485	0	0
Nd-Fe-B	0.03	0.2	0	0	0.007	0.485	0.4	0.5

2.3. Characterization

The phase composition of the samples was analyzed using a Bruker D8 Discover X-ray diffractometer (XRD) (Bruker, Billerica, MA, USA) with Cu kα radiation (λ = 1.5406 Å), a voltage of 40 kV, and a current of 20 mA. In the range of 2θ from 20° to 90°, the XRD patterns were scanned with a step length of 0.02°and a scanning rate of 6°/min. The chemical compositions and morphologies of nanowires with partially dissolved AAO templates were characterized using Quantum 450 FEG scanning electron microscopy (Beijing, China) (SEM) and energy dispersive spectroscopy (EDS). The microstructure of the nanowires was observed using the Tecnai G2 F30 S-TWIN transmission electron microscope (TEM) (Thermo Fisher Scientific, Waltham, MA, USA). The magnetic properties of nanowires in the AAO templates were measured using a Lake Shore 7407 type vibrating sample magnetometer (VSM) in a maximum 15 kOe magnetic field parallel to the long axis of the nanowires.

3. Results

3.1. Phase Analysis

Figure 3a,b show the XRD patterns of Fe-Co, Nd-Fe-B, and Nd-Fe-B/Fe-Co composite nanowire arrays before and after annealing. It can be seen from Figure 3a that the as-deposited Nd-Fe-B/Fe-Co nanowires consist of a Fe_7Co_3 phase. Compared with Fe-Co nanowires, there is a higher amount of amorphous diffuse scattering peaks in the XRD patterns of Nd-Fe-B and Nd-Fe-B/Fe-Co nanowires, which is due to the existence of a Nd atom with a large radius improving the amorphous forming ability of the nanowires.

Figure 3b shows the XRD patterns of nanowires annealed at 600 °C for 2 h in high-purity Ar. In comparison, the diffraction peak intensity for annealed nanowires increases and half-width decreases, indicating that the grains' size and crystallinity improved after annealing [28]. The 600 °C-annealed Nd-Fe-B/Fe-Co nanowires present a complex phase composition containing the newly formed $Nd_2(Fe, Co)_{14}B$, NdB_4, NdB_6, and Fe_7Nd phases as well as the Fe_7Co_3 phase existing in as-deposited nanowires. The formation of the $Nd_2(Fe, Co)_{14}B$ phase illustrates that in addition to forming the Fe_7Co_3 phase with Fe atoms during the alternating electrodeposition process, some Co atoms also enter the $Nd_2Fe_{14}B$ main phase through replacing some Fe atoms during heat treatment. As a result of the substitution of Co (0.126 nm) with a smaller atomic radius for Fe (0.127 nm), lattice contraction occurs in the 2:14:1-type phase according to the Bragg equation, $\lambda = 2d\sin\theta$, leading to the right shift of the 2:14:1-type phase diffraction peak as seen from the example of 204 peaks in the inset of Figure 3b, in which the right shift is approximately 0.4°.

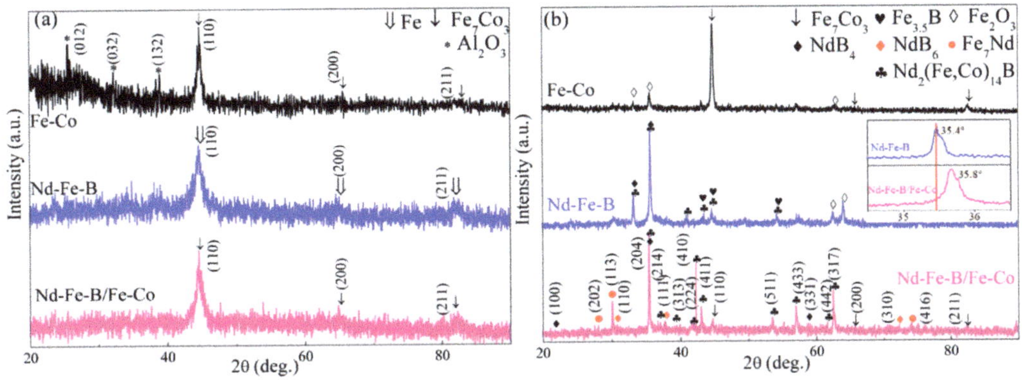

Figure 3. XRD patterns of nanowires with different compositions: (**a**) as-deposited; (**b**) annealed at 600 °C for 2 h. The inserted image is the magnification view of the diffraction peak for the (204) plane at the angle of 34.5°–36.5°.

Compared with Nd-Fe-B nanowires, there are more diffraction peaks of the 2:14:1 type phase in the XRD pattern of Nd-Fe-B/Fe-Co nanowires, which is related to the deposition efficiency of Nd atoms and their structural evolution. (1) Rare earth metals are difficult to reduce due to their low potential, which can be realized through the co-deposition of an RE-TM alloy. The transition metal Fe and Co ions in Fe-Co solutions with similar structural properties and positive electrode potential are easily reduced and deposited into the AAO template. The generation of a large number of Fe and Co atoms promotes dragging more Nd ions into AAO template pores during the alternating electrodeposition process, inducing more Nd atoms to realize co-deposition, thereby improving the deposition efficiency and quantity of Nd atoms.

Assuming that the phase transition process only considers the energy required for phase crystallization without other heat, the phase Gibbs free energy ΔG (i.e., the latent heat of phase transition from amorphous to crystalline) can be considered to be approximately equal to the phase formation enthalpy ΔH. During multiphase structure transformation, it is easier for the phase with smaller ΔG to overcome the energy barrier, then nucleate and grow up, completing phase transition. Based on this, the thermodynamic formula of the Arrhenius Equation can be extended and obtained from the following formulas [29]:

$$\ln k_{2i}/k_{1i} = -\Delta H_i(1/T_2 - 1/T_1)/R \ (i = 1, 2, 3, 4, 5) \tag{1}$$

$$k_{1i} = I_{\max i(843\ \mathrm{K})}/t,\ k_{2i} = I_{\max i(873\ \mathrm{K})}/t \tag{2}$$

$$\Delta G = \Delta H \tag{3}$$

wherein k_{1i} and k_{2i} are the crystallization rates for the phase in nanowires annealed at different temperatures of T_1 (570 °C) and T_2 (600 °C), respectively. $I_{\max i}$ is the maximum diffraction peak intensity of each phase in annealed nanowires. The subscript *i* represents the phase formed during the annealing process, and *R* is the gas constant (8.314 J/(mol·K)).

Figure 4a,b show the XRD patterns of Nd-Fe-B/Fe-Co and Nd-Fe-B nanowires annealed at temperatures of T_1 (570 °C) and T_2 (600 °C), respectively. Table 2 lists the $I_{\max}$ of each phase in nanowires. The ΔG of $Nd_2(Fe, Co)_{14}B$, NdB_4, NdB_6, Fe_7Nd, and Fe_7Co_3 phases in Nd-Fe-B/Fe-Co composite nanowires are calculated using Formulas (1)–(3) to be 29,188.6 J/mol, 90,461.4 J/mol, 84,392.5 J/mol, 82,220.4 J/mol, and −46,742.8 J/mol, respectively. The ΔG (Fe_7Co_3) < 0 indicates that the Fe_7Co_3 phase can spontaneously nucleate and crystallize, consistent with the XRD results shown in Figure 3a, thus confirming that the Fe_7Co_3 phase exists in as-deposited nanowires. The generation of other phases in Nd-Fe-B/Fe-Co nanowires must be completed via absorbing the heat annealing process, corresponding to the positive ΔG values. During the multiphase phase transition process, the phase with a smaller energy barrier will preferentially nucleate and grow, then undergo phase transition. According to the thermodynamic calculation results, it can be inferred that the phase nucleation ordering Nd-Fe-B/Fe-Co nanowires during annealing is roughly as follows: a $Nd_2(Fe, Co)_{14}B$ phase with the smallest phase transition energy barrier is preferentially formed, followed by Fe_7Nd, NdB_6, and NdB_4 phases. It is noted that the ΔG of the $Nd_2(Fe, Co)_{14}B$ phase formed after adding the Co atom is lower than that of the $Nd_2Fe_{14}B$ phase, indicating that the addition of Co reduces the precipitation temperature of the 2:14:1-type phase, thus further promoting the precipitation of more 2:14:1-type hard magnetic phases in composite nanowires, which is consistent with the conclusion proposed by Kojima [30].

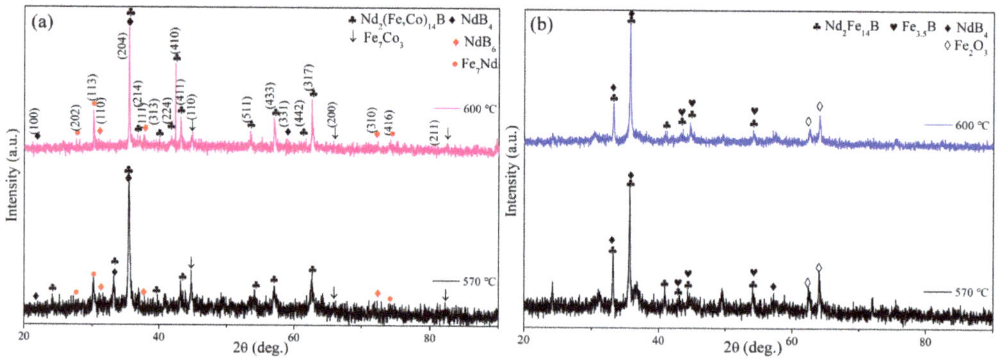

Figure 4. XRD patterns of nanowires with different compositions after annealing at 600 °C and 570 °C: (**a**) Nd-Fe-B/Fe-Co nanowires; (**b**) Nd-Fe-B nanowires.

Table 2. The maximum peak intensities of each phase in Nd-Fe-B/Fe-Co and Nd-Fe-B nanowires after annealing at 600 °C and 570 °C, and the corresponding Gibbs free energy.

Specimen	Phase	I_{max}		$\Delta G = \Delta H$/(J/mol)
		600 °C	570 °C	
Nd-Fe-B/Fe-Co	$Nd_2(Fe, Co)_{14}B$	347	300	29,188.6
	Fe_7Nd	138	91	82,220.4
	Fe_7Co_3	83	104	−46,742.8
	NdB_4	68	44	90,461.4
	NdB_6	78	53	84,392.5
Nd-Fe-B	$Nd_2Fe_{14}B$	271	230	33,264.0
	NdB_4	108	58	125,723.9
	$Fe_{3.5}B$	72	59	43,845.8

The XRD results show that the composite nanowires not only retain the soft magnetic Fe_7Co_3 phase formed in an as-deposited state but also stimulate the generation of greater amounts of the hard magnetic $Nd_2(Fe, Co)_{14}B$ phase during the crystallization process. This improves the comprehensive magnetic properties of the nanowires.

3.2. Morphology and Microstructure

Figure 5 shows the SEM image of Nd-Fe-B/Fe-Co composite nanowire arrays after partial dissociation from the AAO templates using 5 wt. % NaOH solution. As can be seen from Figure 5a, the Nd-Fe-B/Fe-Co composite nanowires are arranged orderly and uniformly throughout the template range, the deposition efficiency is very high, and the template filling rate is nearly 100%. Figure 5b shows the statistical analysis of the nanowires' diameter using Nano Measurer software, illustrating that the average diameter of the Nd-Fe-B/Fe-Co nanowire is 71 nm, which corresponds well to the pore diameter of the AAO template. The lateral-view SEM image shown in Figure 5c displays that the length of the nanowires array is approximately 27.01 μm, and the nanowires stand vertically and parallel to each other under the support of the residual AAO templates. The magnified image on the upper right shows the morphology of the nanowires separated from the template holes, which present as elongated and uniformly cylindrical. SEM images present that Nd-Fe-B/Fe-Co nanowires with uniform size are arranged neatly, parallelly, and in a highly ordered fashion. The nanowire energy spectrum analysis results shown in Figure 5d illustrate that the as-deposited Nd-Fe-B/Fe-Co composite nanowires mainly contain Nd, Fe, Co, and B elements, revealing that the metal cations in the electrolyte used for synthesizing the main phases of the nanowire are successfully reduced to atoms and deposited in AAO template pores during the electrochemical deposition process. Figure 6 shows the SEM line scan result of nanowires to further confirm the compositional homogeneity along the nanowire. Figure 6a demonstrates a low-magnification SEM image of Nd-Fe-B/Fe-Co nanowires, and the distribution of the main component elements Fe, Nd, and Co along the nanowires is shown in Figure 6b. The content ratios of Nd, Fe, and Co elements at different positions of 3 μm, 10 μm, and 18 μm are 1:11.23:3.60, 1:9.73:3.01, and 1:10.02:3.28, respectively, illustrating that the relative content of Nd, Fe, and Co is basically consistent along the nanowires. The intensity of every element fluctuates in a tiny range, indicating that the composition distribution of the main elements in nanowires is uniform along the nanowires.

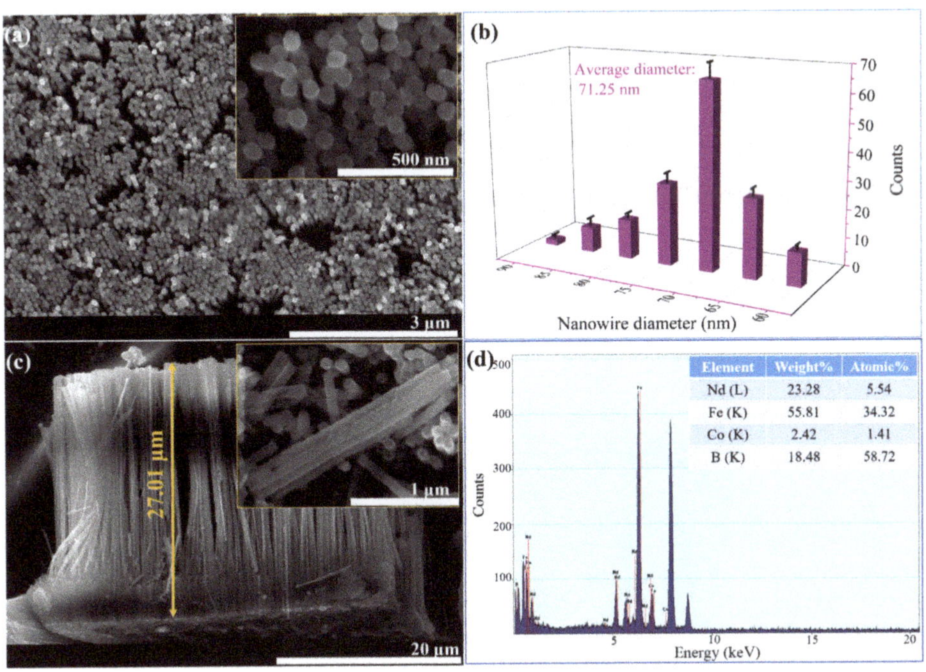

Figure 5. SEM images of Nd-Fe-B/Fe-Co nanowires prepared by means of the alternate electrodeposition method: (**a**) top-view SEM image—the illustration in the upper right corner is the high magnification SEM image; (**b**) the diameter statistical distribution diagram of nanowires; (**c**) cross-sectional SEM image; (**d**) EDS spectrum of nanowires.

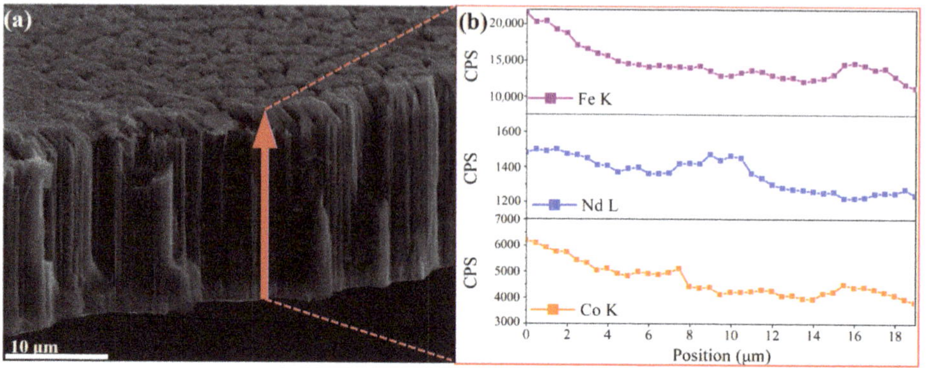

Figure 6. (**a**) SEM image of as-deposited Nd-Fe-B/Fe-Co nanowires; (**b**) the SEM-EDS line scan along the long axis of nanowire (red line in the SEM image).

Figure 7 shows the morphology and lattice structure of deposited Nd-Fe-B/Fe-Co composite nanowires. Figure 7a is the TEM image of the as-deposited nanowires. It can be seen that the dissociated nanowires are uniform elongated cylinders, some of which are agglomerated into bundles, while others are scattered. The illustration in Figure 7a shows that the nanowires possess a uniform diameter of about 70 nm, corresponding with the SEM results. The HRTEM image for zone b of Figure 7a is shown in Figure 7b, and the inserted images in red and blue borders show the Inverse Fast Fourier Transform (IFFT) patterns of regions I and II, respectively. From the IFFT patterns, it can be measured that the d values of regions I and II are 0.2121 nm and 0.2209 nm, both corresponding to the (110) plane of the Fe_7Co_3 phase. The measured values are slightly larger than the interplanar spacing of the bcc Fe_7Co_3 phase listed in standard X-ray crystallography ($d(110)Fe_7Co_3$ = 0.2025 nm). The larger d values indicate that the RE atom of Nd with a larger radius has existed as a solid solution atom in the Fe_7Co_3 phase; therefore, the phase structure of as-deposited Nd-Fe-B/Fe-Co nanowires is the incomplete crystalline Fe_7Co_3 phase with Nd atom doping.

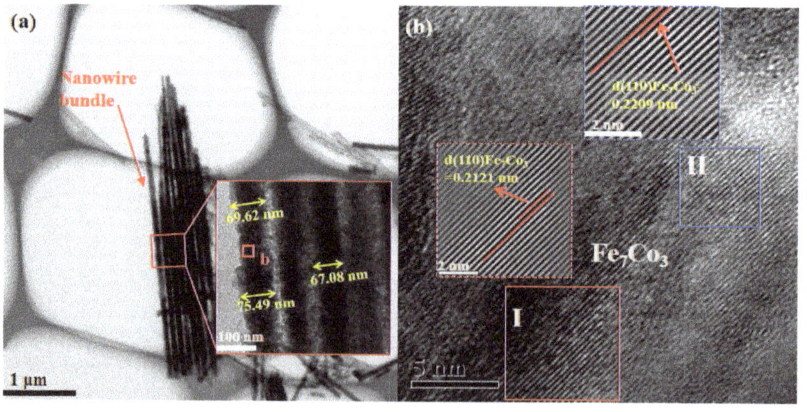

Figure 7. TEM observation and analysis of as-deposited Nd-Fe-B/Fe-Co nanowires: (**a**) TEM image of nanowire bundle, and the inserted image is the magnified view of the selected portion with a red border; (**b**) HRTEM image of zone b in (**a**) and the corresponding IFFT patterns of regions I and II.

Figure 8 displays the TEM analysis of Nd-Fe-B/Fe-Co composite nanowires after annealing at 600 °C for 2 h. Figure 8a is the TEM image for a single nanowire, displaying that the nanowire presents a uniform cylinder-like structure with a smooth outer wall. The selected area electron diffraction (SAED) pattern shown in Figure 8b presents polycrystalline diffraction rings consisting of discrete electron diffraction points, which can be calibrated and analyzed. Based on XRD results, the annealed Nd-Fe-B/Fe Co nanowire is polycrystalline in structure, containing $Nd_2(Fe, Co)_{14}B$, NdB_4, NdB_6, Fe_7Nd and Fe_7Co_3 phases. To confirm the phase composition and distribution of 660 °C-annealed nanowires, zones A–D at different nanowire positions were selected to analyze, and the corresponding results are shown in Figures 9–11.

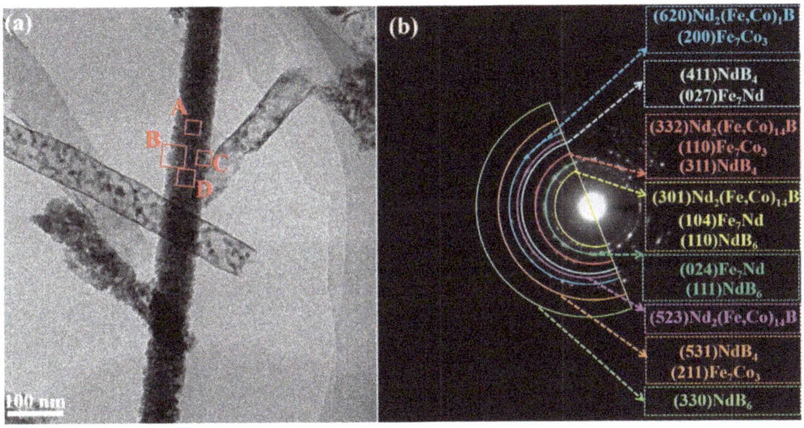

Figure 8. TEM observation and analysis of Nd-Fe-B/Fe-Co nanowires after annealing at 600 °C for 2 h: (**a**) the morphology of a single composite nanowire; (**b**) SAED pattern of the annealed nanowire.

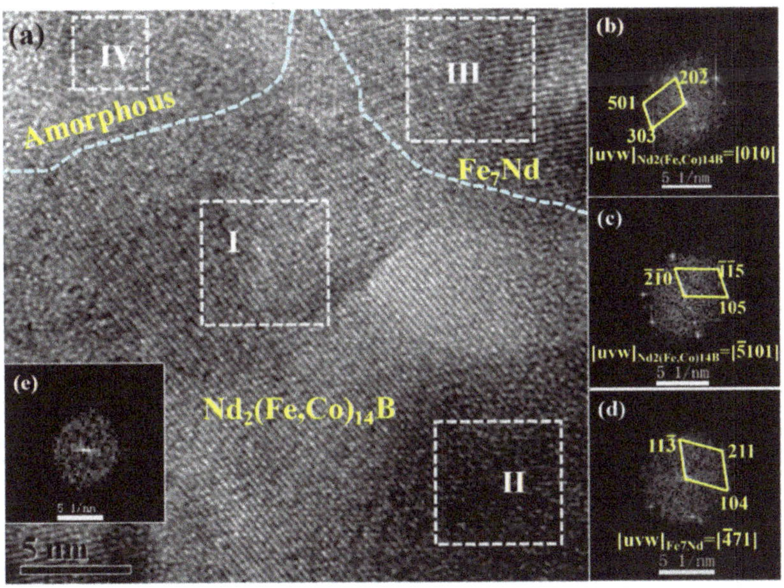

Figure 9. HRTEM analysis of annealed Nd-Fe-B/Fe-Co nanowires: (**a**) HRTEM of zone A in Figure 8a; (**b**–**e**) the corresponding FFT patterns of regions I–IV.

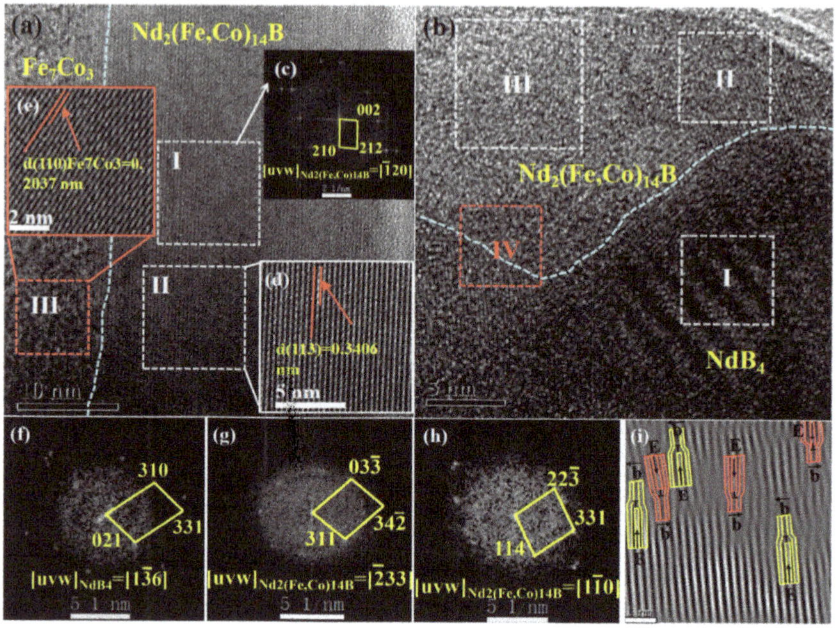

Figure 10. HRTEM analysis of annealed Nd-Fe-B/Fe-Co nanowires: (**a**,**b**) HRTEM images of zones B–C in Figure 8a; (**c**) FFT pattern of the selected region I in (**a**); (**d**,**e**) the magnified view of regions II–III in (**a**); (**f**–**h**) FFT patterns of regions I–III in (**b**); (**i**) the corresponding IFFT pattern of region IV in (**b**).

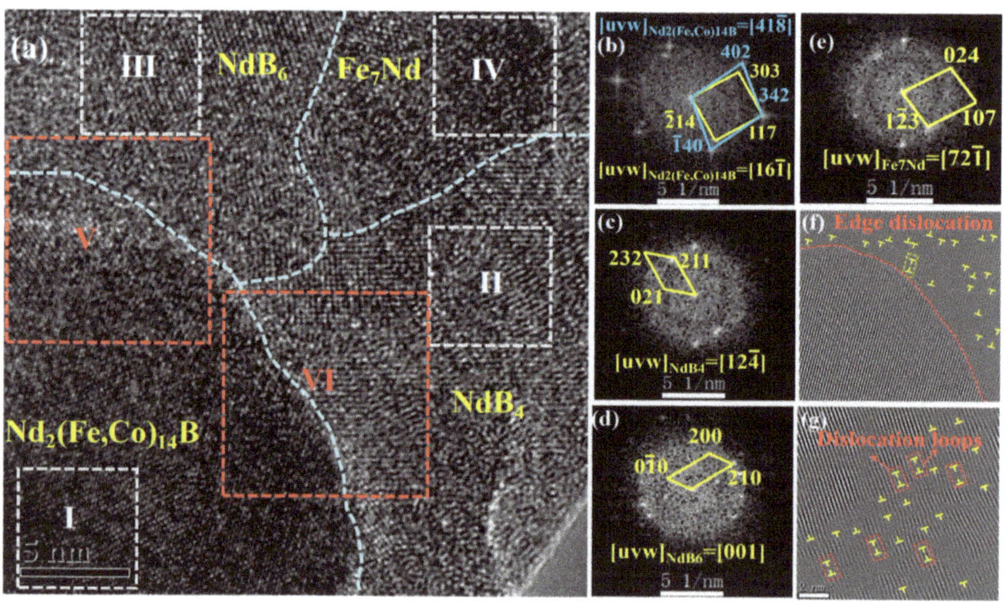

Figure 11. HRTEM analysis of annealed Nd-Fe-B/Fe-Co nanowires: (**a**) HRTEM image for zone D in Figure 8a; (**b**–**e**) FFT patterns of the selected regions I–IV in (**a**); (**f**,**g**) the corresponding IFFT patterns of regions V–VI in (**a**).

The phase analysis results of zone A, which corresponds to the core portion of nanowires, are shown in Figure 9. It can be seen that the core portion of nanowires is a mixed structure consisting of $Nd_2(Fe, Co)_{14}B$, Fe_7Nd phases and a few amorphous phases. The diffraction spots of the FFT patterns for regions I and II shown in Figure 9b,c, respectively correspond to the $(20\bar{2})$, (303), (501) and $(\bar{2}\bar{1}0)$, (105) $(\bar{1}\bar{1}5)$ planes of Nd_2(Fe, Co)$_{14}$B phase with tetragonal structure. The interplanar spacing of the Fe_7Nd phase with rhombohedral structure is about 0.2971 nm, 0.2835 nm, and 0.2742 nm, corresponding to the $(11\bar{3})$, (104), and (211) planes.

Figure 10a,b show the microstructure of zones B and C, which are located in the outer walls of the nanowire. A two-phase structure consisting of Fe_7Co_3 and $Nd_2(Fe, Co)_{14}B$ is observed in Figure 10a. As can be seen from Figure 10b, zone C is composed of NdB_4 and $Nd_2(Fe, Co)_{14}B$ phases with a straightforward phase interface. The IFFT pattern of phase interface region IV shows many edge dislocations, indicating that the soft magnetic NdB_4 exhibits a pinning effect relative to the $Nd_2(Fe, Co)_{14}B$ main phase, increasing resistance to domain wall motion.

The FFT patterns for regions I–IV in Figure 11b–e show that zone D exhibits a multi-phase structure composed of $Nd_2(Fe, Co)_{14}B$, NdB_4, NdB_6, and Fe_7Nd phases with a clear phase interface. As shown in the IFFT patterns of regions V and VI, there are a large number of dislocations and dislocation rings at the phase boundary of $Nd_2(Fe, Co)_{14}B$-NdB_6 and $Nd_2(Fe, Co)_{14}B$-NdB_4, resulting from the pinning effect of NdB_6 and NdB_4 to the (primary) main phase. The existence of the pinning effect will hinder the domain wall motion, making it difficult to twist the domain wall in the magnetization process, thereby facilitating and promoting coercivity.

Overall, the annealed Nd-Fe-B/Fe-Co nanowires comprise $Nd_2(Fe, Co)_{14}B$, Fe_7Co_3, NdB_4, NdB_6, and Fe_7Nd phases, showing a core–shell-like microstructure. Part of the $Nd_2(Fe, Co)_{14}B$ phase and all of the Fe_7Co_3 phase are distributed on the outer wall of the nanowire, and the soft–hard magnetic mixture of $Nd_2(Fe, Co)_{14}B + NdB_4 + NdB_6 + Fe_7Nd$ coexists in the core of nanowire.

According to thermodynamic calculation results in Table 2, the ΔG of $Nd_2(Fe, Co)_{14}B$, NdB_4, NdB_6, Fe_7Nd, and Fe_7Co_3 in the nanowires are 29,188.6 J/mol, 90,461.4 J/mol, 84,392.5 J/mol, 82,220.4 J/mol, and −46,742.8 J/mol, respectively. That is, the Fe_7Co_3 phase prioritizes spontaneous nucleation and crystallization during electrodeposition. Subsequently, $Nd_2(Fe, Co)_{14}B$, Fe_7Nd, NdB_6, and NdB_4 phases are successively formed according to the transition energy barrier for each phase during the heat treatment process.

The phase distribution in nanowires is determined to a great extent by the growth mechanism of the nanowires during electrodeposition (as shown in Figure 12). Prior to electrodeposition, the back of the AAO template is sputtered with a layer of gold film as the conductive electrode. In this process, some gold particles diffuse onto the wall of the template pore, giving it surface absorption energy, promoting the preferential growth of atoms into the tubular structure of the electrode surface and pore wall. Accompanying the formation of the nanotube structure, the surface absorption energy of the nanochannel gradually decreases, making the electric field force the major driving force of electrodeposition. At low current densities, metal atoms pile up from the electrode surface until the nanotubes are completely filled, forming nanowires. Because Fe^{2+} (−0.447 V) and Co^{2+} (−0.28 V) exhibit higher electrode potentials than Nd^{3+} (−2.40 V), Fe^{2+} and Co^{2+} preferentially enter AAO pores under an electric field force and are then reduced on the hole wall, resulting in the following reactions:

$$Fe^{2+} + 2e^{-} = Fe \tag{4}$$

$$Co^{2+} + 2e^{-} = Co \tag{5}$$

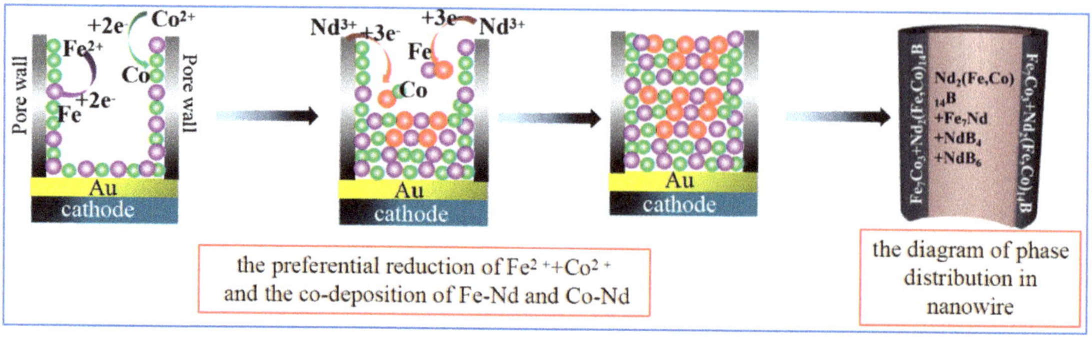

Figure 12. Schematic illustrations for the electrodeposition growth mechanism of Nd-Fe-B/Fe-Co composite nanowires in AAO pores and the phase distribution in nanowires.

After Fe and Co generate the initial nucleus, the Fe-Co phase is formed on the pore wall; then, Nd^{3+} is induced to conduct co-deposition via the following reaction:

$$2Nd^{3+} + 3Fe = 2Nd\ (in\ Nd - Fe) + 3Fe^{2+} \tag{6}$$

$$2Nd^{3+} + 3Co = 2Nd\ (in\ Nd - Co) + 3Co^{2+} \tag{7}$$

Under the combined effect of electric field and surface-absorbed energy, the Fe-Co phase mainly distributes in the outermost layer. Accompanied by heat treatment, a $Nd_2(Fe, Co)_{14}B$ phase with a more minor transition energy barrier first crystallizes at the outer wall, causing the internal components to deviate from 2:14:1. The formation of Fe_7Co_3 and $Nd_2(Fe, Co)_{14}B$ phases reduces Fe and Co content in nanowire, providing conditions for the generation of a Nd-rich phase, leading to the crystallization of a Nd-rich phase in nanowire. Meanwhile, the existence of a Nd-rich phase makes the surrounding components close to 2:14:1 again, which results in a mixed structure composed of a hard magnetic $Nd_2(Fe, Co)_{14}B$ phase and soft magnetic NdB_4 + NdB_6 + Fe_7Nd phases inside the nanowires.

3.3. Magnetic Properties Analysis

Figure 13a,b show the hysteresis loops of Fe-Co, Nd-Fe-B, and Nd-Fe-B/Fe-Co composite nanowire arrays before and after annealing at 600 °C for 2 h, respectively. Samples without removing the AAO template were measured in an external magnetic field parallel to the long axis of the nanowires. Figure 13c,d display the dependences of coercivity (Hc) and saturated magnetization (Ms) on nanowire compositions. The magnetic parameters of coercivity (Hc), remanence (Mr), saturated magnetization (Ms), and remanence ratio (Mr/Ms) are listed in Table 3.

Table 3. The magnetic parameters of nanowires with different compositions.

Sample	State	H_c/Oe	M_r/(emu/g)	M_s/(emu/g)	M_r/M_s
Fe-Co	as-deposited	413.52	15.26	101.29	0.15
	annealed	648.56	10.28	66.28	0.16
Nd-Fe-B	as-deposited	366.70	6.55	77.10	0.09
	annealed	1404.32	15.97	39.74	0.40
Nd-Fe-B/Fe-Co	as-deposited	399.75	6.31	64.01	0.10
	annealed	4203.25	68.51	76.45	0.89

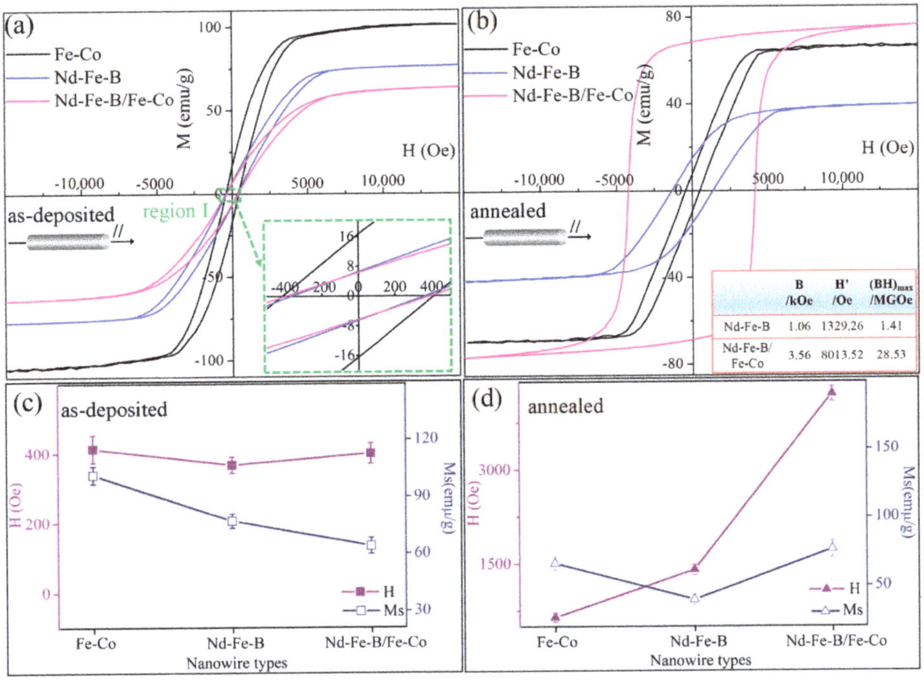

Figure 13. Magnetic hysteresis loops of nanowires with different compositions measured at room temperature under external magnetic field applied direction parallel to the long axis of nanowires: (**a**) as-deposited; (**b**) annealed at 600 °C for 2 h; (**c**) the variation trends of Hc and Ms for as-deposited nanowires with different compositions; (**d**) the variation trends of Hc and Ms for 660 °C-annealed nanowires with different compositions.

Figure 13a displays the hysteresis loops of as-deposited nanowires with different compositions, and the inserted image at the lower right corner is the magnified image of the selected region I. The as-deposited nanowires exhibit prominent soft magnetic properties, with a coercivity of 413.52 Oe for Fe-Co, 366.70 Oe for Nd-Fe-B, and 399.75 Oe for Nd-Fe-B/Fe-Co nanowire arrays. It can be clearly seen from Figure 13c that the Hc for Fe-Co nanowire arrays is slightly higher than that of the other two component nanowires. The Fe-Co nanowires consist of a pure Fe_7Co_3 phase, while the phase composition of as-deposited Nd-Fe-B and Nd-Fe-B/Fe-Co nanowires is incompletely crystallized Fe or Fe_7Co_3 with a few amorphous and microcrystalline phases, resulting in low crystallinity and slightly lower coercivity.

Figure 13b shows that the magnetic properties of Nd-Fe-B/Fe-Co composite nanowires are significantly improved after annealing, with Hc and Mr significantly increasing and Ms slightly decreasing, showing excellent comprehensive magnetic properties of Hc = 4203.25 Oe, Mr = 68.51 emu/g, Ms = 76.45 emu/g, and Mr/Ms = 0.89. It should be noted that compared with Fe-Co and Nd-Fe-B nanowires, the coercivity of composite nanowires is increased without sacrificing the saturation magnetization, as shown in Figure 13d. The excellent magnetic performances of Nd-Fe-B/Fe-Co nanowires are mainly attributed to the following aspects:

(1) The as-deposited Nd-Fe-B/Fe-Co nanowires are soft magnetic Fe_7Co_3 and amorphous phases with low Hc. After annealing, the amorphous phase crystallizes into a polycrystalline structure containing $Nd_2(Fe, Co)_{14}B$, NdB_4, NdB_6, and Fe_7Nd, while the soft magnetic Fe_7Co_3 phase still exists in the nanowires. The generation of the hard magnetic phase $Nd_2(Fe, Co)_{14}B$ with a high anisotropy field hinders the nucleation of

the magnetization reversal domain nucleus and the rotation of the magnetic domain, thereby enhancing the coercivity. At the same time, according to the XRD results and Formula (8) [31,32]:

$$V_i = (I_{\text{max}i}/RIR_i\rho_i)/\left(\sum_i I_{\text{max}i}/RIR_i\rho_i\right)\ (i = 1,\ 2,\ 3) \tag{8}$$

The content of the hard magnetic Nd2(Fe, Co)14B phase in Nd-Fe-B/Fe-Co composite nanowires prepared via alternating electrodeposition can be calculated as a volume fraction of 0.81, which is significantly higher than that of the hard magnetic phase Nd2Fe14B (volume fraction of 0.73) in Nd-Fe-B nanowires. This is due to the presence of Co ions in the alternating deposition process, which promotes more Nd ions in the electrolyte to be pulled into the AAO template and induces co-deposition, increasing the deposition quantity of Nd atoms and providing favorable conditions for the subsequent generation of a greater amount of hard magnetic phases. In the subsequent heat treatment process, some Co atoms enter the 2:14:1-type main phase to replace the Fe atoms, reducing the crystallization formation energy of the 2:14:1-type hard magnetic phase and contributing to precipitating more hard magnetic phases in nanowires. The coercivity is further enhanced by the increased content of the hard magnetic 2:14:1-type phase.

(2) After annealing, the soft and hard magnetic phases intersect with each other in composite nanowires, existing with a large number of edge dislocations at phase interfaces. This causes the domain wall motion in nanowires to be hindered, enhancing the pinning effect between the soft and hard magnetic phases and increasing the pinning resistance, thus increasing the coercivity.

(3) The exchange coupling between the hard magnetic $Nd_2(Fe, Co)_{14}B$ phase and soft magnetic phase of Fe_7Co_3 will prevent the magnetic moments from orienting along the respective easy magnetization directions and promote the improvement of Mr. According to the exchange spring model proposed by Kneller and Hawig, the exchange-correlation length of the soft magnetic phase can be written as [33]:

$$b_{cm} = \pi(A_s/2K_h)^{1/2} \tag{9}$$

wherein A_s is the exchange energy for the soft magnetic phase, and K_h is the magnetic crystal anisotropy constant of the hard magnetic phase [34]. In order to achieve sufficiently strong exchange coupling, the grain size for the soft magnetic phase must be less than $2b_{cm}$. According to Formula (8), where $A(Fe_7Co_3) = 1.67 \times 10^{-11}$ J/m [35], $K(Nd_2Fe_{14}B) = 4.3 \times 10^6$ J/m^3, it can be calculated that the soft magnetic phase size in the Nd-Fe-B/Fe-Co composite magnet is less than 13.8 nm, which in line with the results observed in the TEM image. Therefore, there is a strong exchange coupling effect in soft–hard magnetic phases in annealed nanowires, facilitating to increase Ms. Based on the Stoner Wohlfarth theory, the phenomenon of remanence enhancement will occur in the magnet under the coupling of soft and hard magnetic exchange when Mr/Ms is much higher than 0.5. As the data listed in Table 3, the Mr/Ms of annealed Nd-Fe-B/Fe-Co composite nanowires is as high as 0.89, therefore exhibiting the magnetic characteristics of high remanence and a high magnetic energy product of 28 MGOe under strong exchange coupling.

It can be seen from the magnetic properties analysis that the characteristics of Nd-Fe-B/Fe-Co magnetic nanowires are mainly controlled via the magnetic exchange coupling mechanism with the assistance of the nucleation field and pinning field. The presence of the Fe_7Co_3 phase and the strong exchange coupling of soft and hard magnetic phases increase Hc and produce a remanent magnetic enhancement effect, showing a demagnetization curve similar to that of the hard magnetic phase, and the magnetic energy product is significantly increased. The magnetic performance of Nd-Fe-B magnetic nanowires has been significantly improved in comparison with those obtained in previous research. That is to say, compositing with a soft magnetic Fe-Co alloy is an effectual route to enhance the comprehensive magnetic properties of Nd-Fe-B-based nanowires, rendering it more suitable for application in magnetic storage media.

4. Conclusions

In conclusion, highly ordered Nd-Fe-B/Fe-Co hybrid magnetic nanowires with cylinder-like structures were fabricated in porous AAO templates by means of alternating electrochemical deposition in Nd-Fe-B and Fe-Co aqueous solution. The as-deposited Nd-Fe-B/Fe-Co hybrid nanowires are composed of Nd-atom-doped incompletely crystallized Fe_7Co_3 and amorphous phases, which have typical soft magnetic properties. The annealing treatment recombines the atoms to form a polycrystalline structure containing hard magnetic Nd_2(Fe, Co)$_{14}$B and soft magnetic NdB_4, NdB_6, Fe_7Nd, and Fe_7Co_3 phases. The nucleation sequence of each phase in nanowires is as follows: the Fe_7Co_3 phase prioritizes spontaneous nucleation and crystallization in as-deposited nanowires. Subsequently, the Nd_2(Fe, Co)$_{14}$B phase preferentially nucleates, followed by Fe_7Nd + NdB_6 + NdB_4, in order, during the heat treatment process. The structural characterization revealed that part of the Nd_2(Fe, Co)$_{14}$B and all of the Fe_7Co_3 phase distribute in the outer wall of a nanowire; the soft–hard magnetic mixture of Nd_2(Fe, Co)$_{14}$B + NdB_4 + NdB_6 + Fe_7Nd coexists in the core of nanowire. The coercivity improvement after annealing is mainly due to the formation of a hard magnetic Nd_2(Fe, Co)$_{14}$B phase with a high anisotropy field, hindering the nucleation of the magnetization reversal domain nucleus and the rotation of the domain wall, thereby heightening the nucleation field of the annealed nanowires.

Compared with Nd-Fe-B nanowires, the comprehensive magnetic properties of Nd-Fe-B/Fe-Co composite nanowires are significantly improved with a coercivity of 4203.25 Oe and a saturation magnetization of 76.45 emu/g. The excellent magnetic performances can be attributed to the generation of a greater amount of the hard magnetic Nd_2(Fe, Co)$_{14}$B phase in composite nanowires, the pinning effect between different phases, and the remanence enhancement effect generated by the strong exchange coupling of hard and soft magnetic phase after being combined with Fe-Co alloy, cooperatively resulting in a remanence ratio as high as 0.89 and a high magnetic energy product of 28 MGOe.

Therefore, synthesizing Nd-Fe-B/Fe-Co composite nanowires through alternating electrochemical deposition is an effectual method to improve coercivity without sacrificing remanence. And the RE-Fe-B-based nanowires have a broad application prospect in the field of magnetic memory materials.

Author Contributions: Investigation, calculation, writing—original draft preparation, L.K.; data curation, formal analysis, investigation, W.Y.; supervision, writing, review, editing, C.C.; project administration, funding acquisition, L.Z.; validation, methodology, visualization, resources, F.C. All authors have read and agreed to the published version of the manuscript.

Funding: This research was funded by the "Hebei Province funding project for introducing overseas persons, grant number C20220314" and the "Scientific Research Projects of Colleges and Universities in Hebei Province, China, grant number ZD2022099".

Institutional Review Board Statement: Not applicable.

Informed Consent Statement: Not applicable.

Data Availability Statement: Any further detailed data may be obtained from the authors upon reasonable request.

Conflicts of Interest: The authors declare no conflict of interest.

References

1. Gao, D.Q.; Fu, J.L.; Xu, Y.; Xue, D.S. Preparation and magnetic properties of Nd5Fe95-xBx nanowire arrays. *Mater. Lett.* **2008**, *62*, 3070–3072. [CrossRef]
2. Kim, C.W.; Kim, Y.H.; Pal, U.; Kang, Y.S. Facile synthesis and magnetic phase transformation of Nd-Fe-B nanoclusters by oxygen bridging. *J. Mater. Chem. C* **2013**, *1*, 275–281. [CrossRef]

3. Guo, Z.Z.; Cao, J.W.; Yan, D.M.; Liu, F.F.; Zhang, C.; Guo, J.B.; Fu, Y.D. The influences of annealing on surface morphology and microstructure of Nd-Fe-B thin film. *J. Phys. Conf.* **2020**, *1507*, 102013. [CrossRef]
4. Kobayashi, K.; Urushibata, K.; Matsushita, T.; Sakamoto, S.; Suzuki, S. Magnetic properties and domain structures in NdFeB sintered magnets with Tb additive reacted and diffused from the sample surface. *J. Alloys Compd.* **2014**, *615*, 569–575. [CrossRef]
5. Li, Z.B.; Zhang, Z.X.; Wang, D.S.; Xu, Z.Y.; Li, Y.F.; Zhao, Q.; Zhang, X.F. Magnetic properties, microstructure and magnetization reversal in sintered $Nd_{12.2}Fe_{81.8}B_6$ magnets with the addition of $Tb_{17}Fe_{75}B_8$. *J. Magn. Magn. Mater.* **2019**, *483*, 5–9. [CrossRef]
6. Tan, X.; Parmar, H.; Zhong, Y.; Chaudhary, V.; Ramanujan, R.V. Effect of Dy substitution on the microstructure and magnetic properties of high (BH)max Nd-Dy-Fe-Co-B nanoparticles prepared by microwave processing. *J. Magn. Magn. Mater.* **2019**, *471*, 278–285. [CrossRef]
7. You, C.Y.; Wang, J.W.; Lu, Z.X. Coercivity enhancement of Nd-Fe-B thin film magnets by Dy buffer and capping layers. *J. Magn. Magn. Mater.* **2012**, *324*, 370–1374. [CrossRef]
8. Kim, T.H.; Sasaki, T.T.; Ohkubo, T.; Takada, Y.; Kato, A.; Kaneko, Y.; Hono, K. Microstructure and coercivity of grain boundary diffusion processed Dy-free and Dy-containing Nd-Fe-B sintered magnets. *Acta Mater.* **2019**, *172*, 139–149. [CrossRef]
9. Liu, X.B.; Altounian, Z. The partitioning of Dy and Tb in Nd-Fe-B magnets: A first-principles study. *J. Appl. Phys.* **2012**, *111*, 07A701. [CrossRef]
10. Liu, Q.B.; Tang, X.; Chen, R.J.; Wang, J.Y.; Zong, J.W.; Aru, Y.; Xu, Y.H. Effect of Tb-Fe diffusion on magnetic properties and thermal stability of hot-deformed magnets. *J. Alloys Compd.* **2019**, *773*, 1108–1113. [CrossRef]
11. Rahimi, H.; Ghasemi, A.; Mozaffarinia, R. Coercivity enhancement mechanism in Dy-substituted Nd-Fe-B nanoparticles synthesized by sol-gel base method followed by a reduction-diffusion process. *J. Magn. Magn. Mater.* **2017**, *429*, 182–191. [CrossRef]
12. Zhang, X.C.; Zhao, G.P.; Xia, J.; Yue, M.; Yuan, X.H.; Xie, L.H. Micromagnetic simulation of Sm-Co/α-Fe/Sm-Co trilayers with various angles between easy axes and the film plane. *Chin. Phys. B* **2014**, *23*, 097504. [CrossRef]
13. Sawatzki, S.; Heller, R.; Mickel, C. Largely enhanced energy density in epitaxial SmCo5/Fe/SmCo5 exchange spring trilayers. *J. Appl. Phys.* **2011**, *109*, 123922. [CrossRef]
14. Fan, J.P.; Liang, R.Y.; Bai, Y.H.; Yang, Y.; Sun, J.; Jiang, Y.N.; Wang, F.; Xu, X.H. Magnetic properties and magnetic reversal process of exchange-coupled $Nd_2Fe_{14}B/\alpha\text{-}Fe_{16}N_2$ bilayer. *J. Appl. Phys.* **2016**, *119*, 233902. [CrossRef]
15. Zhang, J.; Takahashi, Y.K.; Gopalan, R.; Hono, K. $Sm(Co,Cu)_5$/Fe exchange spring multilayer films with high energy product. *Appl. Phys. Lett.* **2005**, *86*, 122509. [CrossRef]
16. Finger, R.T.; Kozloweski, G. Microstructure and magnetic properties of Fe-Co alloys. *J. Appl. Phys.* **1997**, *81*, 4110–4111. [CrossRef]
17. Maclaren, J.M.; Schulthess, T.C.; Bulter, W.H.; Mchenry, M. Electronic structure, exchange interaction and curie temperature of Fe-Co. *J. Appl. Phys.* **1999**, *85*, 4833–4835. [CrossRef]
18. Sourmail, T. Near equiatomic FeCo alloys: Constitution, mechanical and magnetic properties. *Prog. Mater Sci.* **2005**, *50*, 816–880. [CrossRef]
19. Khushboo, B.; Prasanta, K.; Pallavi, P.; Ajay, G. Study of magnetic nanowires of amorphous $Co_{20}Fe_{60}B_{20}$ prepared by oblique angle deposition on nanorippled substrate. *J. Magn. Magn. Mater.* **2021**, *529*, 167842.
20. Mansouri, N.; Benbrahim-Cherief, N.; Chainet, E.; Charlot, F.; Encinas, T.; Boudinar, S.; Benfedda, B.; Hamadou, L.; Kadri, A. Electrodeposition of equiatomic FeNi and FeCo nanowires: Structural and magnetic properties. *J. Magn. Magn. Mater.* **2020**, *493*, 165746. [CrossRef]
21. Han, G.; Lu, J.D.; Gao, Y.H. FeCo nanowires deposited in a magnetic field. *J. Magn. Magn. Mater.* **2015**, *393*, 199–203. [CrossRef]
22. Kaniukov, E.Y.; Shumskaya, A.E.; Kozlovskiy, A.L.; Zdorovets, M.V.; Trukhanov, A.V.; Zubar, T.I.; Tishkevich, D.I.; Vinnik, D.A.; Khairetdinova, D.R.; Evstigneeva, S.A.; et al. Structure and magnetic properties of FeCo nanotubes obtained in pores of ion track templates. *Nano Struct. Nano Object* **2021**, *26*, 100691. [CrossRef]
23. Chuanto, M. Magnetic properties of exchange coupled $SmCo_5$/FeCo composite particles synthesized by magnetic self-assembly. *Chem. Phys. Lett.* **2018**, *696*, 31–35.
24. Cui, C.X.; Chen, F.H.; Yang, W.; Li, H.F.; Liu, Q.Z.; Sun, J.B. Electrochemical fabrication, microstructure and magnetic properties of Sm_2Co_{17}/Fe_7Co_3 dual phase nanocomposite. *Mater. Chem. Phys.* **2015**, *160*, 315–320. [CrossRef]
25. Kim, C.S.; Zha, L.; Lie, M.N.; Yang, W.Y.; Yang, J.B. Micromagnetic simulation for optimizing nanocomposite $Nd_2Fe_{14}B/\alpha$-Fe permanent magnets by changing grain size and volume fraction. *J. Magn. Magn. Mater.* **2021**, *523*, 167622. [CrossRef]
26. Ryo, H.S.; Kim, K.G.; Kim, Y.J. An analytic study on coercivity mechanism of exchange coupled $Nd_2Fe_{14}B/\alpha$-Fe nanocomposite magnets. *J. Magn. Magn. Mater.* **2019**, *469*, 531–534. [CrossRef]
27. Choong, J.Y.; Eon, B.P. The effect of magnetic field treatment on the enhanced exchange coupling of a $Nd_2Fe_{14}B/Fe_3B$ magnet. *J. Magn. Magn. Mater.* **2019**, *475*, 554–562.
28. Vorobjova, A.I.; Shimanovich, D.L.; Sycheva, O.A.; Ezovitova, T.I.; Tishkevich, D.I.; Trykhanov, A.V. Studying the thermodynamic properties of composite magnetic material based on anodic alumina. *Russ. Microelectron.* **2019**, *48*, 134–136. [CrossRef]
29. Pang, R.X.; Cui, C.X.; Yang, W.; Kang, L.C. The microstructure and magnetic behaviors of Pr-Fe-B/Fe_7Co_3 dual phase nanowires: As a perpendicular magnetic recording candidate. *Mater. Charact.* **2021**, *180*, 111410. [CrossRef]
30. Kojima, A.; Makino, A.; Inoue, A. Effect of Co addition on the magnetic properties of nanocrystalline Fe-rich Fe-Nb-(Nd,Pr)-B alloys produced by crystallization of an amorphous phase. *Scripta Mater.* **2001**, *44*, 1383–1387. [CrossRef]

31. Pang, R.X.; Cui, C.X.; Yang, W.; Guo, M.F. Fabrication and magnetic properties of Tb-doped multiphase Pr-Tb-Fe-B magnetic nanowire arrays. *Mater. Chem. Phys.* **2021**, *262*, 124299. [CrossRef]
32. Ding, H.W.; Cui, C.X.; Yang, W.; Sun, J.B. Effects of cobalt addition on microstructure and magnetic properties of PrNdFeB/Fe_7Co_3 nanocomposite. *J. Rare Earth.* **2017**, *35*, 468. [CrossRef]
33. Chen, W.; Gao, R.W.; Liu, L.M.; Zhu, M.G.; Han, G.B.; Liu, H.Q.; Li, W. Effective anisotropy, exchange-coupling length and coercivity in $Nd_{8-x}R_xFe_{87.5}B_{4.5}$ (R = Dy, Sm, x = 0–0.6) nanocomposite. *Mater. Sci. Eng. B* **2004**, *110*, 107–110. [CrossRef]
34. Zhao, T.; Xiao, Q.F.; Zhang, Z.D. Effect of magnetocrystalline anisotropy on the magnetic properties of Fe-rich R-Fe-B nanocomposite magnets. *Appl. Phys. Lett.* **1999**, *75*, 2298–2300. [CrossRef]
35. Skomski, R.; Coey, J.M.D. Giant energy product in nanostructured two-phase magnets. *Phys. Rev. B* **1993**, *48*, 812–816. [CrossRef]

materials

MDPI

Article

Fabrication and Properties of a Biodegradable Zn-Ca Composite

Mengsi Zhang [1], Xinyuan Wang [1,†], Shuo Zhang [1,‡], Tiebao Wang [1], Xin Wang [1], Shuiqing Liu [2], Lichen Zhao [1,*] and Chunxiang Cui [1]

1 Key Laboratory for New Type of Functional Materials in Hebei Province, School of Materials Science and Engineering, Hebei University of Technology, Tianjin 300130, China
2 School of Mechanical Engineering, Hebei University of Technology, Tianjin 300130, China
* Correspondence: zhlch@hebut.edu.cn
† Current address: Institute of Metal Research, Chinese Academy of Sciences, Shenyang 110016, China.
‡ Current address: Yangzhou Museum, Yangzhou 225126, China.

Abstract: In recent years, Zn and its alloys have become some of the most promising degradable metals as in vivo implants due to their acceptable biocompatibility and more suitable degradation rate compared with Mg-based and Fe-based alloys. However, the degradation rate of Zn-based materials after implantation in the body for orthopedic applications is relatively slow, leading to long-term retention of the implants after fulfilling their missions. Moreover, the excessive release of Zn^{2+} during the degradation process of Zn-based implants usually leads to high cytotoxicity and delayed osseointegration. To provide a feasible solution to the problem faced by Zn-based implants, a Zn-Ca composite was fabricated by an air pressure infiltration method in this work. The XRD pattern of the composite suggests that the composite is fully composed of Zn-Ca intermetallic compounds. The degradation tests in vitro show that the composite has a much higher degradation rate than pure Zn, and the high Ca content regions in the composite can preferentially degrade as sacrificial anodes. In addition, the composite can efficiently induce Ca-P deposition during immersion tests in Hank's solution. Cytotoxicity tests indicate that L-929 cells exhibit around 82% cell viability (Grade 1) even after being cultured in the 100% extract prepared from the Zn-Ca composite for 1 day and show excellent cell viability.

Keywords: biodegradable metals; Zn-Ca composite; Zn-Ca intermetallic compound; cytotoxicity

1. Introduction

As biodegradable orthopedic implant materials for temporary implantation in vivo, Zn and its alloys have received widespread attention in recent years due to their more suitable degradation rates compared with Mg-based and Fe-based materials. However, recent studies have shown that the degradation rates of Zn-based materials after implantation in the body are relatively slow, leading to long-term retention of the implants after fulfilling their missions [1]. Meanwhile, due to the much lower tolerance values of organisms to Zn compared with that of Mg and Fe [1,2], the excessive release of Zn^{2+} of Zn-based implants during the degradation process can also lead to severe cytotoxicity in vitro and delayed bone osseointegration in vivo [1]. In view of the fact, improving the degradation rate of Zn-based implants while effectively decreasing the cytotoxicity of implants has become a research hotspot in the degradable Zn-based implant field.

Many studies found that when alloying elements such as Li [3–9], Sr [3,10,11], Ca [10, 12–15], Y [16], Mg [16–22], Ti [23–26], Mn [3,12,27–34], Fe [3,35,36], Cu [3,9,23–25,34,37–41], and Ag [3,42–44] are introduced into Zn for alloying treatment, these alloying elements generally react with Zn to form intermetallic compounds containing Zn elements. These compounds not only significantly affect the strength and plastic deformation ability of the Zn alloys but also have a significant impact on their degradation property and cytotoxicity.

Among these alloying elements, Ca is an essential constant element for the human body. The content of Ca in the human body (1200 g) is much higher than that of other major metal elements such as K, Na, and Mg [2]. The dietary average daily intake (ADI, 743 mg) and recommended daily intake (RDI, 800–1200 mg (19–24 years old)) of Ca are also significantly higher than the corresponding values of other alloying elements such as Mg, Fe, Mn, Cu, Li, etc. [2]. Although the LD_{50} (lethal dose, 50%) of $CaCl_2$ obtained via oral supplement of the mouse (2301 mg/kg) is lower than the 5000 mg/kg of $MgCl_2$; the value is also significantly higher than that of other chlorides such as $FeCl_3$, $MnCl_2$, $CuCl_2$, and LiCl [2]. Clearly, when Zn is alloyed with Ca, the dissolved Ca^{2+} during the degradation process of the alloys is difficult to cause toxicity to human cells and tissues. In addition, Li et al. [10] investigated the degradation property of the Zn-1Ca alloy in Hank's solution, and the results showed that the Zn-1Ca alloy has a more negative corrosion potential and faster degradation rate than pure Zn in Hank's solution. Zou et al. [13] investigated the degradation property of Zn-xCa (x = 0.5, 1, 2, 3 wt.%) alloys in Hank's solution, and the results showed that the Zn-xCa alloys also had a more negative corrosion potential and faster degradation rate than pure Zn. Yang et al. [3] also obtained similar experimental results when they investigated the degradation behavior of Zn-xCa (x = 0.1, 0.4, 0.8 wt.%) alloys in simulated body fluids. In addition, the cytotoxicity test results conducted by Yang et al. [3] also suggested that the concentration of Zn^{2+} in the 100% extraction solution of Zn-xCa alloys decreased when the Ca content increased, and the cell activity of the alloys on MC3T3-E1 also increased accordingly.

Although the above experimental results indicate that Zn-Ca compounds in Zn-Ca alloys are beneficial to increasing the degradation rate and cell activity of Zn alloys, there are few studies that report the degradation properties, cytotoxicity, and other relevant properties of Zn-Ca compounds themselves. In view of this fact, this paper used an air compression infiltration method to prepare a Zn-Ca composite composed of Zn-Ca compounds. The microstructures and phase structures of the composite were characterized, and the mechanical properties, biodegradability, and cytotoxicity of the composite were also investigated.

2. Materials and Methods

2.1. Fabrication of a Zn-Ca Composite

Commercially pure Zn ingot (≥99.995%) and Ca particles (≥99.0%) were used as raw materials. Zn-Ca intermetallic compounds having a nominal Zn/Ca atomic ratio of 2 were fabricated by a vacuum induction melting furnace. The obtained compounds were crushed and sieved with 100 and 200 meshes of standard sieves, and Zn-Ca compound powder with a size of around 75–150 μm was then obtained. The SEM images of the obtained Zn-Ca compound powder are shown in Figure 1. The particle size of the compounds is relatively uniform, and these particles have irregular shapes and sharp edges. The XRD pattern of the powder is presented in Figure 2. In addition to the $CaZn_2$ phase, $CaZn_{3.04}$ and CaZn phases were also found in the powder. The presence of $CaZn_{3.04}$ and CaZn phases is due to the high viscosity of the Zn-Ca alloy melt during the preparation of $CaZn_2$, resulting in uneven distribution of Ca in the alloy melt. According to the Zn-Ca phase diagram [45,46], the CaZn and $CaZn_3$ phases are the most likely phases to be formed besides the $CaZn_2$ phase under the current condition.

A Zn-Ca composite was fabricated by an air pressure infiltration method (APIM). The schematic illustration of the APIM setup is depicted in Figure 3. The obtained Zn-Ca compound powder was firstly filled into a steel mold with an inner diameter of 20 mm; the mold was then heated to 405 °C and held at that temperature for at least 60 min. Secondly, the Zn melt with a temperature of 560 °C was cast into the mold (the weight ratio of Zn to Zn-Ca compound powder is around 2.5:1), and the mold was quickly filled with 0.3–0.4 MPa compressed air and held for several minutes. When the mold was cooled to room temperature, the specimen was removed from the mold, and a Zn-Ca composite was obtained.

Figure 1. (**a**) SEM image of the Zn-Ca compound powder, (**b**) is the higher magnification of the powder.

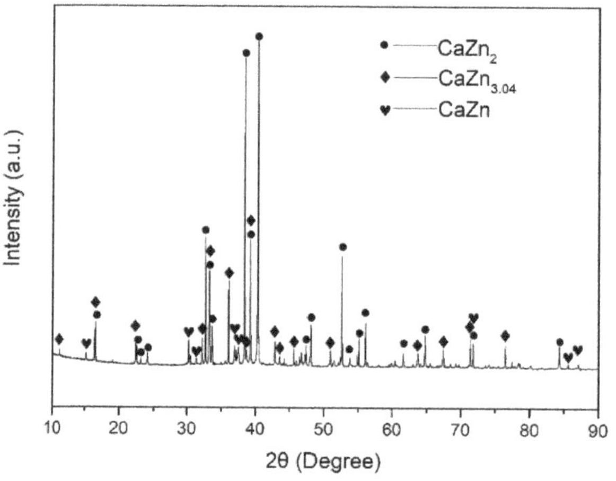

Figure 2. XRD pattern of the obtained Zn-Ca compound powder.

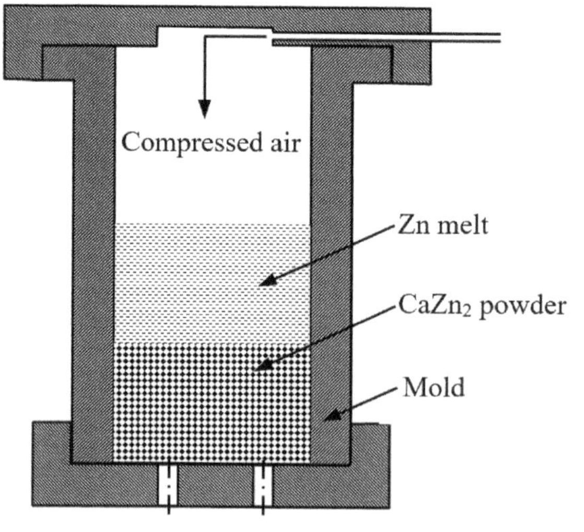

Figure 3. The schematic illustration of the setup for the air pressure infiltration method.

2.2. Microstructures and Phase Structures

The microstructures of specimens were observed by an optical microscope (Axio Imager M2m, Cari Zeiss, Oberkochen, Germany) and a scanning electron microscope (SEM, S-4800, Hitachi, Tokyo, Japan). The phase structures of specimens were characterized by an X-ray diffractometer (XRD, D8 FOCUS, Bruker, Billerica, Germany) with Cu K_α radiation (12°/min).

2.3. Compress Tests

Compressive mechanical properties of the Zn-Ca composite were characterized on a WDW-300 electronic universal testing machine (Zhongzheng, Jinan, China) at room temperature. The specimens for the tests were cut into cylinders with a size of φ4 mm × 8 mm using wire electrical discharge machining. The tests were carried out under displacement control with a crosshead speed of 1 mm/min. At least 5 specimens were tested.

2.4. In Vitro Biodegradation Tests

The potentiodynamic polarization curves of specimens were measured on a CHI660E electrochemical workstation using a standard three-electrode system. The testing specimen, a saturated calomel electrode (SCE), and a graphite rod served as the working electrode, the reference electrode, and the counter electrode, respectively. A Hank's solution was employed as the corrosive medium, whose chemical composition was presented in Table 1. A 1 M HCl solution and 7.4% $NaHCO_3$ solution were used to adjust the pH value of Hank's solution to 7.4 at 37 °C. The testing specimen was first immersed in Hank's solution at 37 °C to measure open circuit potential (OCP). When the OCP was stable, the polarization curve of the specimen was then measured at a scanning rate of 0.5 mV/s.

Table 1. Chemical composition of Hank's solution.

Number	Reagent	Concentration
1	$CaCl_2$	0.14 g/L
2	NaCl	8.00 g/L
3	KCl	0.40 g/L
4	$NaHCO_3$	0.35 g/L
5	Glucose ($C_6H_{12}O_6$)	1.00 g/L
6	$MgSO_4{\cdot}7H_2O$	0.06 g/L
7	KH_2PO_4	0.06 g/L
8	$Na_2HPO_4{\cdot}12H_2O$	0.06 g/L
9	$MgCl_2{\cdot}6H_2O$	0.10 g/L

Immersion tests were also conducted in Hank's solution at 37 °C. The specimens with a size of φ20 mm × 3 mm were immersed in 125 mL of Hank's solution for different periods. After immersion, the specimens were removed from the solutions and gently washed with deionized water and then dried at room temperature. At last, the corrosion products deposited on the specimens were washed with 200 g/L CrO_3 solution.

2.5. Cytotoxicity Tests

An indirect contact method was used to evaluate the cytotoxicity of the Zn-Ca composite. The employed cells were L-929 murine fibroblast cells (Cell Bank of Chinese Academy of Sciences, Shanghai City, China). The detailed process can be found in our previous work [47]. Firstly, the sterilized specimens were soaked in Roswell Park Memorial Institute (RPMI) 1640 medium containing 10% calf serum (0.2 mg/L, Zhejiang Tianhang Biotechnology Co., Ltd., Huzhou City, China) for 24 h at 37 °C. After that, the solution in the extract container was reserved as 100% extract. Secondly, 100 μL of solution having L-929 cells, whose concentration was 2×10^4 cells /mL RPMI 1640 medium containing 10% calf serum, were incubated in a 96-well cell culture plate for 24 h at 37 °C in a humidified atmosphere containing 5% CO_2. After that, 100 μL of 100%, 50%, 25%, 12.5%, and 6.25% extracts were added to the different wells of the culture plate, respectively. The negative control group was added to RPMI 1640 medium containing 10% calf serum, and the positive group was added to 10% dimethyl sulphoxide (DMSO). The culture plate was then incubated in a humidified atmosphere containing 5% CO_2 for 24 h at 37 °C. Then, 10 μL of MTT (thiazolyl blue tetrazolium bromide) (5 mg/mL) were added to different wells and continuously incubated for 4 h. After incubation, 180 μL of DMSO were added to the wells of the plate. The optical density (*OD*) was then measured at 570 nm by a microplate reader, and the cell relative growth rate (*RGR*) was calculated by the following formula.

$$RGR(\%) = \frac{OD_{\text{Specimen}}}{OD_{\text{Control}}}. \quad (1)$$

3. Results and Discussion

3.1. Microstructures and Phase Structures of the Zn-Ca Composite

The optical images of the Zn-Ca composite are shown in Figure 4. It can be seen that the particles with irregular shapes are uniformly distributed in the composite, and a shell layer is wrapped around the outer surface of many particles. The SEM images of the composite are presented in Figure 5. There is indeed a shell layer around the particles. The EDS spectrum result suggests that the atomic ratio of Ca/Zn of the particle (area A in Figure 5a) is 1:2.13, which is very close to 1:2. Therefore, it can be deduced that the particle is likely $CaZn_2$. Comparing the EDS spectrum results of area A, point B, area C, and point D (Figure 5d–g), it can be found that the farther away from the particle, the lower the Ca content in the specimen. It is noted that the atomic ratio of Ca/Zn of point D (Figure 5g) is 1:12.42, which is close to 1:13. Thus, it can be deduced that the phase existing at point D may be $CaZn_{13}$. In addition, the EDS results of points E and F suggest that the composite contains some O element, especially the white particle (point F), which has a higher oxygen content. Although the oxygen content of the specimen cannot be precisely determined by EDS, the results also indicate that the composite inevitably underwent a certain degree of oxidation during the preparation process, and further experiments are needed to investigate the effect of oxides on the properties of the composite. In addition to that, it is found in Figure 5 that the particles in the specimen are sunken compared with the surrounding region. Not only that, the shell layers around the particles also exhibit a relatively light degree of depression. Since the specimen shown in Figure 5 has been corroded by a 4% nitric acid alcohol solution, it can be determined that the particles with the highest Ca content in the specimen are the least corrosion-resistant. As the Ca content decreases, the corrosion resistance of the specimen also increases accordingly. In addition, it is also noted that the composite is not dense, and there are some pores in the specimen.

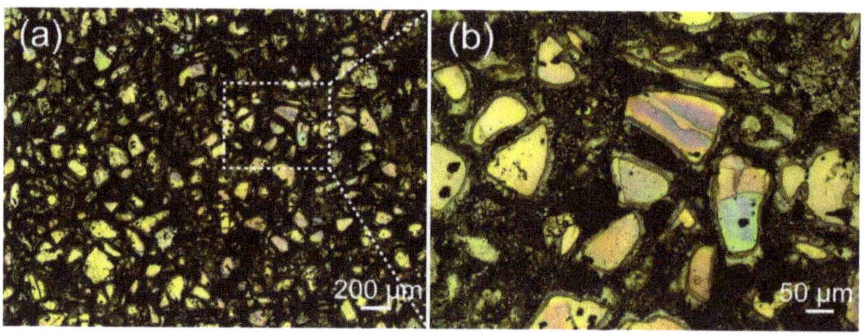

Figure 4. Optical images of the Zn-Ca composite, (**b**) is the higher magnification of the rectangular area in (**a**).

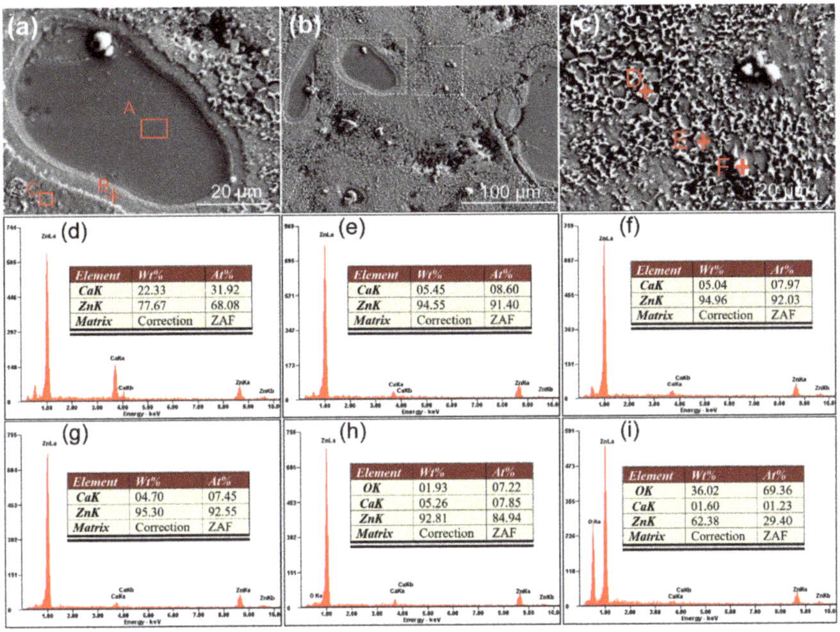

Figure 5. SEM images and EDS results of the Zn-Ca composite, (**a**,**c**) are the higher magnifications of the corresponding rectangular areas in (**b**), (**d**–**i**) are the EDS results of the area A, point B, and area C in (**a**) and points D, E, and F in (**c**), respectively.

The XRD pattern of the obtained Zn-Ca composite is shown in Figure 6, and the inset is a magnification pattern within the range of 32–44 degrees. The detected phases in the composite include $CaZn_{13}$, $CaZn_2$, and $CaZn_5$. No Zn phase is found in the specimen, and the $CaZn_{3.04}$ and CaZn phases existing in the original Zn-Ca compound powder are also not found. Combined with the SEM images and EDS spectra of the Zn-Ca composite in Figure 5, it can be inferred that the XRD result is due to the diffusion of Ca element from the Zn-Ca compound powder to the Zn melt as well as the formation of new phases and the disappearance of old phases during the preparation of the composite. When the Zn melt was pressed into the pores of the Zn-Ca compound powder during the infiltration process, the Ca element in the surface layer of the compound particles inevitably diffused into the surrounding Zn melt under the high-temperature action of the Zn melt. On the basis of Fick's law, the farther away from the particles, the lower the diffused Ca content. The diffused Ca element then reacted with Zn to form new Zn-Ca compounds such as $CaZn_5$ and $CaZn_{13}$ (Figure 5g). The formation of these solid compounds not only consumed a large amount of Zn melt but also increased the resistance of Ca element within $CaZn_2$ particles to diffuse outwards so that the diffusion of Ca element in the core region of the particles had not yet occurred. As a result, the core region of the particles could still retain its original high Ca content (such as $CaZn_2$ shown in Figure 5a,d). On the contrary, the Ca content in the surface layer of these particles was significantly reduced (Figure 5e), and a shell layer with a low Ca content was formed around the particles (Figures 4 and 5a). The disappearance of $CaZn_{3.04}$ and CaZn phases existing in the original Zn-Ca compound powder was obviously due to the diffusion of Ca elements and the occurrence of new reactions during the infiltration process. With the diffusion of the Ca element and the reactions of Ca with Zn, the Zn in the pores of the Zn-Ca compound particles was almost depleted, which ultimately led to the absence of the Zn phase in the XRD pattern of the composite. As a result, the obtained composite is entirely composed of Zn-Ca compounds.

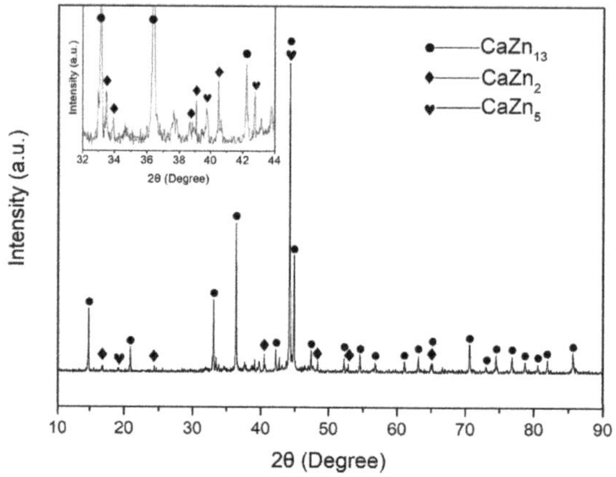

Figure 6. XRD pattern of the Zn-Ca composite.

3.2. *Compressive Mechanical Properties of the Zn-Ca Composite*

The compressive stress–strain curve of the Zn-Ca composite is depicted in Figure 7. Unlike as-cast pure Zn [10,48] and some Zn alloys such as Zn-1Ti [49], Zn-1X (X = Mg, Ca, Sr) [10], Zn-2Mg [50], and Zn-3Cu [50] that have good compressive plasticity, the composite exhibits typical brittle fracture characteristics. Clearly, the result can be attributed to the fact that both the matrix and reinforcing phases of the composite are Zn-Ca compounds. The calculated mean compressive strength of the composite is 281.1 ± 55.3 MPa. The compressive strength value is obviously not high among the Zn alloys that also have brittle fracture characteristics (e.g., Zn-3Mg-1Ti alloy with eutectic structure has a compressive strength of 625.1 MPa [49]), which is due to the fact that the specimen is not dense (Figure 5). Since no extensometer was used during the compressive process, the elastic modulus of the specimen was not calculated.

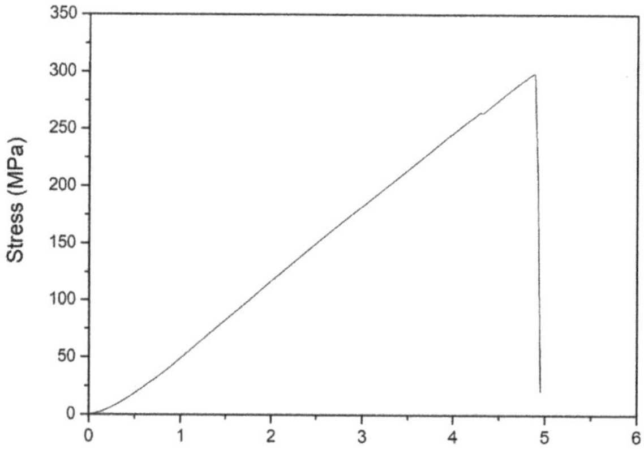

Figure 7. Typical compressive stress–strain curve of the Zn-Ca composite.

3.3. *In Vitro Biodegradable Properties of the Zn-Ca Composite*

The potentiodynamic polarization curve of the Zn-Ca composite is shown in Figure 8. For comparison, the polarization curve of an as-cast pure zinc is also plotted in Figure 8. The corrosion potentials (E_{corr}) and the corrosion current densities (I_{corr}) derived from the polarization curves are listed in Table 2. Clearly, the corrosion potential of the Zn-Ca composite is more negative than that of pure Zn. Moreover, the composite also exhibits a higher degradation rate than pure Zn.

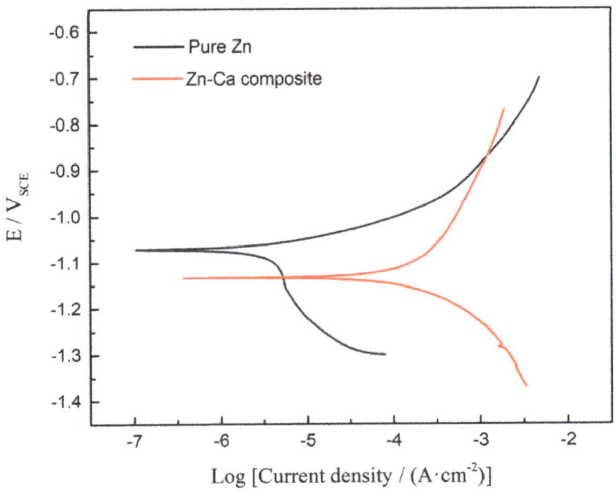

Figure 8. Potentiodynamic polarization curves of the Zn-Ca composite and pure Zn.

Table 2. Corrosion potentials and corrosion current densities derived from the polarization curves.

Specimens	E_{corr} (V_{SCE})	I_{corr} ($\mu A/cm^2$)
Pure Zn	−1.070	3.98
Zn-Ca composite	−1.131	229

The SEM images of the Zn-Ca composite after immersion in Hank's solution for 5 h are shown in Figure 9. A thin layer of corrosion products composed of spherical particles has been deposited on the Zn-Ca composite. Due to the dehydration of the specimen during the drying process after immersion, the corrosion product layer was warped. Nevertheless, it still can be inferred that the surface of the specimen has been completely covered by the corrosion product layer after immersion for only 5 h. Under the warped corrosion product layer, the original scratches on the specimen surface after sanding and polishing can still be seen, although some corrosion product particles were deposited. The EDS spectrum result (Figure 9c) suggests that the corrosion product layer is rich in Zn, O, Ca, and P elements. Figure 10 shows the SEM images of the Zn-Ca composite after immersion in Hank's solution for 10 h. Although the corrosion product layer also cracked, the cracked corrosion product layer did not undergo severe warping. This result is mainly due to the thickening of the corrosion product layer. In addition, the corrosion product layer on the specimen surface has partially detached, and there is still corrosion product deposition on the exposed surface. Unlike after immersion for 5 h (Figure 9), the original scratches on the exposed surface of the partially detached corrosion product layer are no longer obvious. The EDS result of the exposed surface (area A in Figure 10b) suggests that the surface is still rich in Zn, O, Ca, and P elements. However, the EDS spectrum of point B in Figure 10b shows that the corrosion product layer is rich in O, Ca, and P as well as a small amount of Mg, while the Zn content is very low. This result further indicates that the corrosion product layer deposited on the composite surface is already thick enough to prevent the collection of Zn information from the specimen itself after soaking for 10 h.

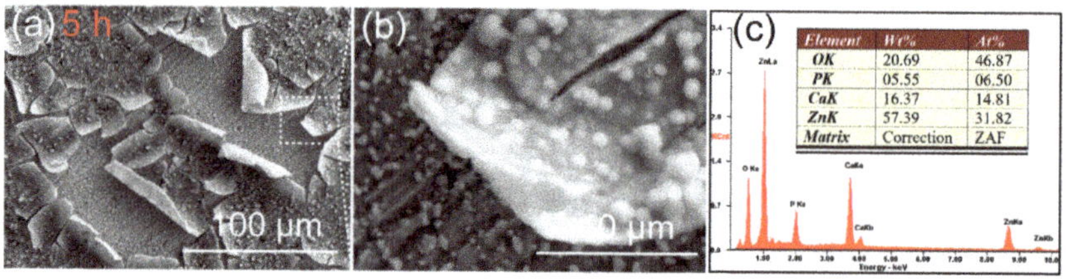

Figure 9. (**a**,**b**) SEM images of the Zn-Ca composite after immersion in Hank's solution for 5 h, (**b**) is the higher magnification of the rectangular area in (**a**), (**c**) is the EDS result of the area shown in (**b**).

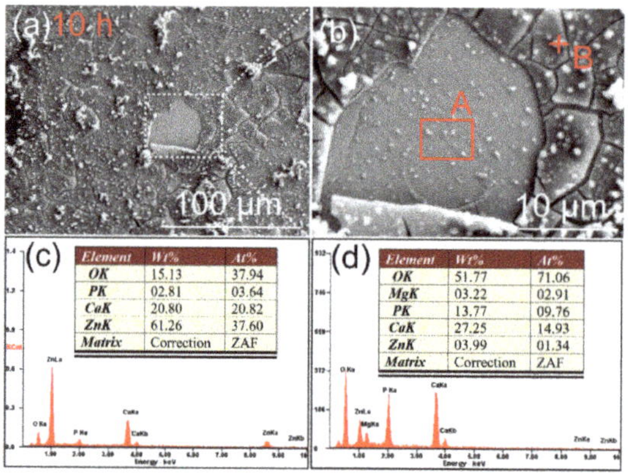

Figure 10. (**a**,**b**) SEM images of the Zn-Ca composite after immersion in Hank's solution for 10 h, (**b**) is high magnification of the white rectangular area in (**a**); (**c**,**d**) are the EDS results of area A and point B in (**b**), respectively.

The SEM images of the Zn-Ca composite after immersion in Hank's solution for 5 h and 10 h and then removal of corrosion products with chromic acid solution are shown in Figure 11. After immersion for only 5 h, obvious localized corrosion was observed on the surface of the specimen. Corrosion pits were mainly found in the core regions of the reinforcing particles with core–shell structures. As seen in Figure 11a–c, the core regions of some particles with core–shell structures were heavily corroded. However, there are also particles where the core regions are not severely corroded (red arrow in Figure 11b). In addition, it should be noted that the pits present between the reinforcing particles (white arrows in Figure 11) were not caused by localized corrosion but were formed during the preparation process of the specimen. After immersion for 10 h, the core regions of many reinforcing particles have corroded off a layer and formed obvious depressions. Unlike the core regions, the shell layers surrounding these regions and their outer regions (Zn-rich region) still did not undergo significant degradation. However, it can be inferred that these Zn-rich regions will also undergo serious corrosion with the extension of immersion time.

The reason for the above phenomena can be attributed to the preferential degradation of the local regions with more negative corrosion potential in the specimen as sacrificial anodes. It is known that the standard electrode potential of Ca is -2.87 V_{SHE}, which is significantly lower than that of Zn (-0.762 V_{SHE}). Therefore, it can be inferred that the electrode potentials of Zn-Ca compounds are also lower than that of Zn. Now, the inference has been confirmed by the polarization curves shown in Figure 8. Similarly, $MgZn_2$ and Mg_2Zn_{11} produced by the reaction of Zn with the element Mg, which also has a more negative standard electrode potential than Zn (-2.372 V_{SHE} vs. -0.762 V_{SHE}), have a more negative corrosion potential than Zn in the corrosion solution [51]. Then, it can be further inferred that the Zn-Ca compound having a higher Ca content, such as $CaZn_2$, will have a more negative corrosion potential. As the Ca content in the compound decreases, the corrosion potential of the compound will be closer to that of Zn. Consequently, corrosion couples will be formed between the $CaZn_2$ core regions and their shell layers, as well as the shell layers and their outer matrix (Ca content is relatively lower) due to the difference in corrosion potential, ultimately leading to the preferential degradation of the core regions that have the most negative corrosion potential. As for the shell layers around the core regions did not show obvious degradation compared with their outer matrix during soaking for 5 h and 10 h; this result is mainly due to the small difference in Ca content between the shell layers and its outer matrix (Figure 5). Clearly, the potential difference between the two electrodes constituting a corrosion couple significantly affects the degradation of the sacrificial anode. When the potential difference between the two electrodes is small, it will take a longer time to observe significant degradation of the sacrificial anode.

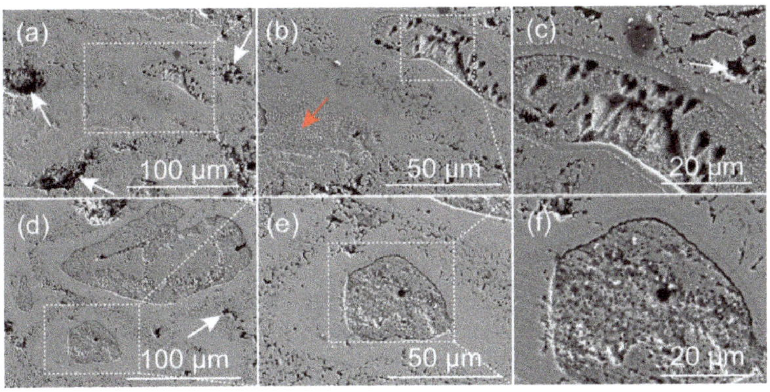

Figure 11. SEM images of the Zn-Ca composite after immersion in Hank's solution for 5 h (**a–c**) and 10 h (**d–f**) and then removal of corrosion products with chromic acid solution.

The preferential degradation of the sacrificial anodes (the core regions of the reinforcing particles) not only led to severe localized corrosion but also made the degradation rate of the Zn-Ca composite significantly higher than that of pure Zn (Figure 8 and Table 2). The corrosion current density of the Zn-Ca composite is almost 57.5 times that of pure Zn. However, the corrosion current densities of the Zn-0.8Ca, Zn-1Ca, and Zn-2Ca alloys measured by Yang et al. [3], Li et al. [10], and Zou et al. [13] are only 1.23, 1.19, and 2.33 times that of pure Zn, respectively. The result is clearly attributed to the fact that the Zn-Ca composite is entirely composed of Zn-Ca compounds.

In addition, the preferential degradation of the sacrificial anodes also accelerated the cathodic reaction (2) occurring on the protected cathodes, thereby significantly increasing the pH value of the soaking solution.

$$2H_2O + O_2 + 4e \rightarrow 4OH^-. \quad (2)$$

Koji et al. [52] measured the pH change in Hank's solution soaking of bulk $CaZn_2$ during immersion tests. After 24 h of immersion, the pH value of the solution was basically stabilized at approximately 8.7. Zhao et al. [53] believed that the OH^- ions produced by the reaction (2) are essential for inducing the deposition of Ca and P elements on biodegradable Zn-based materials. Moreover, the higher the pH of the soaking solution, the more favorable the deposition of calcium phosphates [54]. Although the pH value of Hank's solution is only 8.7, measured by Koji et al. [52], the pH value of Hank's solution soaking of the Zn-Ca composite should possess a higher pH value due to the accelerated reaction (2) on the protected cathodes. The higher the pH value of the soaking solution, the greater the amount of OH^- ions accumulated on the surface of the composite should be. Thus, the compounds containing Ca and P elements are more easily deposited on the surface of the Zn-Ca composite. As a result, a corrosion product layer containing Ca and P elements was deposited on the Zn-Ca composite only after 5 h of immersion in Hank's solution (Figure 9). For the porous Zn scaffold, only some tiny corrosion products were dispersedly deposited on the porous Zn scaffold even after 1 d of immersion in Hank's solution under the same condition [55]. It can be seen that the Zn-Ca composite is far easier to induce Ca and P deposition and exhibits excellent bioactivity.

3.4. Cytotoxicity of the Zn-Ca Composite

The cell viability of L-929 after incubation in the 100%, 50%, 25%, 12.5%, and 6.25% extracts for 1 d is presented in Figure 12. It can be seen that the cell viability of L-929 after culturing in the 100% extract reaches 82.26 (Grade 1) and shows excellent viability. When the 100% extract is diluted to 50%, 25%, 12.5%, and 6.25%, the cell viability of L-929 is higher than 100% (Grade 0). Clearly, these extracts show no cytotoxicity to L-929 cells. Generally, the cytotoxicity of the extracts prepared from biodegradable Zn-based materials is closely related to the concentration of Zn^{2+} in them. Although different cells have different tolerance concentration thresholds to Zn^{2+}, the thresholds are generally very low. For example, Yang et al. [56] reported that the $ZnCl_2$ solution containing 23.9 μg/mL of Zn^{2+} exhibited severe cytotoxicity on MC3T3-E1 cells, while 12.1 μg/mL of Zn^{2+} significantly promoted cell proliferation. He et al. [57] also reported that when the Zn^{2+} concentration in $ZnCl_2$ solution was lower than 19.62 μg/mL, the solution showed good cell viability on MC3T3-E1 cells. However, when the Zn^{2+} concentration increased to 29.43 μg/mL, the cell viability decreased to around 3.7%. For the L-929 cells, the highest safe concentration of Zn^{2+} is only 5.233 μg/mL, as reported by Kubásek et al. [22]. Although Zn metal has a relatively slow degradation rate, the released Zn^{2+} ions during degradation are still excessive to cells. As a result, pure Zn [56,58,59] and other Zn-based implants such as Zn-4Cu alloy [40] and porous Zn scaffold [55] usually exhibit pronounced cytotoxicity in vitro. When these materials with in vitro cytotoxicity are implanted into the body, a layer of fibrous connective tissue is often observed around the implants [56,58,60–62]. The connective tissue layer prevents new bone from directly bonding to the implant, resulting in delayed osteointegration.

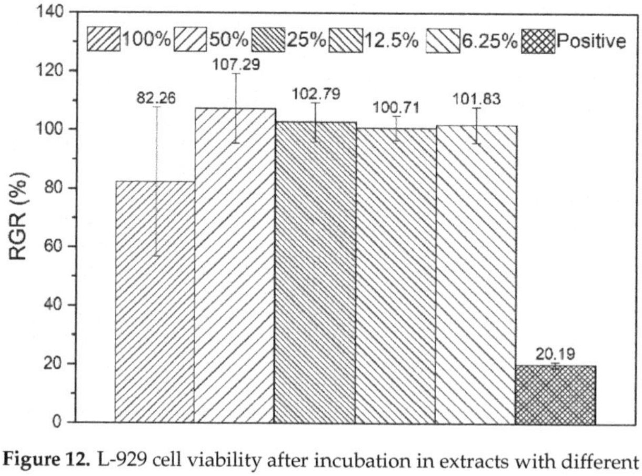

Figure 12. L-929 cell viability after incubation in extracts with different concentrations for 1 day.

Although the Zn^{2+} concentration in the 100% extract prepared from the Zn-Ca composite was not measured, it still can be inferred that the released Zn^{2+} concentration should not be high during the degradation. Clearly, this result is attributed to the protective effect of the sacrificial anodes with high Ca content on the cathodes with high Zn content. In addition, Yang et al. [3] pointed out that the appropriate content of Ca is helpful in eliminating the toxicity of Zn. Therefore, the dissolved Ca^{2+} during the degradation of the

Zn-Ca composite may also play a certain role in improving the cell viability of the material. Further experiments are required to verify the inference.

4. Conclusions

A Zn-Ca composite was successfully fabricated by an air pressure infiltration method using Zn-Ca intermetallic compound powder and Zn ingot. The microstructure, phase structures, compressive mechanical properties, in vitro degradation properties, and cytotoxicity of the composite were investigated. The main conclusions are as follows:

(1) The Zn-Ca composite is fully composed of Zn-Ca intermetallic compounds. The reinforcing phases are Ca-rich particles with core–shell structures, while the matrix phases are rich in Zn elements.
(2) The composite exhibits typically brittle fracture characteristics during compressive tests, and the compressive strength of the composite is around 281.1 MPa.
(3) The core regions of the reinforcing particles degrade preferentially as sacrificial anodes, resulting in severe localized corrosion of the composite during immersion tests. The preferential degradation of the sacrificial anodes also makes the composite have a much faster degradation rate than pure zinc and a better ability to induce Ca and P deposition than pure Zn in Hank's solution.
(4) The preferential degradation of the Ca-rich core regions also provides the composite with excellent cell viability.

In conclusion, the utilization of Ca-rich Zn-Ca compounds as sacrificial anodes accelerates the degradation of the Zn-based implant while also maintaining excellent cell viability. This strategy provides a new pathway for the preparation of degradable Zn-based implants with relatively fast degradation rates as well as good biocompatibility.

Author Contributions: Conceptualization, L.Z., M.Z. and X.W. (Xinyuan Wang); Methodology, L.Z., X.W. (Xinyuan Wang) and S.Z.; Formal analysis, T.W. and X.W. (Xin Wang); Investigation, M.Z., X.W. (Xinyuan Wang) and L.Z.; Data curation, X.W. (Xinyuan Wang), L.Z. and S.Z.; Supervision, L.Z., S.L. and C.C.; Writing—original draft preparation, M.Z., X.W. (Xinyuan Wang), L.Z. and S.Z.; Writing—review and editing, L.Z., M.Z., X.W. (Xin Wang), T.W. and C.C.; Funding acquisition, L.Z. and S.L. All authors have read and agreed to the published version of the manuscript.

Funding: This work was funded by Science Research Project of Hebei Education Department (No. ZD2021034) and Natural Science Foundation of Hebei Province, China (No. E2021202140).

Institutional Review Board Statement: Not applicable.

Informed Consent Statement: Not applicable.

Data Availability Statement: The data presented in the manuscript will be available on request from the corresponding author.

Conflicts of Interest: The authors declare no conflict of interest.

References

1. Yuan, W.; Xia, D.; Wu, S.; Zheng, Y.; Guan, Z.; Rau, J. A review on current research status of the surface modification of Zn-based biodegradable metals. *Bioact. Mater.* **2022**, *7*, 192–216. [CrossRef] [PubMed]
2. Liu, Y.; Zheng, Y.; Chen, X.; Yang, J.; Pan, H.; Chen, D.; Wang, L.; Zhang, J.; Zhu, D.; Wu, S.; et al. Fundamental theory of biodegradable metals—Definition, criteria, and design. *Adv. Funct. Mater.* **2019**, *29*, 1805402. [CrossRef]
3. Yang, H.; Jia, B.; Zhang, Z.; Qu, X.; Li, G.; Lin, W.; Zhu, D.; Dai, K.; Zheng, Y. Alloying design of biodegradable zinc as promising bone implants for load-bearing applications. *Nat. Commun.* **2020**, *11*, 401. [CrossRef] [PubMed]
4. Zhang, Y.; Yan, Y.; Xu, X.; Lu, Y.; Chen, L.; Li, D.; Dai, Y. Investigation on the microstructure, mechanical properties, in vitro degradation behavior and biocompatibility of newly developed Zn-0.8%Li-(Mg, Ag) alloys for guided bone regeneration. *Mater. Sci. Eng. C* **2019**, *99*, 1021–1034. [CrossRef]
5. Li, G.; Zhu, S.; Nie, J.; Zheng, Y.; Sun, Z. Investigating the stress corrosion cracking of a biodegradable Zn-0.8 wt%Li alloy in simulated body fluid. *Bioact. Mater.* **2021**, *6*, 1468–1478. [CrossRef]
6. Zhu, S.; Wu, C.; Li, G.; Zheng, Y.; Nie, J. Microstructure, mechanical properties and creep behaviour of extruded Zn-xLi (x = 0.1, 0.3 and 0.4) alloys for biodegradable vascular stent applications. *Mater. Sci. Eng. A* **2020**, *777*, 139082. [CrossRef]

7. Young, J.; Reddy, R.G. Synthesis, mechanical properties, and in vitro corrosion behavior of biodegradable Zn–Li–Cu alloys. *J. Alloys Compd.* **2020**, *844*, 156257. [CrossRef]
8. Qin, Y.; Yang, H.; Liu, A.; Dai, J.; Wen, P.; Zheng, Y.; Tian, Y.; Li, S.; Wang, X. Processing optimization, mechanical properties, corrosion behavior and cytocompatibility of additively manufactured Zn-0.7Li biodegradable metals. *Acta Biomater.* **2022**, *142*, 388–401. [CrossRef]
9. Gao, Z.; Zhang, X.; Huang, H.; Chen, C.; Jiang, J.; Niu, J.; Dargusch, M.; Yuan, G. Microstructure evolution, mechanical properties and corrosion behavior of biodegradable Zn-2Cu-0.8Li alloy during room temperature drawing. *Mater. Charact.* **2022**, *185*, 111722. [CrossRef]
10. Li, H.F.; Xie, X.H.; Zheng, Y.F.; Cong, Y.; Zhou, F.Y.; Qiu, K.J.; Wang, X.; Chen, S.H.; Huang, L.; Tian, L.; et al. Development of biodegradable Zn-1X binary alloys with nutrient alloying elements Mg, Ca and Sr. *Sci. Rep.* **2015**, *5*, 10719. [CrossRef]
11. Jia, B.; Yang, H.; Zhang, Z.; Qu, X.; Jia, X.; Wu, Q.; Han, Y.; Zheng, Y.; Dai, K. Biodegradable Zn–Sr alloy for bone regeneration in rat femoral condyle defect model: In vitro and in vivo studies. *Bioact. Mater.* **2021**, *6*, 1588–1604. [CrossRef] [PubMed]
12. Shi, Z.; Li, H.; Xu, J.; Gao, X.; Liu, X. Microstructure evolution of a high-strength low-alloy Zn–Mn–Ca alloy through casting, hot extrusion and warm caliber rolling. *Mater. Sci. Eng. A* **2020**, *771*, 138626. [CrossRef]
13. Zou, Y.; Chen, X.; Chen, B. Effects of Ca concentration on degradation behavior of Zn-xCa alloys in Hank's solution. *Mater. Lett.* **2018**, *218*, 193–196. [CrossRef]
14. Balasubramani, N.; Nan, Y.; Venezuela, J.; Dargusch, M. Ultrasonic treatment for the refinement of brittle CaZn13 phases in a biomedical Zn-Cu-Ca alloy. *Mater. Lett.* **2021**, *305*, 130754. [CrossRef]
15. Yang, N.; Balasubramani, N.; Venezuela, J.; Almathami, S.; Wen, C.; Dargusch, M. The influence of Ca and Cu additions on the microstructure, mechanical and degradation properties of Zn–Ca–Cu alloys for absorbable wound closure device applications. *Bioact. Mater.* **2021**, *6*, 1436–1451. [CrossRef]
16. Huang, H.; Liu, H.; Ren, K.; Shi, J.; Ju, J.; Wu, H.; Jiang, J.; Ma, A.; Xue, F.; Bai, J.; et al. Improvement of ductility and work hardening ability in a high strength Zn-Mg-Y alloy via micron-sized and submicron-sized YZn12 particles. *J. Alloys Compd.* **2021**, *877*, 160268. [CrossRef]
17. Liu, H.; Huang, H.; Zhang, Y.; Xu, Y.; Wang, C.; Sun, J.; Jiang, J.; Ma, A.; Xue, F.; Bai, J. Evolution of Mg-Zn second phases during ECAP at different processing temperatures and its impact on mechanical properties of Zn-1.6Mg (wt.%) alloys. *J. Alloys Compd.* **2019**, *811*, 151987. [CrossRef]
18. Huang, H.; Liu, H.; Wang, L.; Yan, K.; Li, Y.; Jiang, J.; Ma, A.; Xue, F.; Bai, J. Revealing the effect of minor Ca and Sr additions on microstructure evolution and mechanical properties of Zn-0.6 Mg alloy during multi-pass equal channel angular pressing. *J. Alloys Compd.* **2020**, *844*, 155923. [CrossRef]
19. Yao, C.; Wang, Z.; See, L.T.; Zhu, T.; Gao, W. Effects of Mg on microstructure and corrosion properties of Zn-Mg alloy. *J. Alloys Compd.* **2014**, *602*, 101–107. [CrossRef]
20. Vojtěch, D.; Kubásek, J.; Šerák, J.; Novák, P. Mechanical and corrosion properties of newly developed biodegradable Zn-based alloys for bone fixation. *Acta Biomater.* **2011**, *7*, 3515–3522. [CrossRef]
21. Kubásek, J.; Pospíšilová, I.; Vojtěch, D.; Jablonská, E.; Ruml, T. Structural, mechanical and cytotoxicity characterization of as-cast biodegradable Zn-xMg (x = 0.8–8.3) alloys. *Mater. Technol.* **2014**, *48*, 623–629.
22. Kubásek, J.; Vojtěch, D.; Jablonská, E.; Pospíšilová, I.; Lipov, J.; Ruml, T. Structure, mechanical characteristics and in vitro degradation, cytotoxicity, genotoxicity and mutagenicity of novel biodegradable Zn-Mg alloys. *Mater. Sci. Eng. C* **2016**, *58*, 24–35. [CrossRef] [PubMed]
23. Zhang, L.; Liu, X.Y.; Huang, H.; Zhan, W. Effects of Ti on microstructure, mechanical properties and biodegradation behavior of Zn-Cu alloy. *Mater. Lett.* **2019**, *244*, 119–122. [CrossRef]
24. Li, L.; Liu, C.; Jiao, H.; Yang, L.; Cao, F.; Wang, X.; Cui, J. Investigation on microstructures, mechanical properties and in vitro corrosion behavior of novel biodegradable Zn-2Cu-0.01Ti-xLi alloys. *J. Alloys Compd.* **2021**, *888*, 161529. [CrossRef]
25. Lin, J.; Tong, X.; Wang, K.; Shi, Z.; Li, Y.; Dargusch, M.; Wen, C. Biodegradable Zn–3Cu and Zn–3Cu–0.2Ti alloys with ultrahigh ductility and antibacterial ability for orthopedic applications. *J. Mater. Sci. Technol.* **2021**, *68*, 76–90. [CrossRef]
26. Wang, K.; Tong, X.; Lin, J.; Wei, A.; Li, Y.; Dargusch, M.; Wen, C. Binary Zn-Ti alloys for orthopedic applications: Corrosion and degradation behaviors, friction and wear performance, and cytotoxicity. *J. Mater. Sci. Technol.* **2021**, *74*, 216–229. [CrossRef]
27. Bagha, P.S.; Khaleghpanah, S.; Sheibani, S.; Khakbiz, M.; Zakeri, A. Characterization of nanostructured biodegradable Zn-Mn alloy synthesized by mechanical alloying. *J. Alloys Compd.* **2018**, *735*, 1319–1327. [CrossRef]
28. Chen, H.; Shi, Z.; Liu, X. Microstructure and mechanical properties of extruded and caliber rolled biodegradable Zn-0.8Mn-0.4Ag alloy with high ductility. *Mater. Sci. Eng. A* **2020**, *770*, 138543. [CrossRef]
29. Huang, H.; Liu, H.; Wang, L.; Ren, K.; Yan, K.; Li, Y.; Jiang, J.; Ma, A.; Xue, F.; Bai, J. Multi-interactions of dislocations and refined microstructure in a high strength and toughness Zn-Mg-Mn alloy. *J. Mater. Res. Technol.* **2020**, *9*, 14116–14121. [CrossRef]
30. Jia, B.; Yang, H.; Han, Y.; Zhang, Z.; Qu, X.; Zhuang, Y.; Wu, Q.; Zheng, Y.; Dai, K. In vitro and in vivo studies of Zn-Mn biodegradable metals designed for orthopedic applications. *Acta Biomater.* **2020**, *108*, 358–372. [CrossRef]
31. Shi, Z.; Yu, J.; Liu, X.; Wang, L. Fabrication and characterization of novel biodegradable Zn-Mn-Cu alloys. *J. Mater. Sci. Technol.* **2018**, *34*, 1008–1015. [CrossRef]
32. Shi, Z.; Li, Z.; Bai, W.; Tuoliken, A.; Yu, J.; Liu, X. (Fe, Mn)Zn_{13} phase and its core-shell structure in novel biodegradable Zn-Mn-Fe alloys. *Mater. Des.* **2019**, *162*, 235–245. [CrossRef]

33. Sun, J.; Zhang, X.; Shi, Z.; Gao, X.; Li, H.; Zhao, F.; Wang, J.; Wang, L. Development of a high-strength Zn-Mn-Mg alloy for ligament reconstruction fixation. *Acta Biomater.* **2021**, *119*, 485–498. [CrossRef] [PubMed]
34. Jiang, J.; Qian, Y.; Huang, H.; Niu, J.; Yuan, G. Biodegradable Zn-Cu-Mn alloy with suitable mechanical performance and in vitro degradation behavior as a promising candidate for vascular stents. *Mater. Sci. Eng. C* **2022**, *133*, 112652. [CrossRef] [PubMed]
35. Shi, Z.; Gao, X.; Chen, H.; Liu, X.; Li, A.; Zhang, H.; Wang, L. Enhancement in mechanical and corrosion resistance properties of a biodegradable Zn-Fe alloy through second phase refinement. *Mater. Sci. Eng. C* **2020**, *116*, 111197. [CrossRef] [PubMed]
36. Lin, J.; Tong, X.; Sun, Q.; Luan, Y.; Zhang, D.; Shi, Z.; Wang, K.; Lin, J.; Li, Y.; Dargusch, M.; et al. Biodegradable ternary Zn–3Ge–0.5X (X = Cu, Mg, and Fe) alloys for orthopedic applications. *Acta Biomater.* **2020**, *115*, 432–446. [CrossRef] [PubMed]
37. Qu, X.; Yang, H.; Jia, B.; Yu, Z.; Zheng, Y.; Dai, K. Biodegradable Zn-Cu alloys show antibacterial activity against MRSA bone infection by inhibiting pathogen adhesion and biofilm formation. *Acta Biomater.* **2020**, *117*, 400–417. [CrossRef]
38. Tang, Z.; Niu, J.; Huang, H.; Zhang, H.; Pei, J.; Ou, J.; Yuan, G. Potential biodegradable Zn-Cu binary alloys developed for cardiovascular implant applications. *J. Mech. Behav. Biomed. Mater.* **2017**, *72*, 182–191. [CrossRef]
39. Mostaed, E.; Ardakani, M.S.; Sikora-Jasinska, M.; Drelich, J.W. Precipitation induced room temperature superplasticity in Zn-Cu alloys. *Mater. Lett.* **2019**, *244*, 203–206. [CrossRef]
40. Niu, J.; Tang, Z.; Huang, H.; Pei, J.; Zhang, H.; Yuan, G.; Ding, W. Research on a Zn-Cu alloy as a biodegradable material for potential vascular stents application. *Mater. Sci. Eng. C* **2016**, *69*, 407–413. [CrossRef]
41. Tang, Z.; Huang, H.; Niu, J.; Zhang, L.; Zhang, H.; Pei, J.; Tan, J.; Yuan, G. Design and characterizations of novel biodegradable Zn-Cu-Mg alloys for potential biodegradable implants. *Mater. Des.* **2017**, *117*, 84–94. [CrossRef]
42. Sikora-Jasinska, M.; Mostaed, E.; Mostaed, A.; Beanland, R.; Mantovani, D.; Vedani, M. Fabrication, mechanical properties and in vitro degradation behavior of newly developed Zn-Ag alloys for degradable implant applications. *Mater. Sci. Eng. C* **2017**, *77*, 1170–1181. [CrossRef] [PubMed]
43. Wątroba, M.; Mech, K.; Bednarczyk, W.; Kawałko, J.; Marciszko-Wiąckowska, M.; Marzec, M.; Shepherd, D.E.T.; Bała, P. Long-term in vitro corrosion behavior of Zn-3Ag and Zn-3Ag-0.5Mg alloys considered for biodegradable implant applications. *Mater. Des.* **2022**, *213*, 110289. [CrossRef]
44. Liu, Z.; Qiu, D.; Wang, F.; Taylor, J.A.; Zhang, M. The grain refining mechanism of cast zinc through silver inoculation. *Acta Mater.* **2014**, *79*, 315–326. [CrossRef]
45. Okamoto, H. Ca-Zn (Calcium-Zinc). *J. Phase Equilib. Diff.* **2013**, *34*, 171. [CrossRef]
46. Messing, A.F.; Adams, M.D.; Steunenberg, P.K. Contribution to the phase diagram calcium-zinc. *Trans. ASM* **1963**, *56*, 345–350.
47. Zhao, L.; Xie, Y.; Zhe, Z.; Wang, X.; Qi, Y.; Wang, T.; Wang, R.; Cui, C. Fabrication and properties of biodegradable ZnO nano-rods/porous Zn scaffolds. *Mater. Charact.* **2018**, *144*, 227–238. [CrossRef]
48. Zhao, L.; Song, Y.; Zhang, Z.; Wang, X.; Wang, T.; Cui, C. Fabrication and investigation on properties of degradable Zn-1Al alloy for biomedical applications. *Mater. Rep.* **2018**, *32*, 1192–1196.
49. Zhang, S.; Yuan, P.; Wang, X.; Wang, T.; Zhao, L.; Cui, C. Fabrication and properties of Zn-3Mg-1Ti alloy as a potential biodegradable implant material. *Materials* **2022**, *15*, 940. [CrossRef]
50. Zhao, L.; Yuan, P.; Zhang, M.; Wang, X.; Qi, Y.; Wang, T.; Cao, B.; Cui, C. Preparation and properties of porous Zn-based scaffolds as biodegradable implants: A review. *J. Mater. Sci.* **2023**, *58*, 8275–8316. [CrossRef]
51. Byun, J.M.; Yu, J.M.; Kim, D.K.; Kim, T.Y.; Jung, W.S.; Kim, Y.D. Corrosion behavior of Mg_2Zn_{11} and $MgZn_2$ single phases. *Korean J. Met. Mater.* **2013**, *51*, 413–419. [CrossRef]
52. Hagihara, K.; Shakudo, S.; Fujii, K.; Nakano, T. Degradation behavior of Ca–Mg–Zn intermetallic compounds for use as biodegradable implant materials. *Mater. Sci. Eng. C* **2014**, *44*, 285–292. [CrossRef] [PubMed]
53. Zhao, L.; Zhang, Z.; Song, Y.; Liu, S.; Qi, Y.; Wang, X.; Wang, Q.; Cui, C. Mechanical properties and in vitro biodegradation of newly developed porous Zn scaffolds for biomedical applications. *Mater. Des.* **2016**, *108*, 136–144. [CrossRef]
54. Okido, M.; Kuroda, K.; Ishikawa, M.; Ichio, R.; Takai, O. Hydroxyapatite coating on titanium by means of thermal substrate method in aqueous solutions. *Solid. State Ionics* **2002**, *151*, 47–52. [CrossRef]
55. Yuan, P.; Zhang, M.; Wang, X.; Qi, Y.; Wang, T.; Zhao, L. Effects of polylactic acid coating on properties of porous Zn scaffolds as degradable materials. *Mater. Charact.* **2023**, *199*, 112852. [CrossRef]
56. Yang, H.; Qu, X.; Lin, W.; Chen, D.; Zhu, D.; Dai, K.; Zheng, Y. Enhanced Osseointegration of Zn-Mg Composites by Tuning the Release of Zn Ions with Sacrificial Mg-Rich Anode Design. *ACS Biomater. Sci. Eng.* **2019**, *5*, 453–467. [CrossRef]
57. He, J.; Fang, J.; Wei, P.; Li, Y.; Guo, H.; Mei, Q.; Ren, F. Cancellous bone-like porous Fe@Zn scaffolds with core-shell-structured skeletons for biodegradable bone implants. *Acta Biomater.* **2021**, *121*, 665–681. [CrossRef]
58. Su, Y.; Yang, H.; Gao, J.; Qin, Y.; Zheng, Y.; Zhu, D. Interfacial zinc phosphate is the key to controlling biocompatibility of metallic zinc implants. *Adv. Sci.* **2019**, *6*, 1900112. [CrossRef]
59. Murni, N.S.; Dambatta, M.S.; Yeap, S.K.; Froemming, G.R.A.; Hermawan, H. Cytotoxicity evaluation of biodegradable Zn-3Mg alloy toward normal human osteoblast cell. *Mater. Sci. Eng. C* **2015**, *49*, 560–566. [CrossRef]
60. Yang, H.; Qu, X.; Lin, W.; Wang, C.; Zhu, D.; Dai, K.; Zheng, Y. In vitro and in vivo studies on zinc-hydroxyapatite composites as novel biodegradable metal matrix composite for orthopedic applications. *Acta Biomater.* **2018**, *71*, 200–214. [CrossRef]

61. He, J.; Li, D.; He, F.; Liu, Y.; Liu, Y.; Zhang, C.; Ren, F.; Ye, Y.; Deng, X.; Yin, D. A study of degradation behaviour and biocompatibility of Zn-Fe alloy prepared by electrodeposition. *Mater. Sci. Eng. C* **2020**, *117*, 111295. [CrossRef] [PubMed]
62. Yuan, W.; Xia, D.; Zheng, Y.; Liu, X.; Wu, S.; Li, B.; Han, Y.; Jia, Z.; Zhu, D.; Ruan, L.; et al. Controllable biodegradation and enhanced osseointegration of ZrO_2-nanofilm coated Zn-Li alloy: In vitro and in vivo studies. *Acta Biomater.* **2020**, *105*, 290–303. [CrossRef] [PubMed]

materials

MDPI

Article

Effect of In Situ NbC-Cr_7C_3@graphene/Fe Nanocomposite Inoculant Modification and Refinement on the Microstructure and Properties of W18Cr4V High-Speed Steel

Lina Bai [1,2,*], Guixing Zheng [2], Lijie Zhang [2], Shuangjin Liu [1,*], Laichun Xu [2], Haowen Zheng [3] and Jie Liu [2]

1 Key Laboratory for New Type of Functional Materials in Hebei Province, School of Materials Science and Technology, Hebei University of Technology, Tianjin 300130, China
2 Department of Basic, Army Military Transportation University, Tianjin 300161, China; jsj_tj@126.com (G.Z.); 18920810661@163.com (L.Z.); lcxu11@163.com (L.X.); jiejiechonger@163.com (J.L.)
3 Department of Automation and Intelligent Science, College of Artificial Intelligence, Nankai University, Tianjin 300350, China; zhw142857@163.com
* Correspondence: blnjjxy@126.com (L.B.); 2003022@hebut.edu.cn (S.L.)

Abstract: A novel graphene-coated nanocrystalline ceramic particle, iron-based composite inoculant was developed in this study to optimize the as-cast microstructure and mechanical properties of W18Cr4V high-speed steel (HSS). The effects of the composite inoculant on the microstructure, crystal structure, and mechanical properties of HSS were analyzed using transmission electron microscopy, scanning electron microscopy, energy dispersive spectroscopy, and X-ray diffraction. The (002) and (020) crystal planes of the Fe_3C and Cr_7C_3 phases, respectively, were collinear at two points in the reciprocal space, indicating a coherent relationship between the Fe_3C and Cr_7C_3 phases in the tempered modified HSS. This contributed to an improved non-uniform nucleation rate and refining of the HSS grains. The mechanical properties of the modified steel exhibited a general improvement. Specifically, the modification treatment enhanced the hardness of HSS from HRC 63.2 to 66.4 and the impact toughness by 48.3%.

Keywords: W18Cr4V high-speed steel; nanocrystalline inoculant; mechanical properties; grain refining

1. Introduction

As the most common tungsten high-speed steel (HSS) grade, known as T1 HSS in the United States and SKH2 HSS in Japan, W18Cr4V plays a key role in industrial production. W18Cr4V belongs to the ledeburitic steel, which has high hardness and wear resistance and is one of the important tool steels. Specifically, W18Cr4V is more affordable than super-hard materials, making it the most suitable material for cutting tools and drills in various machines. Moreover, it is employed in high-temperature bearings that are required in the aviation field, highly temperature-resistant rolls, and high-load molds [1,2]. The microstructure of as-cast W18Cr4V mainly comprises δ eutectoid, martensite, residual austenite, and eutectic martensite [3,4]. The stability of the microstructure and mechanical properties of HSS directly ensure the safety and efficiency of mechanical equipment made from HSS. Current microalloying technologies primarily focus on the effects of different alloying and microalloying elements on steel materials [5–7]. However, W18Cr4V is a high-alloy steel, and incorporating excessive alloying elements can lead to compositional segregation, which is difficult to resolve. Li et al. reported that a modification in GCr15 steel increased impact toughness by 32.02% [8]. Further, Belyanchikov et al. reported that nitrides of Nb, V, Zr, and other elements promoted the precipitation of MC type carbides in the molten HSS [9].

In this study, we developed a novel nanocrystalline composite inoculant to address the severe component segregation problem in W18Cr4V. The preparation cost of the new compound inoculant is lower than that of powder metallurgy HSS. Moreover, it does not alter the cold and hot processing of HSS ingots and is suitable for large-scale industrial production [10–13]. We also investigated the microstructure of W18Cr4V after modification with the new nanocomposite inoculant to study its effects on the mechanical properties of HSS.

2. Experimental Procedure

2.1. Preparation of Composite Inoculants

Based on the thermodynamics and mismatch theory of metal materials, graphene packages of NbC-Cr_7C_3@graphene/Fe were prepared via in situ reaction and mechanical grinding (GN-2, Simoloyer Mill, Oute Vacuum Technology Co., Ltd., Shenyang, China) [14]. The in situ reaction ceramic particles were prepared using a vacuum tube furnace (SK-G06143, Zhonghuan Electric Furnace Co., Ltd., Tianjin, China). The process of preparing Cr_7C_3 and NbC in situ reaction ceramic particles was as follows: chromium powder and graphene were mixed at a mass ratio of 12:1, heated in the vacuum tube furnace to 1160 °C, held at this temperature for 0.5 h, and then cooled in the furnace. Niobium powder and graphene were mixed at a mass ratio of 10:1, heated in the vacuum tube furnace to 1200 °C, held at this temperature for 1 h, and then cooled in the furnace.

The composite inoculants were processed by wet milling using a GN-2 Simoloyer mill at a speed of 700 $rmin^{-1}$ for mechanical grinding. The inner wall size of the ball milling jar is $\phi60 \times 100$ mm and the material is GCr15 bearing steel. Tungsten steel balls with diameters of 2 mm, 5 mm, and 8 mm were added to the ball milling jar, and the corresponding mass ratio of the three kinds of grinding balls was 1:2:5. The high-energy ball milling method employs the rotation or vibration of a ball mill to create a strong impact, grinding, and stirring action on the raw materials. This process effectively transfers kinetic energy to the target material through a stirrer, enhancing energy utilization and material performance. Consequently, it has become an essential method for the preparation of nanomaterials.

2.2. Casting and Performance Testing

W18Cr4V HSS was melted and inoculated in a vacuum induction furnace (ZG-0.025, Alkaline Vacuum Furnace Co., Ltd., Jinzhou, China) with a vacuum degree of 6.67×10^{-3} Pa. W18Cr4V was completely melted, inoculated with 0.5 wt.% composite inoculant, electromagnetically stirred for 2 to 3 min, and poured into a $\phi70$ mm cast iron mold. The W18Cr4V cast and tempered samples were cut into cubes of side length 10 mm. Before observing the microstructure, one surface of the samples was polished with sandpaper, then mechanically polished, and a 4% dilute nitric acid solution was used for corrosion.

The mechanical properties of multiple control test steels, such as hardness, wear resistance, and impact toughness, were also examined. W18Cr4V is commonly used in high-temperature applications. Therefore, it must maintain high hardness at high temperatures. Hardness testing (HR-150A, Rockwell, Dianying Optical Instrument Co., Ltd., Shanghai, China) requires that no fewer than five points on each stress surface of the sample be tested; the overall number of measurements should not be less than 20. The red hardness test involves heating the samples in a box resistance furnace to 600 °C, maintaining them at this temperature for 60 min, and removing them for air cooling; the process must be repeated four times. Impact toughness characterizes the ability of HSS to absorb impact energy when subjected to an impact load. The experiment was conducted using a pendulum impact-testing machine (JB-200, Zhongke Metrology Instrument Co., Ltd., Yangzhou, China) at room temperature. The notchless test samples were processed to dimensions of $10 \times 10 \times 55$ mm. A wear resistance test was conducted using a wear tester (M-200, AVIC Era Instrument Equipment Co., Ltd., Beijing, China). In the comparative experiment, each group consisted of 15 samples. The stains on the surface of the samples ($10 \times 10 \times 12$ mm) were first cleaned with acetone solution. The samples were worn for 90 min at 200 N.

Further, the properties of the new inoculant and its effects on the as-cast and tempered structures of W18Cr4V HSS were systematically studied. The influence of nanocrystalline ceramic particles present in the inoculant on the carbides in the steel and the refinement mechanism were evaluated using metallographic microscopy (Axio Imager M2m, Carl Zeiss., Oberkochen, Germany), scanning electron microscopy (SEM; Hitachi S-4800, Hitachi., Tokyo, Japan), high-resolution transmission electron microscopy (HRTEM; Jeol 2011, JEOL., Ltd., Tokyo, Japan), energy spectrum analysis, X-ray diffraction (XRD; Bruker D8 FOCUS, Brooke Co., Ltd., Saarbrucken, Germany), and Digital Micrograph 3.2 software (AMETEK, Inc., Berwyn, IL, USA).

3. Results and Discussion

3.1. Detection and Analysis of the Nanocomposite Inoculant

The SEM results revealed that the powder mixture comprised NbC ceramic particles, Cr_7C_3 ceramic particles, cast iron powder, and graphene at a mass ratio of 5:5:4.5:0.5 (Figure 1a). Furthermore, flake graphene was observed to be mixed in the metal particle gaps. Figure 1b shows the SEM image obtained from high-energy ball milling of the inoculant raw material for 4 h. The microstructure of the composite inoculant comprised numerous small dark gray, light gray, and bright white particles, along with a small amount of larger particles, some of which had bright white edges. Figure 1c shows the energy spectrum of Area 1 in Figure 1b, indicating the percentages of C, Cr, Nb, and Fe in Area 1 as 51.64 at.%, 44.94 at.%, 0.34 at.%, and 3.09 at.%, respectively. The energy spectrum results for Area 1 indicated that the dark gray particles were mainly chromium carbide; using the XRD analysis results, they were determined to be Cr_7C_3. Figure 1d,e show the energy spectra of selected areas, Spot 1 and Area 2, respectively, from Figure 1b. The bright, white-selected areas of Spot 1 and Area 2 had significantly higher C content, reaching 86.83 at.% and 79.57 at.%, respectively. Moreover, the C content in these selected areas was higher than that required to form carbides, indicating that the excess C in the composite inoculant was graphene, which formed a coating on the surface of the ceramic particles.

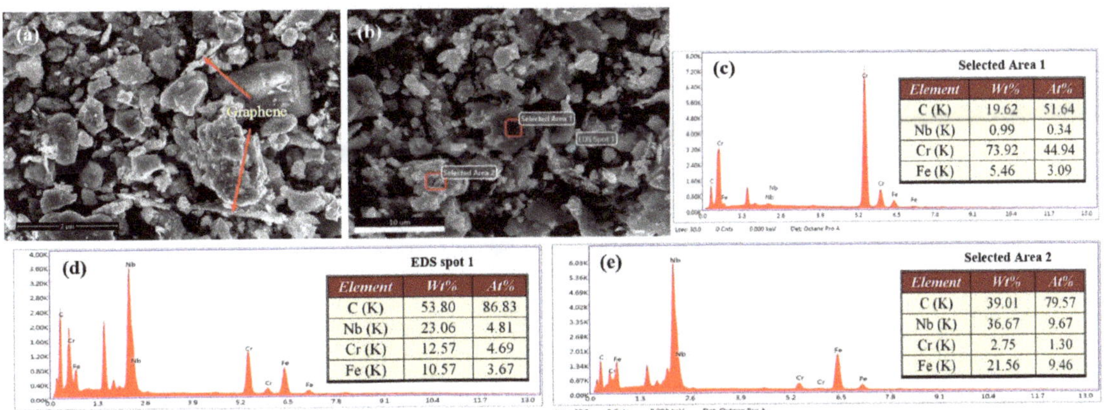

Figure 1. (**a**) SEM image of the composite power before milling; (**b**) SEM micrographs of the composite inoculant for 4 h of ball milling; (**c**) EDS spectra of Area 1 in Figure 1b; (**d**) EDS spectra of Spot 1 in Figure 1b; (**e**) EDS spectra of Area 2 in Figure 1b.

The XRD pattern of the composite inoculant after 10 h of high-energy ball milling is shown in Figure 2a. The main phases in the inoculant were Cr_7C_3, Fe (ferrite), Fe_7C_3, NbC, and graphene. A small amount of the Fe_7C_3 phase was found in the composite inoculant, indicating that prolonged high-energy ball milling increased the atomic potential energy, resulting in Fe replacing a small amount of Cr in the iron powder. Figure 2b shows a TEM image of the NbC-Cr_7C_3@graphene/Fe composite inoculant, revealing that the composite inoculant particles were significantly refined through long-term high-speed rotation and collision with the steel balls. The average composite inoculant grain size

(Figure 2b), measured using Image-Pro software (Figure 2c), was less than 50 nm, with a clear agglomeration of the nanocrystals. Ball milling for 10 h enables the composite inoculant to achieve nanocrystalline morphology that meets the experimental requirements, compared with ball milling for 4 h.

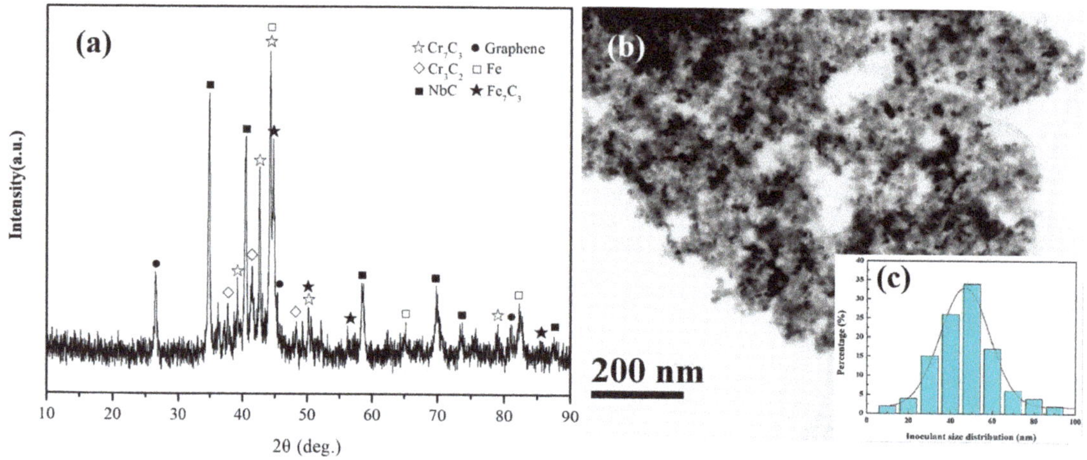

Figure 2. (**a**) XRD patterns of the composite inoculant for 10 h of ball milling; (**b**) TEM image of the inoculant for 10 h of ball milling; (**c**) the inoculant size distribution in Figure 2b.

Figure 3a shows a TEM image of the NbC-Cr_7C_3@graphene/Fe composite inoculant after 10 h of high-energy ball milling, revealing a significant refinement of the inoculant particles, which resulted in the ceramic particles reaching nanocrystal size. The polycrystalline diffraction ring in Figure 3b reveals the SAED pattern of region A in Figure 3a. The main phases in region A, determined through Digital Micrograph software measurement and PDF card analysis, include the NbC, Cr_7C_3, graphene, and $C_{0.08}Fe_{1.92}$ phases.

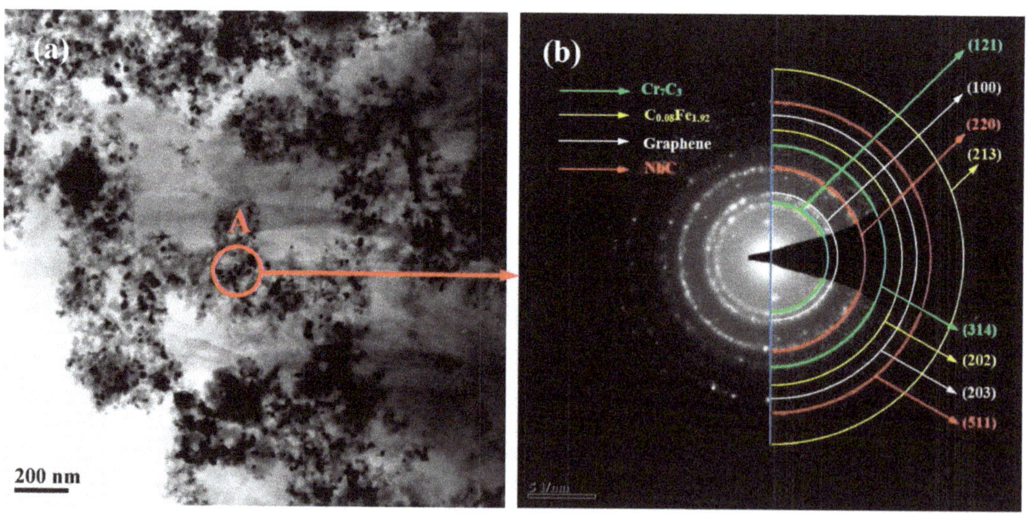

Figure 3. (**a**) TEM low image of the NbC-Cr_7C_3@Graphene/Fe inoculant for 10 h of ball milling; (**b**) SAED pattern of region A in Figure 3a.

To further determine the microstructure and phase composition of the composite inoculant using HRTEM, fast Fourier transform (FFT) analysis was performed on Zones I and II of the images (Figure 4a). Digital Micrograph software was used to measure the crystal plane spacing (Figure 4b); they were found to be 2.12 Å and 2.05 Å, with an angle of 122.36° between crystal planes. According to the calculations and PDF cards, the crystal planes in Zone I were (101) and $(\bar{1}\bar{1}0)$, and the included angle between the crystal planes was 120°. Thus, the phase in Zone I was determined to be $C_{0.12}Fe_{1.88}$ (martensite) through a combination of formula calculations and software measurements. Two sets of reciprocal lattices appear in the FFT of Zone II, as shown in Figure 4c. Zone II was assumed to comprise graphene and martensite via analysis and calculation, indicating that the lamellar structure of graphene was torn during high-energy ball milling. Moreover, the single-layer graphene tightly adhered to the surface of ceramic particles under the strong impact of the grinding ball and ceramic particles, resulting in an overlap of the two phases in Zone II. Furthermore, the orientation difference between the crystal planes of the two phases was 14°31′. The particle phases of the composite inoculant were partitioned in the HRTEM image (Figure 4d), and FFT analysis was performed in the selected area. The FFT diffraction patterns of Zones I and II are shown in Figure 4e,f, respectively. The software analysis and calculation results indicated that Zone I comprised the Cr_7C_3 phase, whereas Zone II comprised the graphene phase. Thus, the experimental results revealed that layered graphene had a coating effect on Cr_7C_3 ceramic particles.

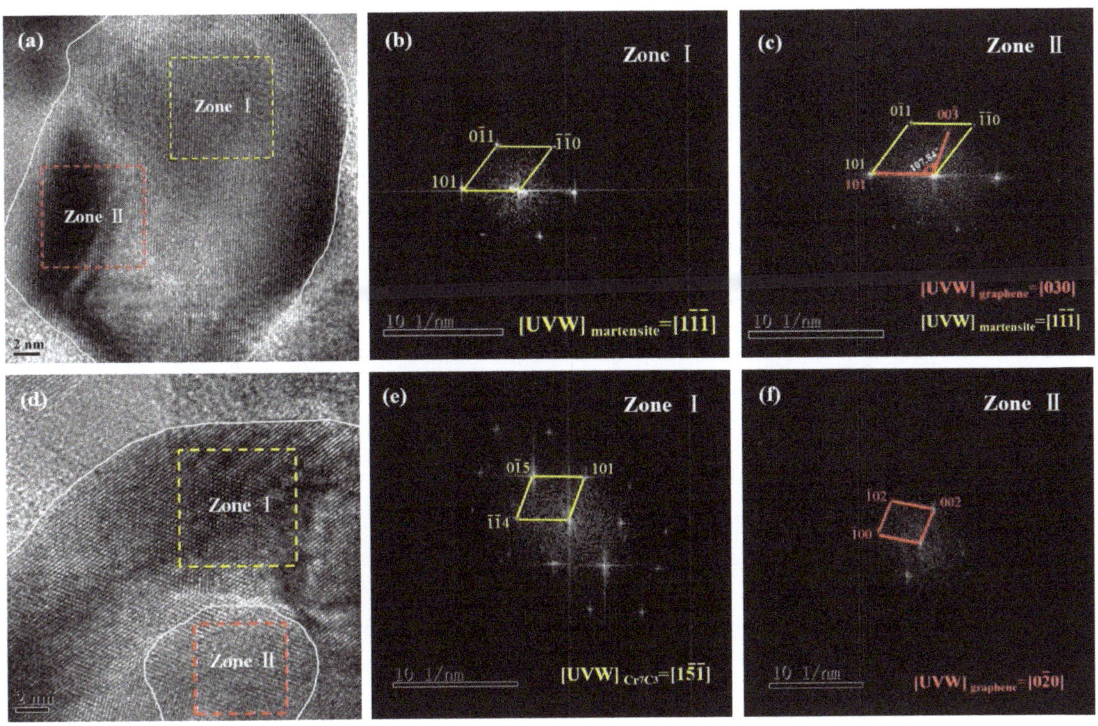

Figure 4. (**a**) HRTEM image of the inoculant; (**b**) FFT of Zone I in Figure 4a; (**c**) FFT of Zone II in Figure 4a; (**d**) HRTEM image of the inoculant; (**e**) FFT of Zone I in Figure 4d; (**f**) FFT of Zone II in Figure 4d.

3.2. *Influence of the Nanocomposite Inoculant on the Microstructure of W18Cr4V HSS*

Metallographic microscopy results reveal that the as-cast microstructure of W18Cr4V comprises fishbone-like eutectic carbides, a black eutectoid at the grain center, and bright white residual austenite distributed along the grain boundaries (Figure 5a,b). The as-cast microstructure of the modified W18Cr4V was significantly refined; the continuous distribution of fishbone-like eutectic carbides was reduced and disconnected, reducing the effect of splitting on the steel matrix, network carbide thickness, and grain size. In the enlarged image of the area (Figure 5a), intragranular carbides appear as strips and are unevenly distributed. However, the enlarged area in Figure 5b shows granular and more uniform intragranular carbides. Furthermore, the amount and size of fishbone carbides distributed along the grain boundaries in the as-cast modified structure were significantly reduced; the distribution of carbides was relatively independent (Figure 5b). According to the Hall–Petch formula [15], the as-cast modified steel has better yield strength. Consequently, the significantly refined carbides were more likely to be crushed during forging. The tempering process for W18Cr4V is complex and requires three treatments. The temperature increase causes the alloy elements to precipitate from the steel matrix. The microstructure of the unmodified tempered W18Cr4V, observed using SEM (Figure 5c), showed coarse granular carbides mostly precipitated at the grain boundaries on the steel matrix. The modified tempered steel exhibited a change in the distribution and morphology of the carbides; the precipitated secondary carbides were smaller and more uniform (Figure 5d). The block carbides, which were distributed on the matrix and enhanced the crystal structure of HSS, were analyzed in a specific region, namely, A, to examine the particle phase (Figure 5e). Figure 5f shows the SAED image of region A. Here, the (111) plane measured using Digital Micrograph software was 6.437 Å, which is close to the calculated crystal plane spacing (6.071 Å). Furthermore, the crystal plane indices of the other two reciprocal points in the reciprocal space were calculated and compared with the reference PDF cards (#65-1347), indicating that region A was the Fe_3W_3C phase. The crystal band axis in this selection area was $\left[0\bar{4}4\right]$ and the three crystal planes were (111), $(3\bar{1}\bar{1})$, and (400).

Massive black carbide particles and a flat noodle martensite structure were observed on a wide range of the dark gray steel matrix in Figure 6a. The image shows that the density of carbides in steel was extremely high and the distance between carbides was less than 1 μm. When martensite makes contact with carbide in steel, its growth is hindered. This effect is similar to the fine-grained effect, because the boundary blocks the growth of martensite. Furthermore, if the grain size is less than the determined value (2–4 μm), the martensite can be dislocation structure or lath martensite [16]. Figure 6b shows an HRTEM image of region B, which is divided into two regions based on the grain boundary trend. The FFT diffraction patterns are shown in Figure 6c,d. Two sets of reciprocal lattices can be observed in the reciprocal space, as shown in Figure 6c. According to the parallelogram rule, the $(00\bar{2})$ crystal plane of the Fe_3C phase and the (020) crystal plane of the Cr_7C_3 phase were collinear at two points in the reciprocal space. Consequently, the $(00\bar{2})$ and (020) crystal planes can be determined to be parallel and coplanar in the positive space, indicating that the Fe_3C and Cr_7C_3 phases have a coherent relationship in the tempered modified HSS [17]. The FFT diffraction patterns of Zone I are shown in Figure 6d. Using software and formulas to calculate the crystal plane spacing and angle between crystal planes, Zone I was determined to comprise the Cr_7C_3 phase; the crystal band axis of this selected region was $\left[20\bar{2}\right]$. The three crystal plane indices were (101), (020), and (121), respectively. The inverse FFT pattern of Zone II in Figure 6b is shown in Figure 6e, indicating edge dislocations at the grain boundary. These dislocations can prevent the sliding of the grain plane and improve the impact toughness of HSS. Thus, the new compound inoculant can promote heterogeneous nucleation, increase the content of primary carbides in steel, hinder the growth of martensite, and facilitate the formation of cryptocrystalline martensite.

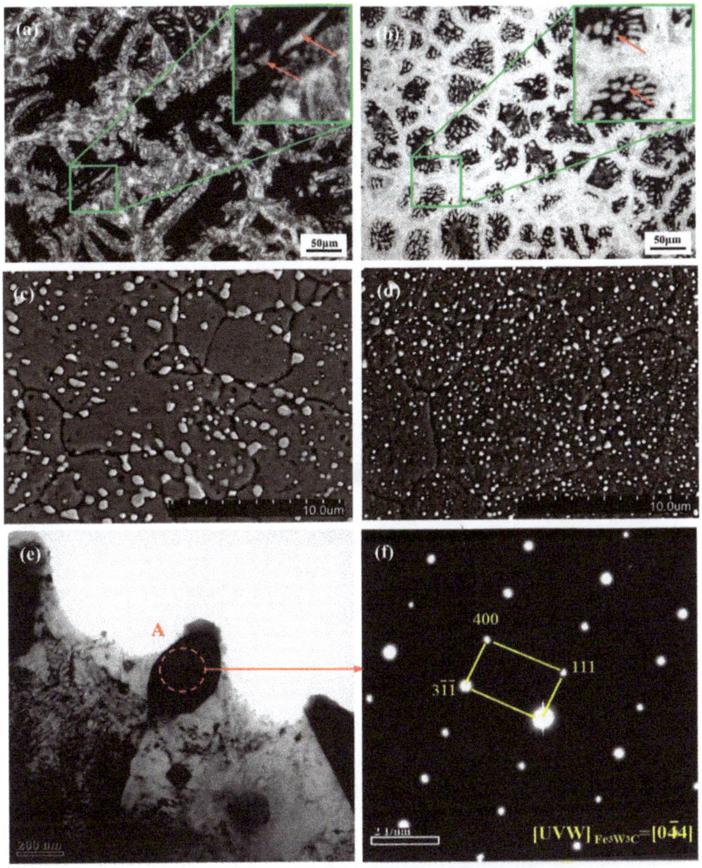

Figure 5. Microstructure of the as-cast HSS: (**a**) optical micrographs of the unmodified HSS; (**b**) optical micrographs of the modified HSS. SEM images of the tempered HSS in the states of (**c**) microstructure of the unmodified HSS; (**d**) microstructure of the modified HSS. TEM images of the tempered modified HSS: (**e**) TEM lower image of the tempered modified HSS; (**f**) SADE pattern of region A in Figure 5e.

3.3. *Effect of the Nanocomposite Inoculant Modification on the Mechanical Properties of W18Cr4V HSS*

The hardness and red hardness of the tempered W18Cr4V samples were measured and are shown in Figure 7a. The values in the graph are the arithmetic mean values of multiple measurements. The results revealed that the NbC-Cr_7C_3@graphene/Fe composite inoculant affected the mechanical properties of the modified HSS. The hardness of the tempered unmodified HSS was HRC (Hardness Rockwell C Scale) 63.2 and the red hardness was HRC 60.9. In addition, the hardness of the tempered modified HSS was HRC 66.4 and the red hardness was HRC 63.1. The novel composite inoculant had an impact on hardness and red hardness, resulting in a 5.1% and 3.6% increase in hardness and red hardness, respectively.

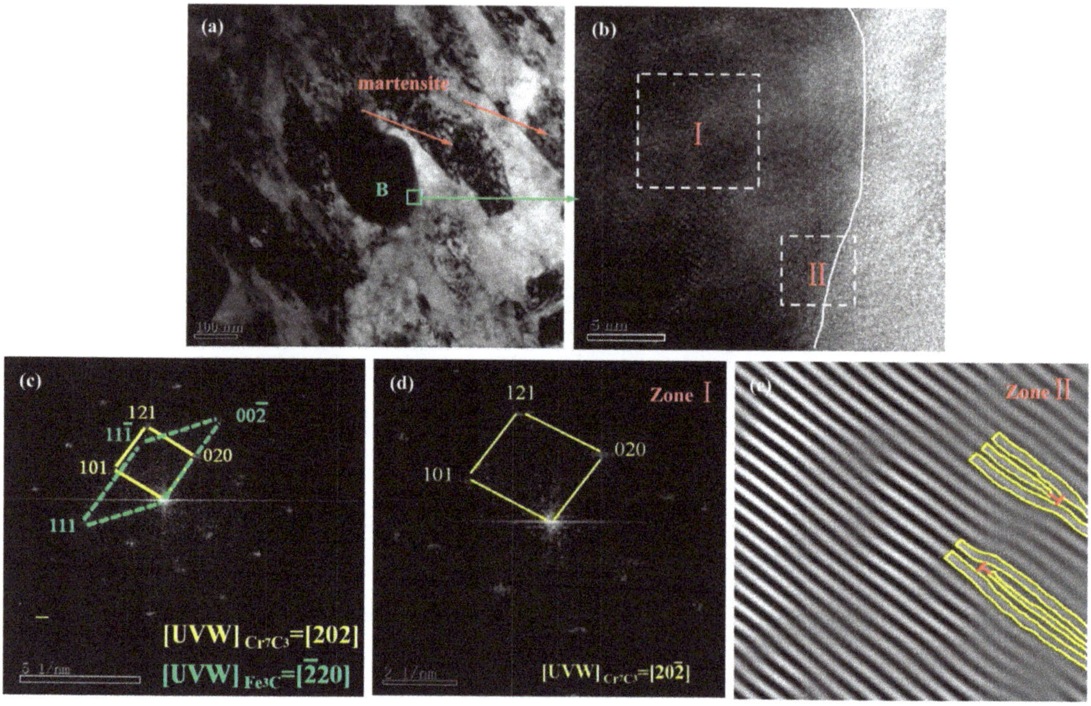

Figure 6. (**a**) TEM lower image showing martensitic structure and secondary carbide dispersion; (**b**) HRTEM images of region B in Figure 6a; (**c**) FFT of Figure 6b; (**d**) FFT of Zone I in Figure 6b; (**e**) inverse FFT of Zone II in Figure 6b.

W18Cr4V is a brittle material; hence, a small pendulum (0–150 J) was used for the impact test. The results of the impact toughness tests are shown in Table 1. The impact toughness of the modified steel increased significantly from 0.153 to 0.227 MJ/m^2 (Figure 7b). The modification treatment increased the impact toughness of W18Cr4V steel by 48.3%. The fracture morphology of the unmodified HSS was relatively shallow, as observed via SEM. Multiple continuously distributed secondary cracks were observed (Figure 7c), indicating a small bonding force between the test sample grains. The cracks propagated smoothly after being impacted by external forces, and the structure fractured along the grain plane. The morphology of the modified HSS exhibited river-like fluctuations, with many "dimples" of varying sizes; no secondary cracks were observed (Figure 7d). This indicates that the fracture of the sample occurred gradually in layers when subjected to impact force, showing increased resistance and good toughness. Owing to the refinement of the grain size of the modified HSS, the precipitation phase and impurities at the grain boundaries were also reduced [18,19]. The ceramic particle phase in the NbC-Cr_7C_3@graphene/Fe composite inoculant also reduced the size of the secondary carbides and increased the amount of cryptocrystalline martensite. These factors enabled the samples to absorb more deformation before fracture, resulting in better impact toughness.

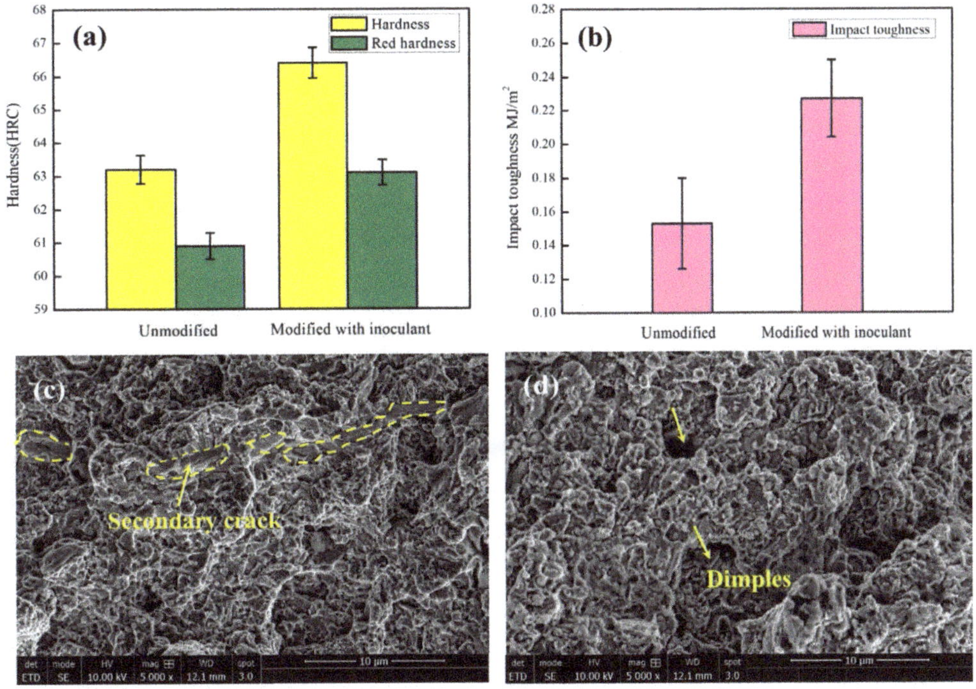

Figure 7. (**a**) Hardness and red hardness comparison of the tempered unmodified and modified HSS; (**b**) comparison chart of impact toughness of the unmodified and modified HSS; (**c**) SEM images of fracture morphologies of the tempered unmodified HSS; (**d**) SEM images of fracture morphologies of the tempered modified HSS.

Table 1. Impact energy and impact toughness of the tempered W18Cr4V HSS.

Sample Category	Impact Energy /MJ(A_k)	Cross-Sectional Area/m^2 (F)	Impact Toughness/(MJ/m^2) (a_k)
Unmodified	$(1.48 \pm 0.19) \times 10^{-5}$	$(9.68 \pm 0.47) \times 10^{-5}$	0.153 ± 0.027
Modified	$(2.21 \pm 0.23) \times 10^{-5}$	$(9.72 \pm 0.36) \times 10^{-5}$	0.227 ± 0.023

A wear resistance test was conducted using a wear tester which was worn for 90 min at 200 N. The microscopic morphology of the unmodified HSS samples exhibited deep parallel groove marks and numerous black areas on the entire surface that peeled off easily (Figure 8a). The microstructures of the modified HSS samples revealed that the groove marks on the worn surface became lighter and denser, and the black areas were significantly reduced (Figure 8b). During the friction test, adhesive wear occurred; the load applied by the testing machine to the sample caused plastic deformation on the surface of the sample. Plastic deformation causes the wear ring to come into contact with the surface of the sample under pressure, and the binding of atoms between the adhesive surfaces produces black areas. The modification treatment increased the equilibrium carbon content and carbon saturation in the steel, significantly improving the red hardness. Consequently, the black areas on the surface of the sample were reduced. During the friction test process, abrasive wear also occurred. Continuous sliding of the wear ring caused the surfaces of the samples to detach and form debris, gradually forming groove marks. However, the ceramic particles in the composite inoculant were evenly distributed in the steel matrix, improving the hardness and refining the microstructure of the steel. This resulted in shallower groove marks on the surfaces of the modified samples. The frictional wear experimental results are listed in Table 2 and show that the wear mass decreased from 10.3 to 6.1 mg. The microstructure of the unmodified steel sample was observed using SEM (Figure 8c), indicating that the martensite structure was broken, and the depth continued to increase, leading to peeling of the sample surface. However, the surface of the modified steel sample exhibited slight wear, a large light-gray area, less obvious peeling, and a lighter broken layer of the martensite structure. Moreover, dark wear pits were occasionally observed (Figure 8d).

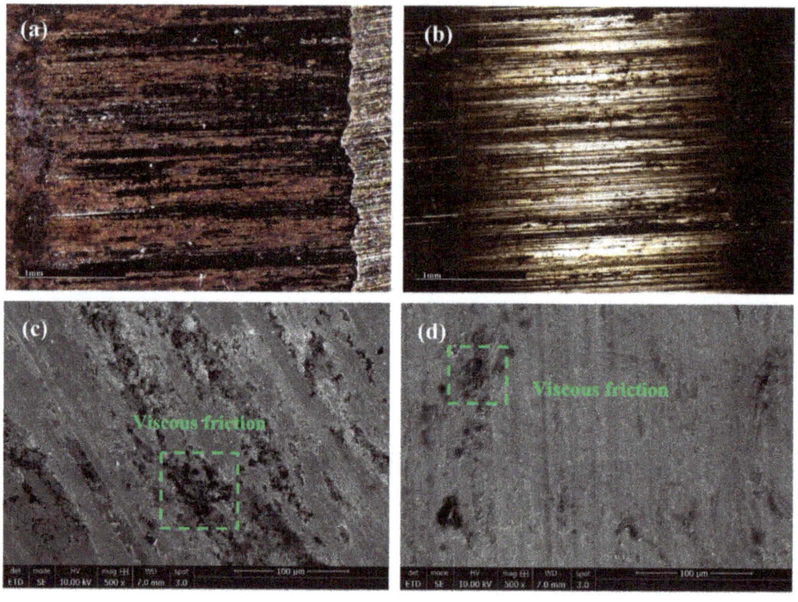

Figure 8. Metallographic diagram of wear surface of the tempered HSS: (**a**) unmodified HSS; (**b**) modified HSS. SEM images of wear surface of the tempered HSS in the states of (**c**) worn surface of the unmodified HSS; (**d**) worn surface of the modified HSS.

Table 2. Frictional wear experiments of the tempered W18Cr4V HSS.

State of HSS	Initial Mass/g	Residual Mass/g	Wear Mass/g
Uninoculated	10.3275	10.3172	0.0103
Inoculated	9.0885	9.0824	0.0061

Sexton et al. found that the wear resistance mechanism of steel materials is related to their hardness, where high hardness makes them susceptible to abrasive wear and low hardness makes them prone to adhesive wear [20]. Archard et al. derived a theory on the dependence of experiments on the load. They found that the wear rate of materials is directly proportional to the load and inversely proportional to the hardness [21]. Based on these research results, the composite inoculant not only increases the content of carbides in HSS but also improves the hardness and wear resistance of HSS, which is consistent with the experimental results of Sexton et al. During the friction and wear tests, the contact area between the test steel and wear ring increased, forming numerous dislocations and particle phases on the worn surface [22–24], which increased the hardness of the modified steel. The improvement in HSS red hardness resulting from modification was related to the content of carbides, with the ceramic particles coated with graphene in the composite inoculant increasing the content of alloying compounds in HSS. The alloying elements entered austenite during quenching. The higher the equilibrium carbon content and carbon saturation in steel, the higher is its quenching temperature, resulting in more carbides being incorporated into austenite. In addition, as the second-phase particle reinforcement, the high-melting-point ceramic particles in the inoculant also increased the hardness and stability of the modified steel.

3.4. Discussion

In the early 1970s, Bramfitt found through experimental results that TiN can reduce the nucleation undercooling of δ-Fe [25], and a two-dimensional mismatch theory was proposed based on crystal structure and data fitting. The specific formula is as follows:

$$\delta_{(hkl)_n}^{(hkl)_s} = \sum_{i=1}^{3} \frac{\left| d_{[uvw]_s^i} \cos\theta - d_{[uvw]_n^i} \right|}{3d_{[uvw]_s^i}} \times 100\% \quad (1)$$

where $(hkl)_n$ is the low-index crystal plane of the nucleation phase, $[hkl]_n$ is the low-index crystal plane of the matrix phase, and $[uvw]_n$ and $d_{[uvw]n}$ are the crystal plane index of the nucleation phase and the atomic spacing in the direction of the crystal plane index, respectively. $[uvw]_s$ and $d_{[uvw]s}$ are the crystal plane index of the base phase and the atomic spacing in the direction of the crystal plane index, respectively, and θ represents the angle between these two crystal belt axes. According to the mismatch calculation formula, the planar mismatch degree of NbC and δ-Fe is 8.72% and the planar mismatch degree of Cr_7C_3 and δ-Fe is 6.14%. The obtained mismatch degrees were less than 12%, making it a potential heterogeneous nucleation core for the steel matrix.

The NbC-Cr_7C_3@graphene/Fe composite inoculant was an iron-based composite inoculant, with a specific gravity very close to that of W18Cr4V metal liquid. The composite inoculants did not exhibit layering in molten steel and had a smaller wetting angle than other types of inoculants. The nanocrystalline ceramic particles in the composite inoculant were dispersed in the molten HSS, providing good nucleation support for the solidification of steel and considerably improving the non-uniform nucleation rate of the modified W18Cr4V HSS. The inoculation and refinement mechanisms of the compound inoculants in the experimental HSS are shown in Figure 9. This experiment utilized in situ reactions to prepare ceramic particle phases, mainly through precise temperature control and chemical reactions with appropriate elemental ratios. The TEM results showed that the grain size of the graphene-coated composite inoculant reached the nanoscale, and the composite inoculant had high surface energy and surface binding energy. During the fabrication process, when the composite inoculant is added to the steel liquid, the nanocrystals can easily combine with the alloy elements in the W18Cr4V steel, thereby enhancing the non-uniform nucleation rate. HRTEM analysis of the inoculant reveals that the layered structure of graphene was torn during high-energy ball milling. Under the strong collision between the grinding ball and the ceramic particles, the single-layer graphene tightly adhered to the surface of the ceramic particles, resulting in an overlap between the two phases, with a difference in crystal orientation of 14°31′. The distribution pattern of the ceramic particle phase in the HSS matrix, crystal structure, existing morphology, and the combination of the original alloying elements in the steel have a significant impact on the W18Cr4V steel. A large number of nanocrystalline ceramic particles in the composite inoculant caused more carbides to be generated during the solidification process of W18Cr4V HSS, and the grain size of the carbides was refined. NbC and Cr_7C_3 ceramic particles can also serve as second-phase particles to enhance intragranular strength, causing HSS to have higher hardness and better stability after modification treatment. Metallographic microscopy results revealed that the grains of the modified steel were refined, and the continuous distribution of fishbone-like eutectic carbides was disconnected. TEM results of the tempered modified HSS indicated that a large range of dark gray steel matrices were distributed with massive black Cr_7C_3 particles and flat noodle martensite. According to the TEM image, the density of carbides in the steel was extremely high, and the distance between carbides was less than 1 μm. The theory of the fine-grained effect indicates that Cr_7C_3 ceramic particles can effectively hinder the growth of carbides and martensite in steel, and an increase in alloy compounds promotes the transformation of austenite into martensite in steel, facilitating the formation of cryptocrystalline martensite. The inverse FFT pattern near the grain boundaries in HRTEM shows edge dislocations at the grain boundary. These dislocations have high strain energy, which is conducive to the precipitation of secondary carbides and can improve the mechanical properties of the modified steel. The morphology of the modified HSS exhibited a river-like fluctuation, with numerous "dimples" of varying sizes. Furthermore, no secondary cracks were observed, indicating that the fracture of the sample was gradually layered when subjected to impact force, with increased resistance and good toughness. The ceramic particle phase in the NbC-Cr_7C_3@graphene/Fe composite inoculant reduced the size of the secondary carbides and increased the amount of cryptocrystalline martensite. These factors enabled the samples to absorb more deformation before fracture, resulting in better impact toughness.

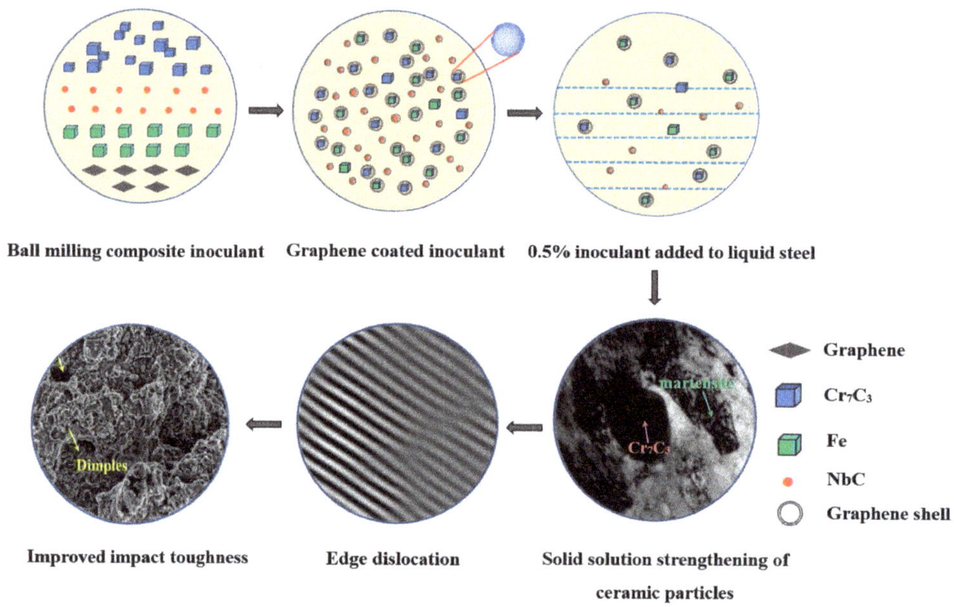

Figure 9. Inoculation and refinement mechanism of the NbC-Cr_7C_3@Graphene/Fe compound inoculants on the experimental HSS.

4. Conclusions

This study investigated the effect of NbC-Cr_7C_3@graphene/Fe composite inoculants on the microstructure and mechanical properties of W18Cr4V HSS during the modification process. The conclusions are as follows:

1. According to the XRD patterns and Jade software analysis, the main phases in the composite inoculant were Cr_7C_3, Fe (ferrite), Fe_7C_3, NbC, and graphene. The average grain size was less than 50 nm, and significant agglomeration of nanocrystals was observed in the TEM image. Furthermore, single-layer graphene tightly covered the surface of the ceramic particles under strong collisions between the grinding ball and ceramic particles, resulting in an overlap of the two phases in the selected area.
2. The as-cast microstructure of the modified W18Cr4V was significantly refined, with the continuous distribution of fishbone-like eutectic carbides reduced and disconnected. This reduction decreased the effect of splitting on the steel matrix. Furthermore, the SEM images of the tempered modified HSS indicate changes in the distribution and morphology of the carbides, showing smaller and more uniform secondary carbides.

3. The (00$\bar{2}$) crystal plane of the Fe_3C phase and the (020) crystal plane of the Cr_7C_3 phase were collinear at two points in the reciprocal space. This indicates that the Fe_3C and Cr_7C_3 phases had a coherent relationship in the tempered modified HSS. The HRTEM results indicate that the ceramic particles provide a nucleation substrate for the carbides in W18Cr4V, improving the non-uniform nucleation rate.
4. The nanocrystalline composite inoculant had an impact on the mechanical properties of W18Cr4V. The hardness of the tempered modified HSS increased from HRC 63.2 to 66.4 and the wear mass decreased from 10.3 to 6.1 mg. Moreover, the modification treatment enhanced the impact toughness of the steel by 48.3%.

Author Contributions: Conceptualization, L.B.; formal analysis, L.B., H.Z. and S.L.; investigation, L.B., S.L., L.X. and G.Z.; writing—original draft preparation, L.B., G.Z. and L.Z.; writing—review and editing, J.L., L.B. and L.X.; data curation, L.B., L.Z., J.L. and H.Z. The authors would also like to thank Chunxiang Cui, Lichen Zhao, and Xin Wang for their formal analysis and funding of these experiments. All authors have read and agreed to the published version of the manuscript.

Funding: This research was funded by Natural Science Foundation of Hebei, grant number E2009000102.

Institutional Review Board Statement: Not applicable.

Informed Consent Statement: Not applicable.

Data Availability Statement: No new data were created or analyzed in this study. Data sharing is not applicable to this article.

Conflicts of Interest: The authors declare no conflicts of interest.

References

1. Hossam, H.; Asiful, H.S.; Hany, S.A.; Ibrahim, A.A. Study on the Microstructure of Vanadium-Modified Tungsten High-Speed Steel-Coded SAE-AISI T1 Steel. *Adv. Mater. Sci. Eng.* **2022**, *2022*, 3469305. [CrossRef]
2. Mihai, O.C.; Mihai, B.; Sorin, C.; Ioana, A.G.; Mariana, I. Sulfonitrocarburizing of High-Speed Steel Cutting Tools: Kinetics and Performances. *Materials* **2021**, *14*, 7779. [CrossRef]
3. Chaus, A.S.; Sahul, M.; Sobota, R. Development of the Chromium-Rich High-Speed Steel: As-Cast Microstructure. *Defect Diffus. Forum.* **2020**, *400*, 59–64. [CrossRef]
4. Zhao, J.Y.; Zhang, Q.X.; Bai, L.N. The effect of Amorphous Nanocrystalline Inoculant on Structures and Properties of High Speed Steel. *Mater. Res. Express.* **2017**, *4*, 066507. [CrossRef]
5. Casellas, D.; Caro, J.; Molas, S. Fracture Toughness of Carbides in Tool Steels Evaluated by Nanoindentation. *Acta Mater.* **2007**, *55*, 4277–4286. [CrossRef]
6. Kheirandish, S.; Karim, K.Y.; Mirdamadi, S. Mechanical properties of M7 High Speed Cast Steel Modified with Niobium. *ISIJ Int.* **2007**, *37*, 721–725. [CrossRef]
7. Lu, L.; Hou, L.G.; Zhang, J.X. Improved the Microstructures and Properties of M3:2 High-Speed Steel by Spray Forming and Niobium Alloying. *Mater. Charact.* **2016**, *117*, 1–8. [CrossRef]
8. Ning, L.; Cui, C.X. Structure and Properties of GCr15 Modified by Multiphase Ceramic Nanoparticles/Fe-C Composite Inoculants. *Mater. Sci. Eng. A* **2018**, *738*, 63–74. [CrossRef]
9. Belyanchikov, L.N. New High-Nitrogen Corrosionresistant Tool and High-Speed Steel. *Russ. Metall.* **2009**, *2008*, 761. [CrossRef]
10. Fang, L.; Wang, T.B. Effects of Amorphous and Nanocrystalline Inoculants on the Micro-Structure and Mechanical Behavior of Fe-Si-Mn Steels. *Adv. Mater. Res.* **2011**, *415–417*, 146–150. [CrossRef]
11. Upadhyaya, R.; Kumar, S.K. Effect of Some Inoculants on the Structure and Properties of Thin Wall Ductile Iron. *Mater. Today* **2018**, *5*, 3595–3601. [CrossRef]
12. Jovicevic-Klug, P.; Podgornik, B. Comparative Study of Conventional and Deep Cryogenic Treatment of AISI M3:2 (EN1.3395) High-Speed Steel. *J. Mater. Res. Technol.* **2020**, *9*, 13118–13127. [CrossRef]
13. Bai, L.N.; Cui, C.X.; Wang, T.B.; Cui, S.; Han, X.H.; Kang, L.C. Preparation of in situ Cr@Graphene/Fe Nanocomposite Inoculant and its Refining Effect on Microstructure and Properties of W18Cr4V High-Speed Steel. *Mater. Sci. Technol.* **2021**, *37*, 224–236. [CrossRef]
14. Decker, S.; Kruger, L. Influence of High-Energy Ball Milling on Mg-PSZ-Reinforced TRIP Steel-Matrix Composites Synthesized by FAST/SPS. *Mater. Sci. Eng. A* **2019**, *761*, 137974. [CrossRef]
15. Ong, C.Y.A.; Blackwood, D.J.; Li, Y. The Effects of W Content on Solid-Solution Strengthening and the Critical Hall-Petch Grain Size in Ni-W Alloy. *Surf. Coat. Technol.* **2019**, *357*, 23–27. [CrossRef]
16. Wang, Y.J.; Sun, J.J.; Jiang, T. A Low-Alloy High-Carbon Martensite Steel with 2.6 GPa Tensile Strength and Good Ductility. *Acta Mater.* **2018**, *158*, 247–256. [CrossRef]

17. Kaneko, K.; Takuya, M.; Yasuhito, K. Formation of Core-Shell Type Structure in Duplex Martensitic Steel. *Scripta Mater.* **2021**, *193*, 112–116. [CrossRef]
18. Su, G.; Gao, X.; Yan, T. Intercritical Tempering Enables Nanoscale Austenite/ε-Martensite Formation in Low-C Medium-Mn Steel: A Pathway to Control Mechanical Properties. *Mater. Sci. Eng. A* **2018**, *736*, 417–430. [CrossRef]
19. Chang, K.K.; Yong, C.K.; Park, J.I. Effects of Alloying Elements on Microstructure, Hardness, and Fracture Toughness of Centrifugally Cast High-Speed Steel Rolls. *Metall. Mater. Trans. A* **2005**, *36*, 87–97. [CrossRef]
20. Sexton, M.D.; Fischer, T.E. The Mild Wear of 52100 Steel. *Wear* **1984**, *96*, 17–30. [CrossRef]
21. Archard, J.F. Contact and Rubbing of Flat Surfaces. *J. Appl. Phys.* **1953**, *24*, 981–988. [CrossRef]
22. Liu, X.; Xiao, L.; Wei, C. Effect of Multi-Directional Forging and Annealing on Abrasive Wear Behavior in a Medium Carbon Low Alloy Steel. *Tribol. Int.* **2018**, *119*, 608–613. [CrossRef]
23. Toth, L.; Nyikes, Z.; Umesh, M. Incresing Tool Steel Surface Wear Resistance by Surface Treatment. *Műsz. Tudományos Közl.* **2020**, *13*, 170–173. [CrossRef]
24. Shah, M.; Bakshi, S.D. Three-Body Abrasive Wear of Carbide-Free Bainite, Martensite and Bainite-Martensite Structure of Similar Hardness. *Wear* **2018**, *402–403*, 207–215. [CrossRef]
25. Bramfitt, B.L. The Effect of Carbide and Nitride Additions on the Heterogeneous Nucleation Behavior of Liquid Iron. *Metall. Trans.* **1970**, *7*, 1987–1995. [CrossRef]

materials

MDPI

Article

Grain Refinement Caused by Larger Particles in Laser 3D Printing of AISI 304L Stainless Steel

Xuhuai Zhang * and Xing Lu *

School of Material Science and Engineering, Dalian Jiaotong University, Xi'an Road Street, Shahekou District, Dalian 116028, China
* Correspondence: xhzhang2025@126.com (X.Z.); lu@djtu.edu.cn (X.L.)

Abstract: Additively manufactured (AM) stainless steel has attracted a lot of attention for its competitive performance advantages over parts prepared by traditional methods. However, the influence of the powder characteristic of AISI 304L stainless steel on the laser 3D printing (3DP) process has yet to be clarified. In this research, the effect of the particle size of atomized AISI 304L stainless steel powder on 3DP of a powder-fed laser was studied, the grain morphology of different printed samples was analyzed by electron backscatter diffraction (EBSD) technology, and the mechanical properties were investigated via quasi-static tensile experiments. It was found that the use of small particles made the columnar crystal area mix with fine equiaxed grains in single-pass laser melting deposition, resulting in an obvious grain refinement effect. However, in multi-pass deposition, samples deposited with small particles exhibited more significant grain coarsening and anisotropy effects than those of larger particles, resulting in a significant reduction in plasticity. This can be attributed to the grain merging and growth mechanism caused by the thermal shock effect of multi-pass laser deposition, which is controlled by the grain configuration formed in the early-pass deposition. The results show that the use of powder particles greater than 50 μm is of great significance for improving the quality of AISI 304L stainless steel 3DP products.

Keywords: laser 3D printing; stainless steel; particle size; grain refinement; mechanical property

1. Introduction

Additive manufacturing (AM), also known as three dimensional printing (3DP), is widely recognized as a revolutionary component processing technology that can enhance the competitiveness of the manufacturing industry [1]. In terms of metal 3D printing, powder bed fusion and directed energy deposition are the two most commonly used AM technologies in industrial applications, both of which utilize metal powders as raw materials. It is widely accepted that the quality of powder used in AM is a key factor affecting process performance and final component quality [2]. Therefore, only by reducing the cost of producing metal powders and fully understanding the impact of powders on process performance can the widespread application of additive manufacturing technology in modern industry be promoted more quickly.

The effect of raw powder on AM includes two different categories, the chemical aspect of the powder surface and the physical properties of the powder itself. The chemical aspect refers to the formation of oxidizing surfaces or other impurities on the surface of the powder, which are introduced into the AM part to reduce its quality [3]. Physical properties refer to the characteristics of flowability, particle size distribution (PSD), internal porosity, etc., which are critical to the quality of the part. Since different AM equipment uses different powder supply and dispensing methods, the optimal powder characteristics are also different.

The PSD of used powder significantly affects the grain size and the microstructure formed during laser 3DP. For particles with a diameter of less than 30 μm, the particles tend to form agglomerates due to the gravitational action dominated by the adhesion force, resulting in an increase in the surface roughness of the deposited layer [4]. At the same time, small particle size powders provide a larger surface area to facilitate heat conduction but, due to their poor flowability, they are prone to irregular build-up during laser powder bed fusion (LPBF), which in turn affects the quality of the final part [5]. For large-size powders, the remaining large-size powders can significantly reduce the surface roughness of the deposited layer when particles with a diameter of less than 60 μm are removed [4]. In addition, large particle size powders can improve the uniformity of powder spreading to a certain extent due to their better fluidity and lower agglomeration tendency, thereby reducing porosity and optimizing grain structure [6].

AISI 304L stainless steel (304L SS), also known as ultra-low carbon stainless steel, has been widely used in nuclear reactor components, chemical processing equipment, and the petroleum industry due to its excellent corrosion resistance, high oxidation resistance, and excellent mechanical properties [7]. In addition, 304L SS is a promising candidate for oxide dispersion-reinforced steel matrices, which are expected to achieve excellent comprehensive properties such as good mechanical properties, high corrosion resistance, and resistance to high-temperature oxidation [8]. Recently, the AM of 304L SS parts has also attracted a lot of attention, and most of the current literature reports focus on the effect of processing parameters on mechanical properties and corrosion resistance [9–13]. For example, Mahmood et al. found that laser scanning speed and laser power were inversely proportional to the average grain size, while there was a direct relationship between the powder feed rate and the average grain size [14]. However, there are few studies on the relationship among powder properties, mechanical properties, and the microstructure of 304L SS stainless steel. Therefore, it is important to establish a link between the properties of 304L SS powder and the performance of the AM final part.

In this research, we systematically studied the effects of the PSD of commercial 304L SS stainless steel powders on the microstructure and mechanical properties of the 3DP samples. The grain refinement behavior of single-pass laser deposition and grain roughening behavior in multi-pass laser deposition were discussed.

2. Materials and Methods

Commercial 304L SS stainless steel powder prepared by nitrogen atomization (Zhuohang Metal, Xingtai, China) was used as the 3DP raw material. In order to investigate the effect of powder particle size on the microstructure and properties of the 3DP samples, the as-received powder was carefully sieved using stainless steel sieves with a mesh size of 150–320. Powders with mesh sizes of 150–170, 170–230, 230–270, and 270–320 were collected and named as Sample A, Sample B, Sample C, and Sample D, respectively. A laser particle size analyzer (Haoyu, HYL-2080, Dandong, China) was used to test the PSD of the sieved powder samples, using deionized water as a dispersant, with a testing time of 30 s.

A powder-fed laser melting deposition (also 3DP) device was used to prepare different samples. The schematic diagram of the equipment is shown in Figure 1, and the printing process parameters are shown in Table 1. The cut 304L SS steel plates with a size of about 50 mm × 50 mm × 10 mm were used as the 3DP substrates. For single-pass laser deposition, a sample with a length of about 20 mm was deposited on the substrate. For multi-pass laser deposition samples, a 42-pass process was used, stacked with 6 layers up to 13 mm in height and 45 mm in length.

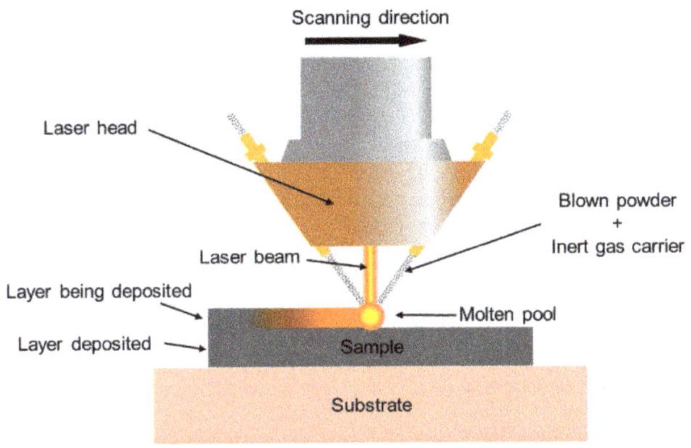

Figure 1. Schematic diagram of the powder feeding laser 3D printing device.

Table 1. The process parameters of powder-fed laser 3D printing selected for this study.

Processing Parameter	Value
Laser power (P)	1100 W
Laser scan speed (v)	480 mm/min
Layer height	1.5 mm
Power feed rate	9.3 g/min
Dwell time between each fill run	30 s
Laser beam diameter (D)	3 mm
Linear energy density (E = P/vD)	91.6 J/mm^2

A wire cutting machine was used to cut the printed sample into different sizes for structural and property analyses. A metallurgical microscope (Zeiss, Axio Observer, Munich, Germany) was used to observe the microstructure of the cross-section of the sample. Tensile tests of the printed samples were performed on a servohydraulic testing machine (Instron 5982, Boston, MA, USA) at a strain rate of 5×10^{-4} s^{-1}. Thin plate samples with typical dog-bone shape, with effective test area dimensions of 1 mm $\times$ 1 mm $\times$ 7 mm, were used for tensile experiments, following the ASTM E8 standard. At least 5 repeated tensile experiments were performed on each sample to obtain reliable data. The fracture surface was examined using a field emission gun scanning electron microscope (Fe-SEM, Zeiss Gemini SEM 360, Oberkochen, Germany). In order to analyze the grain morphology, electron backscatter diffraction (EBSD) images of different samples were taken with an SEM (Zeiss Jena, Supra 55 SEM, Jena, Germany). The foil samples for EBSD analysis were cut from the cross-section of the deposited samples, and subsequently mechanically polished, then followed by electro-polishing under a voltage of 15 V for ~20 s at room temperature. The working distance used for the EBSD measurement was about 13 mm, the scanning step size was 5 μm, the accelerating voltage was 20.00 kV, and the tilt angle of the sample was 70.00°.

3. Results

Figure 2 displays the SEM images and the analysis results of various powder samples using a laser particle size analyzer. After passing through some special stainless steel mesh screens of 150, 170, 230, 270, and 320 mesh sizes, the particle size of 304L SS stainless steel powder becomes remarkably uniform. The average diameter (D50) values of samples A,

B, C, and D, measured using the laser particle size analyzer, were 118.8, 85, 65.28, and 49.18 μm, respectively. Meanwhile, the specific surface areas were 21.09, 27.81, 37.82, and 58.75 m^2/kg, respectively. Notably, the specific surface area of Sample D is 278% of that of Sample A, suggesting a potentially higher level of oxide impurities adsorbed on the powder surface.

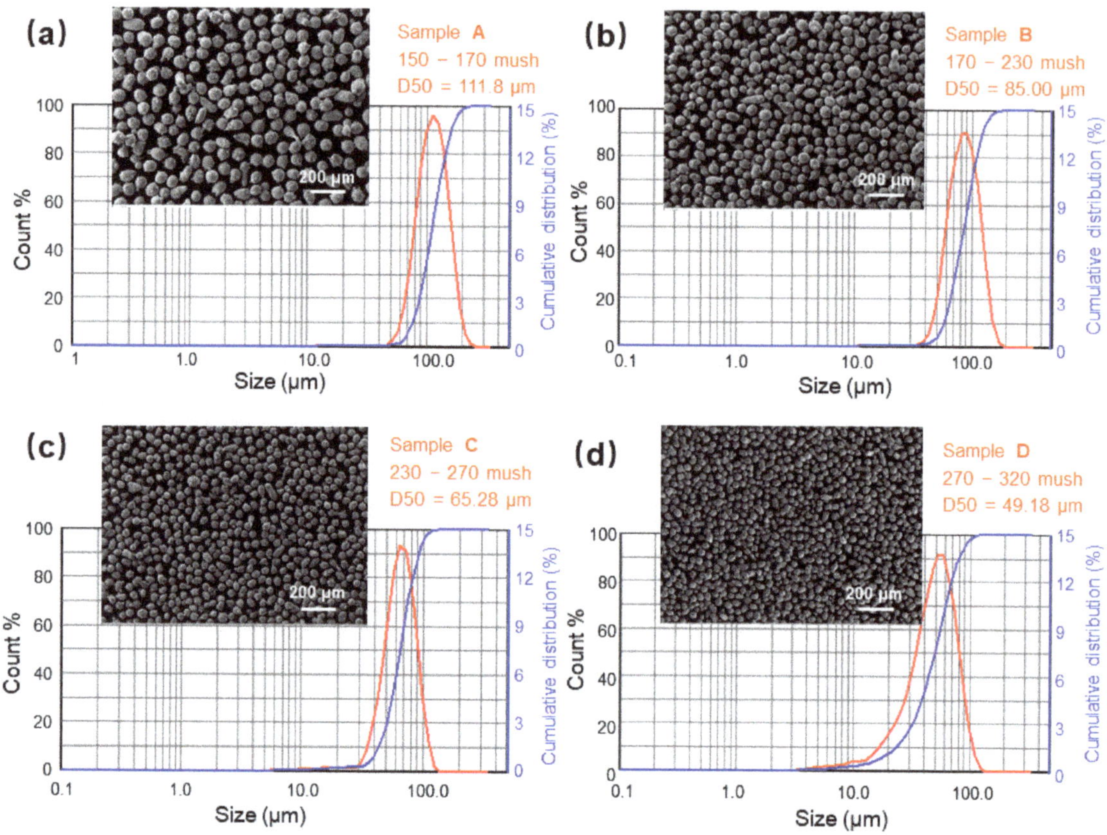

Figure 2. SEM images and particle size distribution curves of different particle samples: (**a–d**) samples A, B, C, and D, respectively.

The microstructure of laser 3DP materials is primarily formed through the solidification of the laser melt pool. Therefore, the size, shape, and cooling conditions of the laser melt pool are important factors that must be considered. Figure 3a–d show the metallographic images of the cross-section of the deposited samples by the single-pass laser deposition using raw powders with different sizes. It shows that the shape of the molten pool formed by laser heating is close to oval, and its width and depth exhibit obvious dependence on the particle size of the powder. As the particle size decreases, the width of the melt pool increases from about 1932 μm to 2005 μm, and the depth also increases from about 1095 μm to 1175 μm. It is suggested that the smaller the powder size, the larger the melt pool formed. In our opinion, it is related to the change in the number of particles and surface area caused by the change in the particle size of the raw powder. According to the geometric characteristics of spherical particles, the relationship between the particle size d and the number of particles N under the same total volume V_{total} is as follows:

$$N = \frac{6V_{total}}{\pi d^3} \tag{1}$$

Assuming that the powder volume is 1 m^3, the number of particles are 1367, 3111, 6869, and 16064 when the particle sizes are 111.8 μm, 85 μm, 65.28 μm, and 49.18 μm, respectively. It can be seen that the reduction in particle size by half and the increase in the number of particles by more than 10 times will significantly increase the total surface area of particles. In this study, Sample D has more particles and greater specific surface area under the condition of consistent powder delivery, which will melt at a faster rate in laser heating to undergo superheating, resulting in a larger melt pool.

Figure 3f shows a schematic of the laser melt pool in this study. The cooling conditions at the bottom and top of the melt pool are different; the bottom cools quickly and has directionality, while the top cools quickly but lacks directionality. As a result, columnar grains typically form at the bottom of the melt pool, while equiaxed grains mainly form at the top and the center, as shown in Figure 3e.

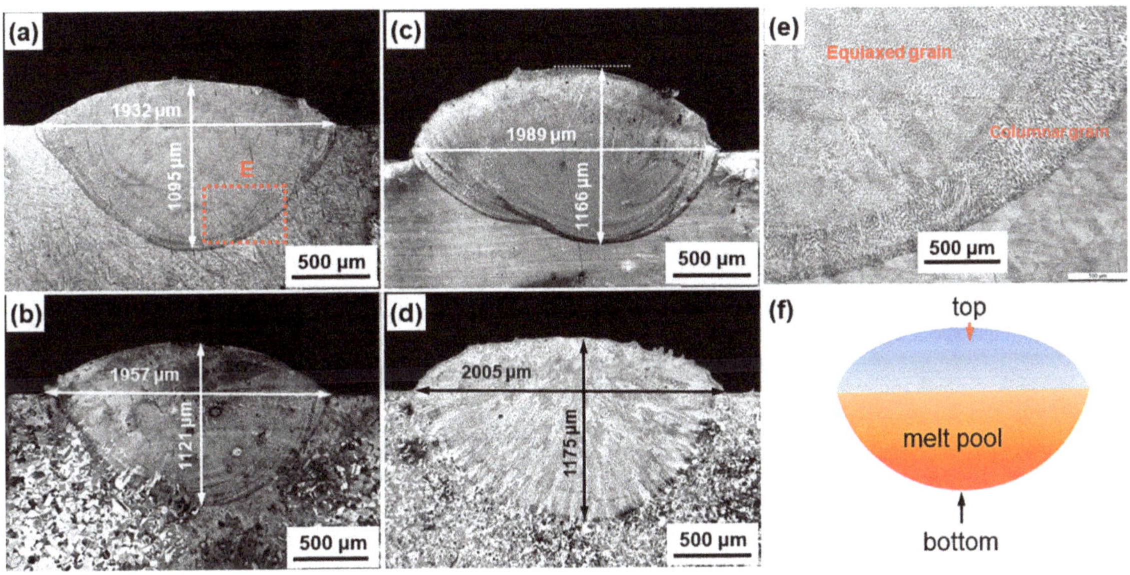

Figure 3. Metallographic photographs of cross-sections of single-pass laser-deposited samples using raw materials with different particle sizes of: (**a**–**d**) powder A–D; (**e**) enlarged view of the dashed box E area in Figure 3a; (**f**) schematic diagram showing the shape of the molten pool.

The metallographic photographs shown in Figure 3a–d also show the presence of casting defects. After staining, the distribution of defects such as cast pores in different samples can be more clearly observed, as shown in the insets in Figure 4. Obviously, samples prepared from small-sized powders contain a higher number of defects, with defect densities of up to 1.74 cm^{-2} in the cross-section (Sample D). With the increase in particle size, the defect density decreases significantly, and when the particle size is 111.8 μm (Sample A), the defect density is only 0.34 cm^{-2}, which is 80% smaller than that of Sample D. It shows that the use of larger particles can effectively reduce the defects of 3DP products.

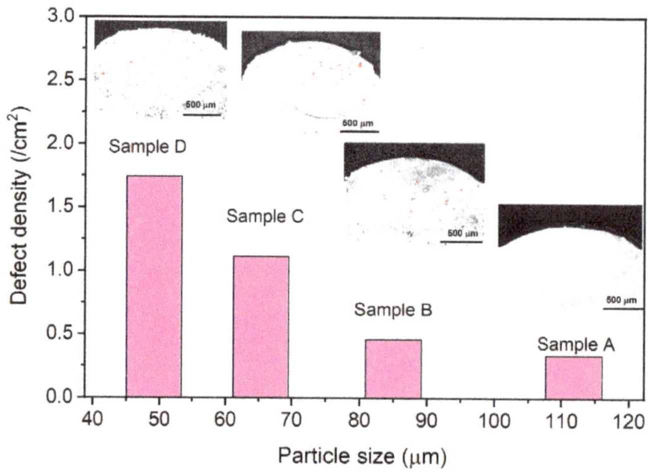

Figure 4. Powder particle size vs. defect density in single-pass laser 3D printing samples, with the insets showing the defect distribution.

To more clearly analyze the effect of particle size on the solidification behavior of the laser melt pool, typical electron backscatter diffraction (EBSD) technology was used to analyze the grain morphology of the samples prepared using different powders after a single pass of printing, as shown in Figure 5a–d.

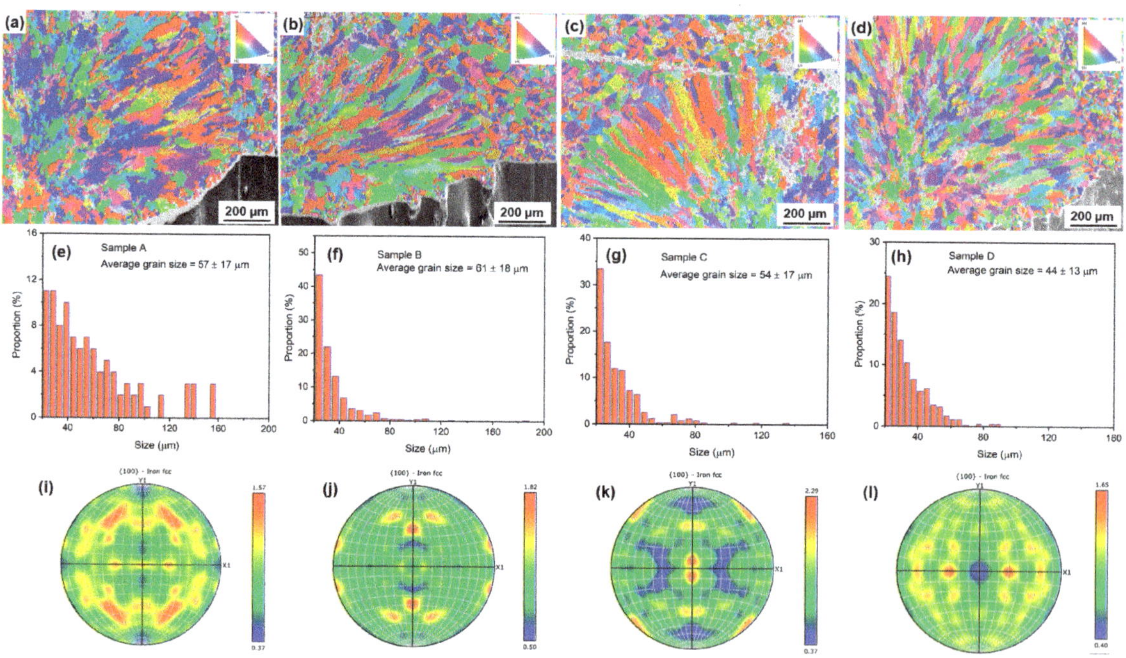

Figure 5. EBSD analyses of samples deposited by single-pass laser deposition using powders A, B, C, and D, respectively: (**a**–**d**) EBSD images; (**e**–**h**) grain size distributions; (**i**–**l**) ⟨100⟩ pole diagrams of FCC Fe.

In general, with the decrease in the average particle size of the raw powder, the proportion of equiaxed grains in a single melt channel gradually increases, and the epitaxial growth of columnar grains to the equiaxed grain region in the melt channel of the finest powder is obvious. Figure 5d,h show the grain features of a sample prepared with a size of ~44 μm, where the diameter of the columnar grains is significantly smaller than that of the samples prepared from larger particle size raw powder (Figure 5e–g). However, there is no significant change in the length of the columnar grains in the samples printed with different sizes of raw powder. In addition, there is a transition zone between the equiaxed grains at the top of the melt pool and the columnar grains at the bottom, which is mixed with columnar grains and equiaxed grains. Interestingly, when the powder particle size is minimized, this transition zone significantly increases, and even a certain amount of equiaxed crystals are mixed throughout the entire columnar grain region. This may be related to the special flowability and the melting behavior of small-sized particles.

Figure 5i–l show the pole plots of different samples by single-pass laser deposition. It shows that there is no obvious preferred orientation in the <100> direction for all the samples, which is related to the underdeveloped columnar grains.

In laser 3DP, the component parts are stacked layer by layer, and the part that is formed first is generally subjected to repeated thermal shocks caused by the melt pool of the later laser heating, causing some changes in the microstructure. Figure 6a shows an optical photograph of the sample prepared for this study, which is based on a thick stainless steel plate with good heat dissipation. In a cross-section perpendicular to the substrate (as shown in the white box B in Figure 6a), there is typically a stacked structure formed by a multi-pass laser melt pool, with columnar and equiaxed grains crossed together, as shown in Figure 6b. The depth of the melt pool produced by the multi-pass laser heating is theoretically the same as that of the single-pass laser heating due to the same amount of powder fed and laser power. However, we observed that the depth decreases after repeated melting and heating in multi-pass laser deposition, ranging from 163 μm to 311 μm, which is about ~1/3 of that of the single-pass melt pool, as shown in Figure 5b. It means that the top of the solidified melt pool is melted 2/3 by the later multi-pass laser heating, i.e., the equiaxed crystal area is essentially re-melted. At the same time, the un-melted and residual part of the solidified melt pool is also subjected to the high-temperature heating and annealing process close to the melting point, which will inevitably have a great impact on the microstructure and mechanical properties.

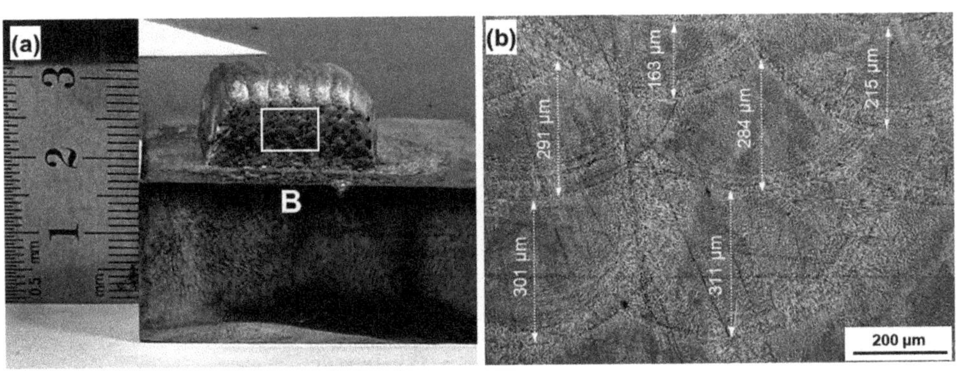

Figure 6. (**a**) Typical macroscopic optical photograph of the bulk samples deposited through multi-pass laser deposition using A powder; (**b**) metallographic photograph of the sample taken from position B indicated by the white square in Figure 6a.

The bottom layer, close to the substrate, was subjected to the highest number of thermal shocks. Therefore, we took EBSD images of the samples at this location for the four samples, as shown in Figure 7a–d. Compared to the single-pass laser deposition sample shown in Figure 5a–d, the columnar grain diameter at the bottom of the multi-pass deposition sample is significantly larger. This is due to the fact that in the multi-pass deposition process, the part that solidifies first is repeatedly heated, causing repeated grain growth.

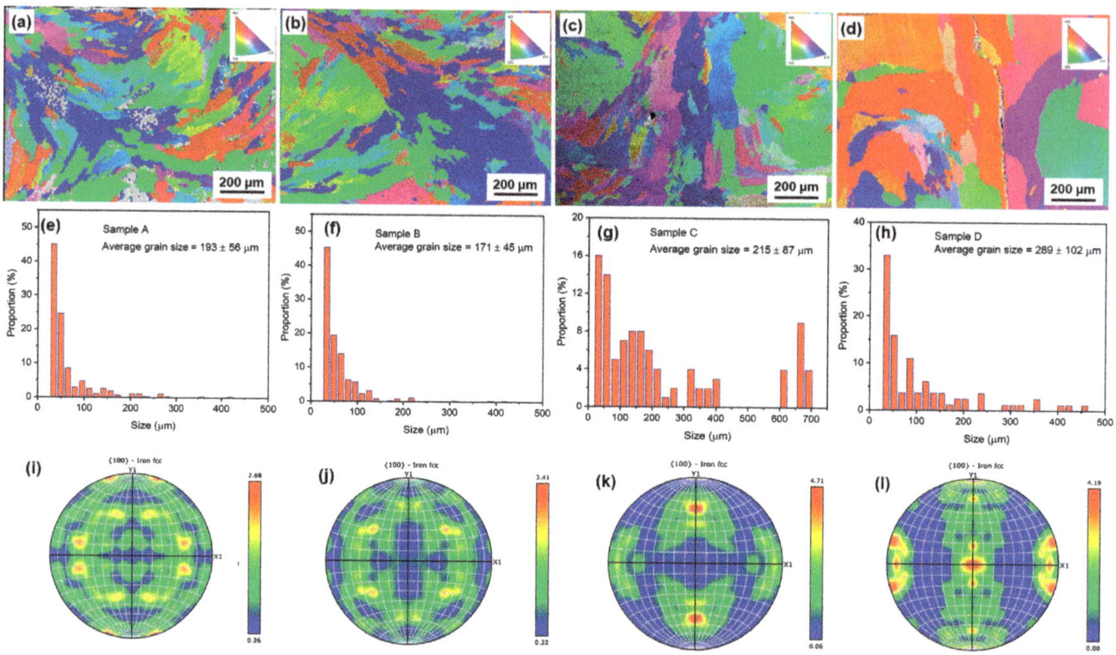

Figure 7. EBSD analyses of the bottoms of bulk samples deposited through multiple passes using powders of different particle sizes, namely samples A, B, C, and D: (**a**–**d**) EBSD images; (**e**–**h**) grain size distributions; (**i**–**l**) ⟨100⟩ pole diagrams of FCC Fe.

Figure 7a–d also show the effect of the particle size of the powder feedstock on the micromorphology of the columnar crystal region at the bottom of the multi-pass deposited sample. Of particular note, it can be found that the diameter of the columnar grains in the multi-pass sedimentation samples increases significantly as the particle size decreases (Figure 7e–h). For laser-deposited samples with a particle size of less than 50 μm, the diameter of the columnar crystals grows to more than a few hundred microns. It not only shows that the influence of laser thermal shock on the microstructure of 3DP stainless steel is too large to be ignored, but also proves that the effect of raw material particle size on the grain size of multi-pass laser deposition samples is exactly the opposite of that of single-pass deposition.

Figure 7i–l show the pole plots of different samples. For samples deposited with large-sized particles, there is no obvious preferred orientation in the <100> direction (Figure 7i–j). However, for samples C and D with smaller particle sizes, the preferred orientation in the <100> direction is significantly strengthened, which can be attributed to the appearance of coarse columnar grains.

Figure 8 shows the tensile stress–strain curves of bulk samples prepared from powders with different particle sizes, and the corresponding mechanical properties are shown in Table 2. All the samples in the present study show good plasticity and better tensile strength than the samples obtained by conventional manufacturing methods. For example, the yield strength ($\sigma_{0.2}$, at 0.2% offset) of Sample A prepared with large particles is measured to be 335 ± 5 MPa, which is about 1.6 times higher than that of the dislocation-free 304L SS stainless steel sample (210 MPa) [15]. It is well known that the concept of stacking fault energy is important in austenitic steels, which influences the plastic deformation mechanisms such as transformation-induced plasticity, twinning-induced plasticity, and microband-induced plasticity [16–18]. Laser 3DP has ultra-fast heating and cooling rates, which can lead to a higher presence of dislocations and residual stresses in the solidified microstructure, thereby increasing the stacking fault energy and contributing to the enhancement of strength. However, with the decreased particle size, the plasticity is reduced. We believe this phenomenon relates to the complex microstructure of the 3DP sample after repeated thermal shocks. The use of small-particle raw materials generally brings more physicochemical and mechanical defects and more uneven grains to the printed product, which brings certain interference to the plastic deformation, resulting in the plasticity of the sample being significantly smaller than that of the large-particle sample.

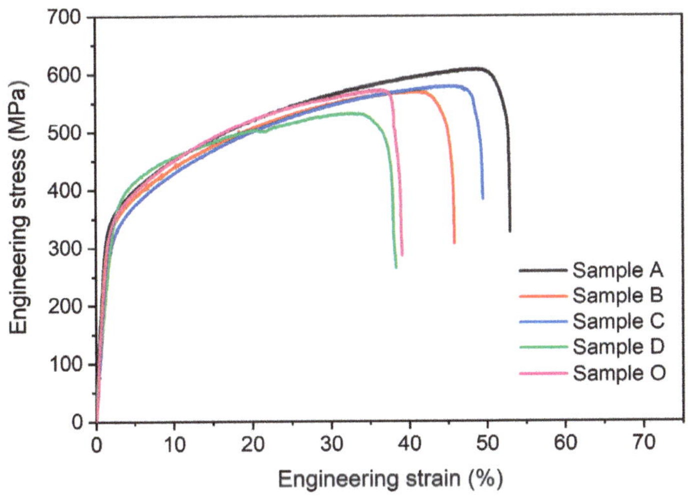

Figure 8. Tensile stress–strain curves of the samples prepared using powders of different particle sizes.

Table 2. Mechanical properties of different samples using raw particles with varying particle sizes.

Sample Name	Particle Size (μm)	Yield Strength (MPa)	Tensile Strength (MPa)	Tensile Plasticity (%)	Single-Pass Hv (MPa)	Multi-Pass Hv (MPa)
A	118.8	335 ± 5	610 ± 15	51.2 ± 1.6	228 ± 2	209 ± 7
B	85	315 ± 9	570 ± 12	44.2 ± 2.2	230 ± 4	189 ± 3
C	65.28	307 ± 12	580 ± 10	47.2 ± 3.5	226 ± 5	184 ± 4
D	49.18	365 ± 13	532 ± 18	36.8 ± 4.8	223 ± 4	170 ± 3
O	82.7	322 ± 7	572 ± 16	37.9 ± 3.9	224 ± 8	191 ± 4

In addition, the 3DP samples with the widest range of particle size distribution before sieving were also subjected to tensile testing for comparison. Interestingly, its strength is similar to that of the large particle sample (samples A and B), while its plasticity is similar to that of the small particle sample (sample D).

304L SS stainless steel is a typical ductile metal material that usually has good plasticity, and its fracture exhibits toughness characteristics. Figure 9a–d compare the fracture morphology of Sample A–D, both of which have typical dimple characteristics, suggesting that they share ductile fracture behavior. For dimple fractures, a larger dimple generally represents better plasticity in ductile metals [19]. This is because the dimples are essentially deformation holes formed on the fracture surface of ductile materials. When subjected to greater plastic deformation, the holes grow bigger to form a larger size dimple. Herein, Figure 9a–c show typical larger dimple features, which are consistent with the better plasticities of Sample A–C. For Sample D (Figure 9d), the dimple size is significantly smaller than those of Sample A–C, and even smaller than that of Sample O with the full particle size range (Figure 9e, un-sieved powder), indicating that the 3DP product with small particles has a relatively brittle deformation behavior. It is important to note that the coarse columnar crystals contained in Sample D are often the source of cracks, as shown in Figure 9f. This is due to the fact that the heat conduction of the fine raw powder is not smooth and the heating is uneven, and some of the un-melted raw material particles are accumulated at the interface front and frozen in the 3DP part. These un-melted particles, along with another casting defect that tends to segregate at the interface, are prone to crack initiation and are responsible for the deterioration in plasticity.

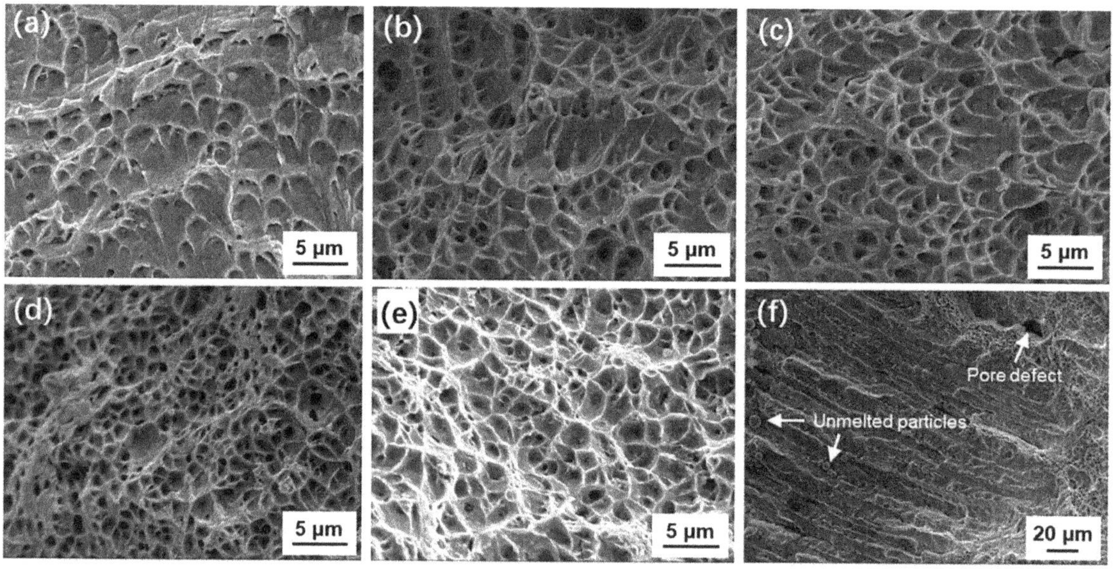

Figure 9. (**a–e**) SEM images of the fracture surfaces of Sample A–O; (**f**) typical SEM image of the fracture surface of columnar crystal region in Sample D.

Table 2 also lists the Vickers hardness test results of different samples. The Vickers hardness values of single-pass deposition samples using different particle sizes of raw materials are 223–230 MPa, with small differences, which is consistent with the grain size results. After multiple deposition processes, the hardness value of the sample significantly decreased compared to that of single-pass deposition processes, and the hardness of the deposited sample gradually decreased as the particle size of the raw material decreased. This is because the smaller the raw material particles, the smaller the grain size deposited in a single pass, resulting in more obvious grain growth during the multi-pass laser melting process, thereby causing a decrease in hardness. In addition, it is known that there is a strong correlation between hardness and the fatigue strength of materials [20]:

$$\sigma_w = 1.6\ \mathrm{HV} \pm 0.1\ \mathrm{HV} \quad (2)$$

where σ_w is the fatigue limit, and HV is Vickers hardness. By substituting the hardness data of the multi-pass samples in Table 2 into Equation (2), the fatigue limits of samples A, B, C, D, and O can be evaluated to be 334.1, 302.2, 294.9, 272.2, and 306.1 MPa, respectively.

It is demonstrated that using larger particles is beneficial for obtaining 3DP products with high fatigue performance.

4. Discussion

The influence of the particle size of raw material powder on the quality of parts and components is one of the key process parameters of laser 3D printing technology. In addition, scan speed and laser power are also key parameters to control grain size and porosity. For example, with a fixed laser power, decreasing the scanning speed allows more particles to bind and form a more contiguous region, thus reducing porosity [21]. However, if the laser power is reduced or the scanning speed is increased, there may be incomplete or partially melted particles in the melt pool that act as new nuclei to promote grain refinement [22].

Under the condition of fixed scanning speed and laser power, it was found that the size of the powder particles also showed a significant adjustment effect on the morphology and size of the solidified grains. The columnar crystal region is mixed with a large number of small equiaxed grains, which can be attributed to the following aspects. Firstly, small particles generally show a higher tendency to agglomerate, which not only reduces the fluidity of the powder but also causes the particles inside the agglomerated clusters to melt incompletely, thus acting as a natural crystal nucleus and refining the grains. Secondly, the specific surface area of small particles is large and the laser absorption rate is high, which may lead to local over melting or porosity. Thirdly, the agglomeration of small particles causes uniformity of particle size distribution, which affects the consistency of the melt pool and increases spatter and defects. Fourthly, small particles have large differences in heat conductivity, and the instability of the melt pool induces faster cooling rates and greater supercooling, which refines the grains and may lead to higher residual stresses.

Another issue that needs to be clarified is the mechanism of grain growth, especially columnar grains, in multi-pass laser deposition. The driving force for grain growth in the first solidified part under the thermal shock of the later laser pass may come from the residual stress formed by the rapid cooling of the laser forming itself and the high interfacial energy provided by the initial microstructure. For residual stresses, Sample D using small particle feedstock has a more complex grain morphology, which usually means that it has greater residual stress. As a result, Sample D has a greater tendency to grow during multi-pass laser deposition. For the interfacial energy, we believe that the interfacial energy provided by different grain configurations in this research plays a major role in the growth of columnar grains.

When the lattice orientation difference between adjacent columnar grains is small, their grain boundaries easily form a co-lattice or near-co-lattice relationship, so they can grow through the grain coarsening mechanism. As shown in Figure 10a, if the difference in atomic orientation between grains A and B is very small, the A/B interface must consist of sparse dislocations. When the temperature rises to a certain range due to thermal shock, the dislocations on the A/B interface can migrate elsewhere through slip and climbing, causing the grain boundaries between grain A and B to disappear and merge into one big columnar grain C. Obviously, this growth mode depends on the crystal orientation of adjacent grains, and the randomness is large, so it results in uneven local growth.

In the case of Figure 5d, the presence of a large number of grain boundaries results in high interfacial energy of the whole system. In particular, some small equiaxed grains are embedded inside the columnar crystals, as shown in Figure 10b. When the temperature rises to a high level due to thermal shock, the curved grain boundaries convex into the columnar grain are flattened under the action of interfacial tension, resulting in the growth of the columnar grain. Its growth form is carried out in the way that large grains swallow small grains, so it is called the grain phagocytosis mechanism, which corresponds to the uniform growth/coarsening of columnar grains.

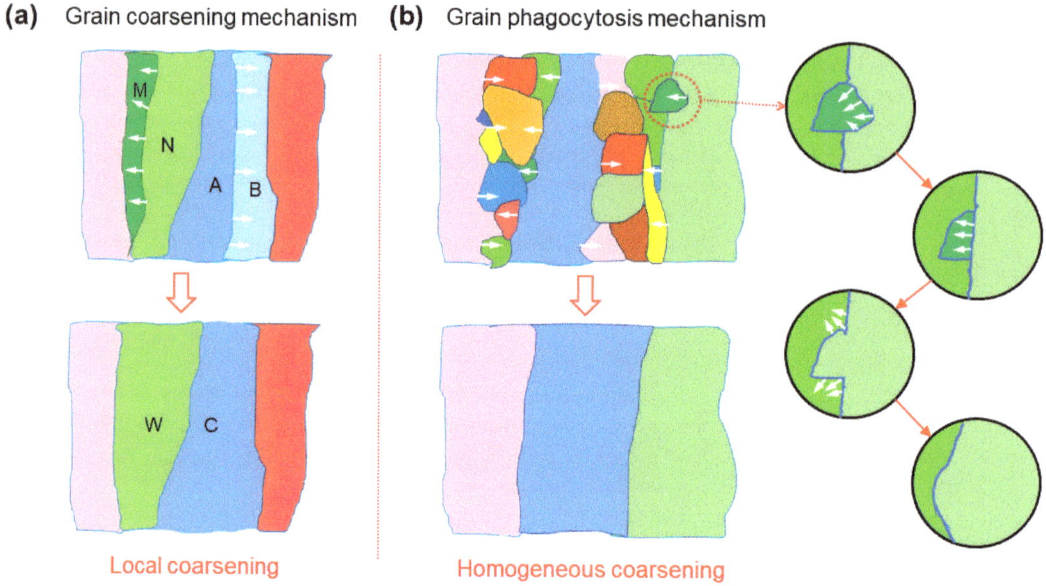

Figure 10. (**a**) Schematic diagram describing the grain coarsening mechanism, in which two grains with similar orientations colored in different shades of blue (A, B) and green (M, N) preferentially merge and grow (C and W); (**b**) schematic diagram describing the grain phagocytosis mechanism derived from the interface behavior between grains with significant size differences.

In this study, an interesting phenomenon was found, that is, the use of fine particles could refine the microstructure of single-pass laser deposition but caused the coarsening of multi-pass laser deposition microstructure. It fully demonstrates the complexity of the influence of powder particle size distribution on the laser melting deposition process. Therefore, the findings of this study are of great significance for improving the performance of laser 3D printing or laser cladding coating products.

5. Conclusions

In the present study, the effects of powder particle size on the microstructure and mechanical properties of laser 3DP 304L SS stainless steel were studied for powder feeding laser melting deposition technology. The results show that the powder particle size has a significant effect on the grain micromorphology of laser 3DP 304L SS stainless steel. With decreases in powder particle size, the microstructure of single-pass laser 3DP tends to be refined, but the coarsening of columnar crystals is induced in the microstructure of multi-pass laser 3DP. This is related to the repeated thermal shocks caused by multi-pass laser melting. The cyclic temperature rises, and complex grain boundary configurations play important roles in grain engulfment and growth. The results of this study clarify the effect of powder particle size on the grain refinement effect of laser 3DP 304L SS stainless steel, deepening the understanding of the solidification process of laser 3DP metal.

Author Contributions: Methodology, formal analysis, investigation, data curation, writing—original draft preparation, X.Z.; conceptualization, writing—review and editing, supervision, project administration, X.L. All authors have read and agreed to the published version of the manuscript.

Funding: This research received no external funding.

Institutional Review Board Statement: Not applicable.

Informed Consent Statement: Not applicable.

Data Availability Statement: Data presented are original and not inappropriately selected, manipulated, enhanced, or fabricated. Any reader interested in this study can contact the author to obtain the original data.

Conflicts of Interest: The authors declare no conflicts of interest.

References

1. Ngo, T.D.; Kashani, A.; Imbalzano, G.; Nguyen, K.T.Q.; Hui, D. Additive manufacturing (3D printing): A review of materials, methods, applications and challenges. *Compos. Part B Eng.* **2018**, *143*, 172–196. [CrossRef]
2. Ren, P.; Ouyang, Y.; Mu, J.; Luo, S.; Tang, Z.; Wu, Y.; Alex Leung, C.L.; Oliveira, J.P.; Zou, Y.; Wang, H.; et al. Metal powder atomization preparation, modification, and reuse for additive manufacturing: A review. *Prog. Mater. Sci.* **2025**, *152*, 101449. [CrossRef]
3. Swayne, M.; Perumal, G.; Padmanaban, D.B.; Mariotti, D.; Brabazon, D. Exploring the impact of laser surface oxidation parameters on surface chemistry and corrosion behaviour of AISI 316L stainless steel. *Appl. Surf. Sci. Adv.* **2024**, *22*, 100622. [CrossRef]
4. Parteli, E.J.R.; Pöschel, T. Particle-based simulation of powder application in additive manufacturing. *Powder Technol.* **2016**, *288*, 96–102. [CrossRef]
5. Jamil, I.R.; Mustaquim, A.M.; Islam, M.; Thakur, M.S.H.; Hasan, M.N. Effect of Powder Bed Fusion Process Parameters on Microstructural and Mechanical Properties of FeCrNi MEA: An Atomistic Study. *Met. Mater. Int.* **2023**, *29*, 659–673. [CrossRef]
6. Roy, S.; Xiao, H.; Angelidakis, V.; Pöschel, T. Combined thermal and particle shape effects on powder spreading in additive manufacturing via discrete element simulations. *Powder Technol.* **2024**, *445*, 120099. [CrossRef]
7. Vaijayanth, S.; Koppula, S.; Sateesh, N.; Subbiah, R. Effect of RF carbonitriding and improvement of mechanical properties on AISI 304 stainless steel—A review. *Mater. Today Proc.* **2020**, *27*, 1655–1659. [CrossRef]
8. Wang, M.; Zhou, Z.; Sun, H.; Hu, H.; Li, S. Microstructural observation and tensile properties of ODS-304 austenitic steel. *Mater. Sci. Eng. A* **2013**, *559*, 287–292. [CrossRef]
9. Zhang, H.; Li, C.; Yao, G.; Shi, Y.; Zhang, Y. Effect of annealing treatment on microstructure evolution and deformation behavior of 304 L stainless steel made by laser powder bed fusion. *Int. J. Plast.* **2022**, *155*, 103335. [CrossRef]
10. Zhang, H.; Xu, M.; Kumar, P.; Li, C.; Liu, Z.; Zhang, Y. Fatigue life prediction model and entropy generation of 304L stainless steel fabricated by selective laser melting. *J. Mater. Process. Technol.* **2021**, *297*, 117279. [CrossRef]
11. Zhang, H.-y.; Zheng, L.-w.; Shi, Q.-x.; Liang, W.; Hu, J. Anisotropy in mechanical property and hydrogen embrittlement resistance of selective laser melted 304 austenitic stainless steel. *Mater. Lett.* **2023**, *339*, 134104. [CrossRef]
12. Zhang, S.; Jia, M.; Wang, W.; Hou, J.; Kuang, W. The effects of heat treatment and surface state on the corrosion resistance of laser powder bed fusion 304L stainless steel in 3.5 wt% NaCl solution. *J. Mater. Res. Technol.* **2024**, *29*, 5620–5632. [CrossRef]
13. Zhang, S.; Wang, S.; Feng, X.; Ning, Z.; Hou, J.; Kuang, W. Insights into the stress corrosion cracking resistance of a selective laser melted 304L stainless steel in high-temperature hydrogenated water. *Acta Mater.* **2023**, *244*, 118561. [CrossRef]
14. Mahmood, M.A.; Popescu, A.C.; Oane, M.; Chioibasu, D.; Popescu-Pelin, G.; Ristoscu, C.; Mihailescu, I.N. Grain refinement and mechanical properties for AISI304 stainless steel single-tracks by laser melting deposition: Mathematical modelling versus experimental results. *Results Phys.* **2021**, *22*, 103880. [CrossRef]
15. Pan, Q.; Guo, S.; Cui, F.; Jing, L.; Lu, L. Superior Strength and Ductility of 304 Austenitic Stainless Steel with Gradient Dislocations. *Nanomaterials* **2021**, *11*, 2613. [CrossRef]
16. Köhnen, P.; Ewald, S.; Schleifenbaum, J.H.; Belyakov, A.; Haase, C. Controlling microstructure and mechanical properties of additively manufactured high-strength steels by tailored solidification. *Addit. Manuf.* **2020**, *35*, 101389. [CrossRef]
17. Kies, F.; Köhnen, P.; Wilms, M.B.; Brasche, F.; Pradeep, K.G.; Schwedt, A.; Richter, S.; Weisheit, A.; Schleifenbaum, J.H.; Haase, C. Design of high-manganese steels for additive manufacturing applications with energy-absorption functionality. *Mater. Des.* **2018**, *160*, 1250–1264. [CrossRef]
18. Rodriguez, J.S.; Duran, J.F.; Aguilar, Y.; Perez Alcazar, G.A.; Souza, R.M.; Zambrano, O.A. Effect of Aluminum Content on the Dynamic Recrystallization of $Fe_{18}Mn_xAl_{0.74}C$ Steels During Hot-Forging Treatments. *Metall. Mater. Trans. A* **2022**, *53*, 2961–2976. [CrossRef]
19. Liu, S.; Wang, X.; Cui, C.; Zhao, L.; Li, N.; Zhang, Z.; Ding, J.; Sha, D. Enhanced grain refinement of in situ CeB6/Al composite inoculant on pure aluminum by microstructure control. *J. Alloy Compd.* **2017**, *701*, 926–934. [CrossRef]
20. Wu, H.; Hamada, S.; Noguchi, H. Fatigue strength prediction for inhomogeneous face-centered cubic metal based on Vickers hardness. *Int. J. Fatigue* **2013**, *48*, 48–54. [CrossRef]

21. Yang, Y.; Ragnvaldsen, O.; Bai, Y.; Yi, M.; Xu, B.-X. 3D non-isothermal phase-field simulation of microstructure evolution during selective laser sintering. *npj Comput. Mater.* **2019**, *5*, 81. [CrossRef]
22. Kang, S.-G.; Gainov, R.; Heußen, D.; Bieler, S.; Sun, Z.; Weinberg, K.; Dehm, G.; Ramachandramoorthy, R. Green laser powder bed fusion based fabrication and rate-dependent mechanical properties of copper lattices. *Mater. Des.* **2023**, *231*, 112023. [CrossRef]

materials

MDPI

Article

Effects of PPA Reinforcement and Sintering Parameters on the Densification and Hardness Properties of Al/Mg/PPA Composites

Osarue Osaruene Edosa [1], Francis Kunzi Tekweme [1], Peter A. Olubambi [2] and Kapil Gupta [1,*]

[1] Department of Mechanical and Industrial Engineering Technology, University of Johannesburg, Johannesburg 2028, South Africa; osaruene@gmail.com (O.O.E.); ftekweme@uj.ac.za (F.K.T.)

[2] Centre for Nanoengineering and Advanced Materials, School of Mining, Metallurgy and Chemical Engineering, University of Johannesburg, Johannesburg 2028, South Africa; polubambi@uj.ac.za

* Correspondence: kgupta@uj.ac.za

Abstract: The utilization of agricultural wastes in composite fabrication leads to attaining sustainability in manufacturing. This study investigates the use of plantain peel ash (PPA) as a reinforcement to fabricate Al/Mg/PPA composites using ball milling and spark plasma sintering (SPS) technology. The impact of PPA weight fraction and SPS parameters on Al/Mg/PPA composites' densification and hardness were analyzed. Microstructural characterization revealed that the PPA reinforcement was uniformly distributed in the aluminum matrix with no considerable microstructural defects. The relative densities of the composites were higher at elevated sintering temperatures, with composites displaying reduced porosity as the sintering temperature rose. The composites also exhibited the highest micro-hardness of 77 HV, improving 86.89% over the sintered aluminum matrix (base material). The Analysis of Variance (ANOVA) results revealed that the sintering temperature and reinforcement material significantly influenced the relative density (RD) of the sintered composites, while the reinforcement material significantly influenced the micro-hardness. Conclusively, the composite samples made using agricultural waste derivatives possess good mechanical properties and are suitable for various industrial applications.

Keywords: composites; hardness; relative density; agro-waste ash; SEM; sintering; XRD

1. Introduction

Agricultural waste reinforcements have continued to be the focus of research in aluminum matrix composite (AMC) because of their potential to enhance composites' physical, mechanical, and tribological properties [1–3]. As reported in the literature, synthetic and agro-industrial waste derivatives are the two main types of ceramic reinforcements [3,4]. Commonly used synthetic reinforcements include SiC, Al_2O_3, AlB_2, B_4C, MgO, TiC, WC, and TiB_2 [2,5], while commonly used agro-industrial waste reinforcements include fly ash, rice husk ash, palm kernel shell ash, wheat husk ash, corn cob ash, sugarcane bagasse, and bamboo leaf ash [4,6,7].

The types of reinforcements used and the processing methods significantly impact the quality and performance of composites. As an alternative to synthetic reinforcement, agricultural waste reinforcement has gained much interest [3,8]. The high acquisition cost and limited availability are some of the drawbacks of synthetic reinforcements.

Using agricultural waste as reinforcement is especially important for promoting a circular economy, which will positively impact the environment by reducing waste and pollution and using fewer natural resources. This is especially true when it comes to achieving zero-waste-to-landfill goals [9]. The percentage chemical composition of agricultural wastes (Al_2O_3, MgO, SiO_2, K_2O, ZnO, MnO, Fe_2O_3, CaO, P_2O_5, and TiO_2), which varies from plant to plant, can be linked to their performance as reinforcements. According to previous studies, AMC's microstructure and properties significantly improved due to the performance of agricultural waste reinforcements [5,10]. Agricultural waste reinforcements have lower economic costs and density advantages. One intriguing feature of agricultural waste ashes is that their percentage chemical composition changes with heat treatment temperatures. For instance, Abshalomu et al. [10] found that when the heat treatment temperature increased from 400 to 600 °C, the percentage concentration of SiO_2 in cashew nut ash increased from 44 ± 2% to 75 ± 2%. Using agricultural waste derivatives as the only reinforcement presents several challenges, including reduced strength, ductility, and fracture toughness [3,8]. However, a significant improvement was seen when agricultural waste reinforcements were combined with synthetic reinforcements [8,11]. It has been proposed that agricultural waste reinforcements supplement synthetic reinforcements because of their inadequacy. Therefore, combining agricultural waste reinforcements with synthetic reinforcements is one efficient way to use them to synthesize AMC [11,12].

The properties of the aluminum matrix and reinforcement define the attributes of the AMC. The properties of AMC depend on the concentration and characteristics (the chemical composition, size, and shape) of the reinforcement. The automotive and aerospace industries, among others, have found it necessary to employ AMC due to its appealing qualities. When employed, AMC has excellent performance and fuel efficiency, among other advantages [13]. AMC reinforced with agricultural waste also has benefits such as low density and coefficient of thermal expansion. The sustainability criteria for any synthesized material include affordability, environmental friendliness, and societal advantages [14]. The uninterrupted fabrication of sustainable AMC will depend on the constant availability of reinforcement materials to avoid supply shortages due to the growing demand for AMC's products. Utilizing green and renewable agricultural waste derivatives as reinforcement materials is one approach to assure sustainability in the manufacturing of AMC. Ikele et al. [12] fabricated hybrid reinforced AMC using SiC and palm kernel shell ash (PKSA). According to the authors, the best value of the mechanical property of the composites was obtained when the weight ratio of SiC to PKSA reinforcements was 1:1, indicating that agricultural waste reinforcement is a great addition to synthetic reinforcement.

Another factor that may affect the quality of AMC is the process of fabrication. There have been several documented methods for producing AMC [3,15,16]. The two most frequently used methods are powder metallurgy (PM) and stir casting. This study explores the capability of the PM process in producing high-quality AMC. The main objective of PM technology for fabricating AMC has been to improve the densification and microstructural features, which ultimately enhances the mechanical properties such as hardness, strength, and fracture toughness of the composites [15,17,18]. In this study, AMC reinforced with plantain peel ash is fabricated using ball milling and spark plasma sintering (SPS). A detailed analysis and discussion of the effects of ball milling parameters, sintering parameters, and reinforcement concentration on the fabricated AMC's microstructure, densification, and hardness have been conducted.

2. Materials and Methods

2.1. Starting Materials

The starting materials for synthesizing the AMC samples for this investigation include pure Al powder (75 μm), Plantain peel ash (70 μm), and Mg powder (100 μm). The Al matrix, purchased from Zimco Aluminum Company, Benoni, South Africa, was reinforced with 0, 5, 10, 15, and 20 wt% PPA. The 2 wt% Mg powder, purchased from Sigma-Aldrich, Johannesburg, South Africa, was introduced into the Al/PPA mixes to improve the wettability [19]. The SEM image and SEM-EDS analysis of the pure Al particles are shown in Figure 1. From the figure, it can be seen that mixed sizes, oval, irregular, and flaky shapes characterize the aluminum powder particles. The processing of the plantain peel wastes to obtain ash is presented in Section 2.2.

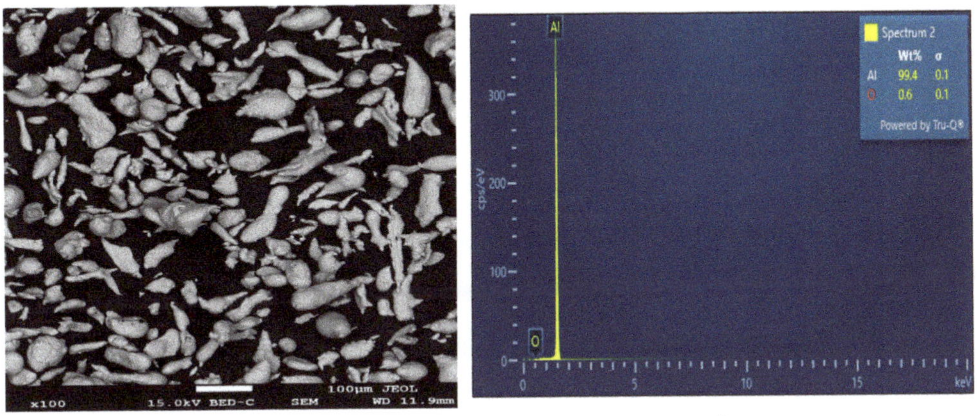

(a) **(b)**

Figure 1. (**a**) SEM image; (**b**) SEM-EDS analysis of pure aluminum powder particles.

2.2. Preparation of Plantain Peel Ash Reinforcement

The plantain peel wastes were sourced from Harper, Liberia, and processed to obtain ash following the procedure described in the literature [8,12]. The processing stages include: (i) sourcing the wastes and drying them in the sun for about two weeks to reduce the moisture content; (ii) burning the dried plantain peel in an incinerator to remove organic material and produce ash residue; (iii) heat treating the ash residue at an average temperature of 550 °C using a furnace to remove the black carbon and other volatile materials. This makes it fit for use as ceramic reinforcement, see Figure 2b. The heat-treated ash was then pulverized and sieved to obtain fine powder particles. Characterization of the ash was done using an SEM-EDS machine (JEOL JSM-7900F, JEOL, Peabody, MA, USA) to determine its elemental composition. The result obtained includes 79.42% K, 4.70% Fe, 4.46% P, 4.18% Si, 1.45% Al, and 1.26% Mg, as illustrated in Figure 2a. The density of the PPA was determined to be 2.34 g/cm^3 using a gas pycnometer (BetterPyc 380, Bettersize, Noi, Vietnam).

2.3. Design of Experiment Using the Taguchi Method

The Taguchi technique is selected to design the experiments in the present study. It uses Orthogonal Arrays (OAs) to systematically vary and test each parameter at different levels (intervals). The orthogonal array provides a set of well-balanced control parameters with settings to achieve the best performance characteristics. The control parameters for this study include reinforcement weight fraction, sintering temperature, sintering pressure, heating rate, and soaking time. The design of experiments for this study uses ($L_{25}5^5$) Taguchi orthogonal array with five parameters tested at five levels, resulting in 25 experimental runs (samples).

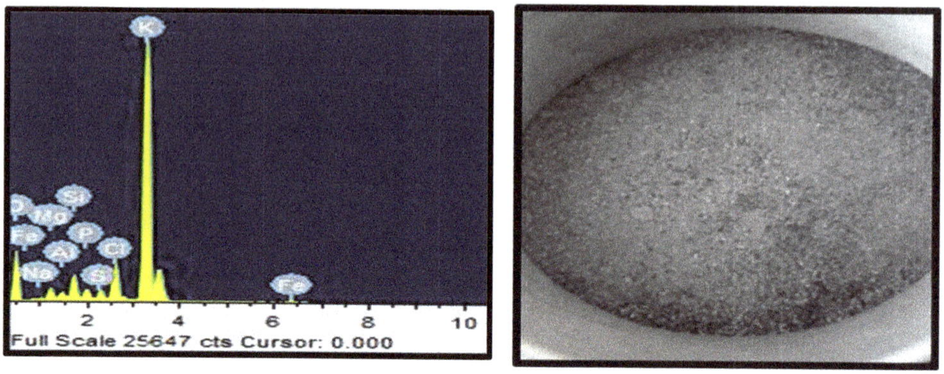

(a) (b)

Figure 2. Plantain peel ash (**a**) EDS analysis, (**b**) powder particles.

2.4. Al/Mg/PPA Composite Sample Production

Al/Mg/PPA samples were fabricated using the PM method (Ball milling and SPS). The first stage in the processing sequence of the PM process involves powder mixing using a high-energy ball mill (PM 100, Retsch, Haan, Germany). The powder mixes were made according to the ratio presented in Table 1 and charged in a chromium steel vial with a ball-to-powder ratio (BPR) of 5:1 under normal atmospheric conditions. The powder mixtures were milled for 90 min at an average speed of 200 rpm. To prevent excessive cold welding and powder particle agglomeration, 1 wt% of propanol was added to the powder mixture. After every 30 min of milling, the milling machine was turned off for 10 min to prevent the milling vial's temperature from rising [18,20]. The compaction and sintering of the milled powders were done using the SPS machine (8604 HHPD-25, FCT Systeme GmbH, Frankenblick, Germany), having a graphite die, a maximum compaction force of 250 KN, a working temperature of 2200 °C, a maximum sintering current of 8000 A, and a vacuum atmosphere. The die walls were covered with graphite sheets before filling with nine grams of the powder mixture. The graphite sheet prevents the sintered samples from sticking to the die walls, thereby facilitating their removal from the die. After the sintering time was completed, the sintering current was switched off to allow the samples to cool to room temperature. The graphite sheets, which formed additional layers to the sintered samples, were removed using a knife. The dimensions of the fabricated composites after the graphite sheet layers have been removed are 20 mm in diameter and 12 mm in thickness. The sintering parameters for this study are presented in Table 2. Figure 3 illustrates the flow chart for the fabrication of Al/Mg/PPA Composites.

Table 1. Sample designation and their theoretical densities.

Sample Designation	Composition	Theoretical Density (g/cm^3)
A1	98Al:2Mg	2.6816
A2	93Al:2Mg:5PPA	2.6636
A3	88Al:2Mg:10PPA	2.6476
A4	83Al:2Mg:15PPA	2.6276
A5	78Al:2Mg:20PPA	2.6096

Table 2. Sintering parameters for Al/Mg/PPA composites.

Parameters	Levels				
	1	2	3	4	5
Sintering temperature (ST) °C	430	460	490	520	550
Holding (sintering) time (HT) min	2	4	6	8	10
Sintering pressure (SP) MPa	12	22	32	42	52
Heating rate (HR) °C/min	50	100	150	200	250

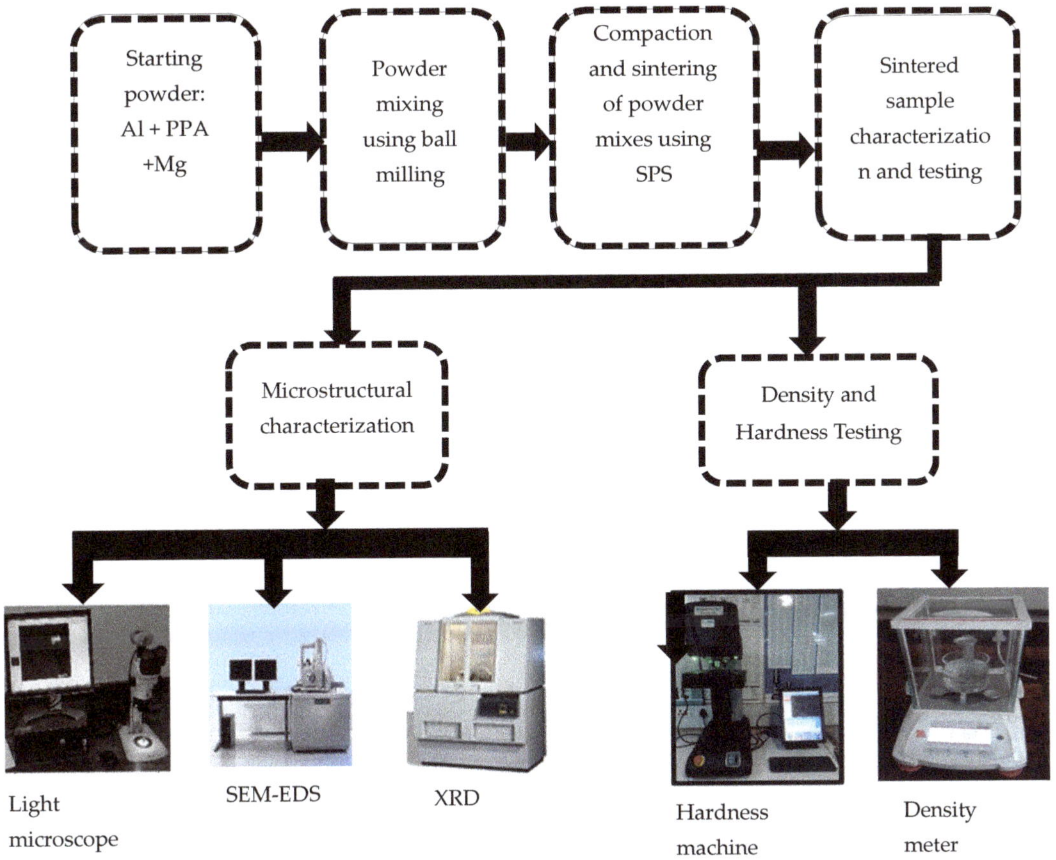

Figure 3. The flow chart for the fabrication of Al/Mg/PPA composites.

2.5. *Characterization of Samples*

The microstructure of the fabricated Al/Mg/PPA samples was characterized using XRD (Empyrean, Worcestershire, UK), light microscope (ZEISS, Jena, Germany), and SEM-EDS, following the standard procedure [19]. Sample preparation was by grinding and polishing using abrasive papers and polishing cloths mounted on the grinding and polishing machine (ATM Saphir 500, ATM Qness GmbH, Mammelzen, Germany) to have a smooth and mirror-like surface finish. Etching of each sample to make the grain boundaries more visible took 30 s using Keller reagent (190 mL distilled water + 5 mL HNO_3 + 3 mL HCl + 2 mL HF). The etching process was consecutively halted by immersing the samples in alcohol and distilled water.

2.6. Density Measurement

The densities of the developed Al/Mg/PPA samples were determined using a digital density meter under ambient temperature and pressure, following the ASTM B962-13 standard, according to Archimedes principles. The density meter has an accuracy of 0.0002 g/cm^3 and an average value of five measurements was reported. The density (ρ_c), theoretical density (ρ_{th}), relative density (RD), and the porosity of Al/Mg/PPA samples were determined using Equations (1)–(4).

$$\rho_c = \frac{w_a}{w_a - w_w} \times \rho_w \quad (1)$$

$$\rho_{th} = \rho_{Al} wt_{Al} + \rho_{Mg}\, wt_{Mg} + \rho_{PPA} wt_{PPA} \quad (2)$$

$$RD = \frac{\rho_a}{\rho_{th}} \times 100 \quad (3)$$

$$Porosity\ (\varnothing) = 1 - \frac{\rho_a}{\rho_{th}} \quad (4)$$

where ρ_c = density of the fabricated composites, w_a = weight in the air, w_w = weight in water, ρ_w = density of water, ρ_{th} is the theoretical density of the composites, ρ_{Al} = theoretical density of the Al matrix (2.7 g/cm^3), wt_{Al} = Weight fraction of Al matrix, ρ_{Mg} = theoretical density of the Mg (1.74 g/cm^3), wt_{Mg} = weight fraction of magnesium, ρ_{PPA} = theoretical density of the PPA reinforcement (2.34 g/cm^3), wt_{PPA} = weight fraction of PPA, and RD = relative density.

Applying the rule of mixture expressed in Equation (2), the evaluated theoretical densities of the fabricated Al/Mg/PPA samples are presented in Table 1.

2.7. Hardness Testing

Hardness testing is a non-destructive test carried out by applying a constant load through a diamond indenter on the polished surface of the specimen to create an indentation on the surface, which is then measured to determine hardness. The hardness of the developed Al/Mg/PPA samples was tested following ASTM E 92 Standard using a Vickers' hardness testing machine with a capacity of 100 kgf (Innovatest Falcon 500, Innovatest, Maastricht, The Netherland, equipped with a diamond pyramid indenter). A constant load of 100 gf was applied for 10 s to the polished sample surfaces, and an average value of five trials was recorded. The RD, hardness, and standard deviation for hardness of the sintered samples are presented in Table 3.

Table 3. RD and hardness results of Al/Mg/PPA samples.

Exp. Runs	Input Processing Parameters					Results			
	ST (°C)	HT (min)	'SP' (MPa)	PPA wt%	HR (°C/min)	Av. Exp. Density (g)	RD (%)	Average Hardness (HV)	Standard Dev. for Hardness
	Experimental Values								
1	430	2	12	20	50	2.3864	91.45	55.1	7.1
2	430	4	22	15	100	2.4531	93.36	56.9	3.8
3	430	6	32	10	150	2.3989	90.61	61.9	7.6
4	430	8	42	5	200	2.4250	91.00	68.9	4.6
5	430	10	52	0	250	2.3853	88.95	72.6	7.3
6	460	2	22	10	200	2.4660	93.14	54.5	7.0

Table 3. *Cont.*

Exp. Runs	Input Processing Parameters					Results			
	ST (°C)	HT (min)	'SP' (MPa)	PPA wt%	HR (°C/min)	Av. Exp. Density (g)	RD (%)	Average Hardness (HV)	Standard Dev. for Hardness
	Experimental Values								
7	460	4	32	5	250	2.5380	95.30	63.0	7.9
8	460	6	42	0	50	2.5021	93.31	61.5	1.6
9	460	8	52	20	100	2.4323	93.21	53.3	2.5
10	460	10	12	15	150	2.5280	96.21	58.8	5.3
11	490	2	32	0	100	2.4644	91.90	64.7	4.4
12	490	4	42	20	150	2.4592	94.24	53.5	4.5
13	490	6	52	15	200	2.4759	94.23	77.0	6.2
14	490	8	12	10	250	2.4750	93.48	52.1	7.3
15	490	10	22	5	50	2.4938	93.63	63.5	5.9
16	520	2	42	15	250	2.6026	99.05	71.5	7.0
17	520	4	52	10	50	2.5243	95.34	55.7	5.5
18	520	6	12	5	100	2.5159	94.45	61.7	5.4
19	520	8	22	0	150	2.5059	93.45	74.5	3.2
20	520	10	32	20	200	2.5168	96.44	56.1	6.2
21	550	2	52	5	150	2.5107	94.26	70.4	8.7
22	550	4	12	0	200	2.5685	95.78	75.8	4.9
23	550	6	22	20	250	2.5580	98.02	52.7	3.1
24	550	8	32	15	50	2.5074	95.43	67.5	4.8
25	550	10	42	10	100	2.5421	96.02	56.2	4.5
Al	550	4	12	100	200	2.6901	99.63	41.2	3.3

3. Results and Discussion

3.1. Optimal Input Parameters for RD

Optimizing the processing parameters for RD results in the best value of output. Using Minitab 20 and the larger-the-better approach expressed in Equation (5), the results of optimization are presented in Tables 4 and 5, and Figure 4, respectively.

$$SN_L = -10log\left(\frac{1}{n}\sum_{i=1}^{n}\frac{1}{y_i^2}\right) \quad (5)$$

where y_i = response, n = number of trials for y_i.

Table 4. Response for signal-to-noise (SN) ratios for RD.

Level	Temperature	Holding Time	Pressure	Material	Heating Rate
1	39.19	39.46	39.49	39.52	39.45
2	39.48	39.54	39.49	39.61	39.44
3	39.42	39.47	39.45	39.43	39.44
4	39.62	39.40	39.53	39.44	39.47
5	39.64	39.48	39.39	39.34	39.54
Delta	0.45	0.14	0.14	0.28	0.11
Rank	1	4	3	2	5

Table 5. Analysis of Variance for RD.

Source	DF	Adj SS	Adj MS	F-Value	p-Value
Regression	5	85.147	17.029	8.05	0.000
Temperature	1	62.362	62.362	29.49	0.000
Holding time	1	0.414	0.414	0.20	0.663
Pressure	1	1.528	1.528	0.72	0.406
Material	1	17.500	17.500	8.28	0.010
Heating rate	1	3.344	3.344	1.58	0.224
Error	19	40.175	2.115		
Total	24	125.322			

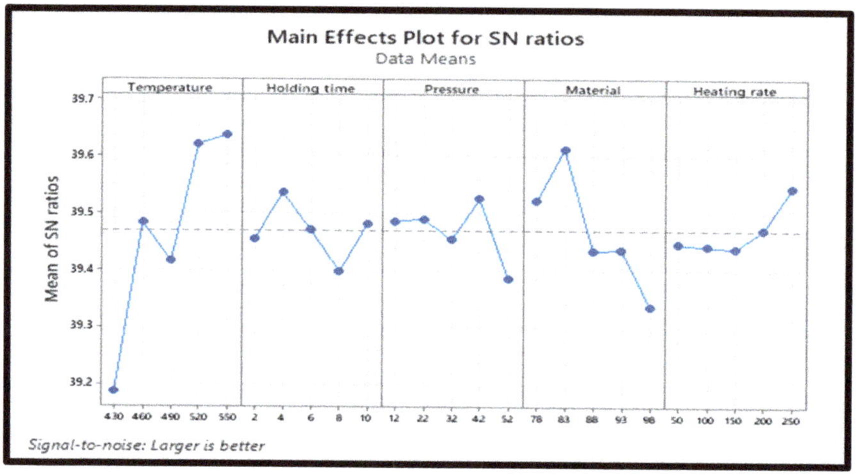

Figure 4. Main effect plot of signal-to-noise ratios for RD.

The results of the Taguchi analysis presented in Tables 4 and 5, and Figure 4 show that the sintering temperature strongly impacted RD (rank 1), followed by the PPA reinforcement (rank 2). The heating rate had the least effect on the RD. As illustrated in Figure 4, the optimal processing parameters for the RD of Al/Mg/PPA samples include a sintering temperature = 550 °C, holding time = 4 min, compaction pressure = 42 MPa, PPA concentration = 15 wt% PPA, and a heating rate = 250 °C. The effects of each input parameter on the RD and their significance level were determined using ANOVA.

As illustrated in Table 5, the F-value and p-value show the factor's significance level. As indicated by the F-value column, the factor with the highest F-value has the strongest influence and vice versa. Similarly, from the p-value column, factors having values less than 0.05 ($p < 0.05$) are statistically significant. As shown in Table 5, temperature and material are the only parameters that have a statistically significant influence on RD. The remaining parameters, though, contributed to the enhanced densification of the composites but in a smaller magnitude.

3.2. *Influence of Input Parameters and Microstructure on Relative Density (RD) and Porosity*

The porosities of the fabricated Al/Mg/PPA composites were estimated using Equation (4). Porosity in composites can be reduced but cannot be eliminated. As the RD decreases, the porosity of the Al/Mg/PPA composites rises, as illustrated in Figure 5. The main factors influencing the porosity levels of the fabricated composites include the reinforcement characteristics and the processing parameters [21,22].

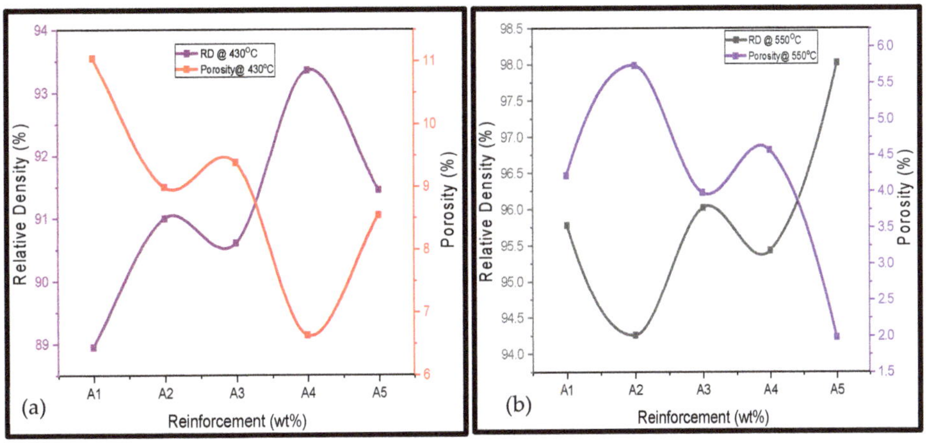

Figure 5. Relationship between RD and porosity of Al/Mg/PPA composites sintered at (**a**) 430 °C and (**b**) 550 °C.

According to Table 3 and Figures 5–7, higher sintering temperatures are required to achieve higher RDs. As seen from Table 3, the lowest sintering temperature of 430 °C produced the lowest RD (88.95%) of the sintered samples, while the highest RD of 99.05% was obtained at a sintering temperature of 520 °C. Therefore, it can be inferred that the sintering temperature is an important factor in the densification of the Al/Mg/PPA composites [23]. Powder particles are transported by diffusion. According to Equation (6), sintering AMC at high temperatures increases the particles' thermal energy and diffusion rate [22,24]. A higher diffusion rate depends on the sintering temperature and holding (sintering) time and is essential for greater consolidation of the composites to create a denser structure. For instance, the microstructure of the composite shown in Figure 6a,b was sintered using a temperature of 430 °C. Poor grain boundaries and pores are evident. Comparing the microstructures of Figure 6a,b with the one of Figure 6c, it can be seen that the microstructural features of Figure 6c is more enhanced with improved grain boundaries and few pores. This is because the microstructure of Figure 6c is sintered at higher temperature of 520 °C. Increased porosity is a function of poor grain boundaries and an increased number of pores. This explains why the RDs of the Al/Mg/PPA samples increased with the rise in the sintering temperature, as illustrated in Figure 7. Furthermore, the SPS machine's higher heating rate necessitated a shorter holding (sintering) time of 4 min, as indicated by Figure 4. Therefore, sintering at higher temperatures and extended holding (sintering) times may result in grain growth (coarse grains), consequently increasing porosity.

$$D = D_0\, exp\left(\frac{-Q}{RT}\right) \tag{6}$$

where D = diffusion coefficient, D_0 = constant, Q = activation energy, R = Boltzmann constant, and T = temperature.

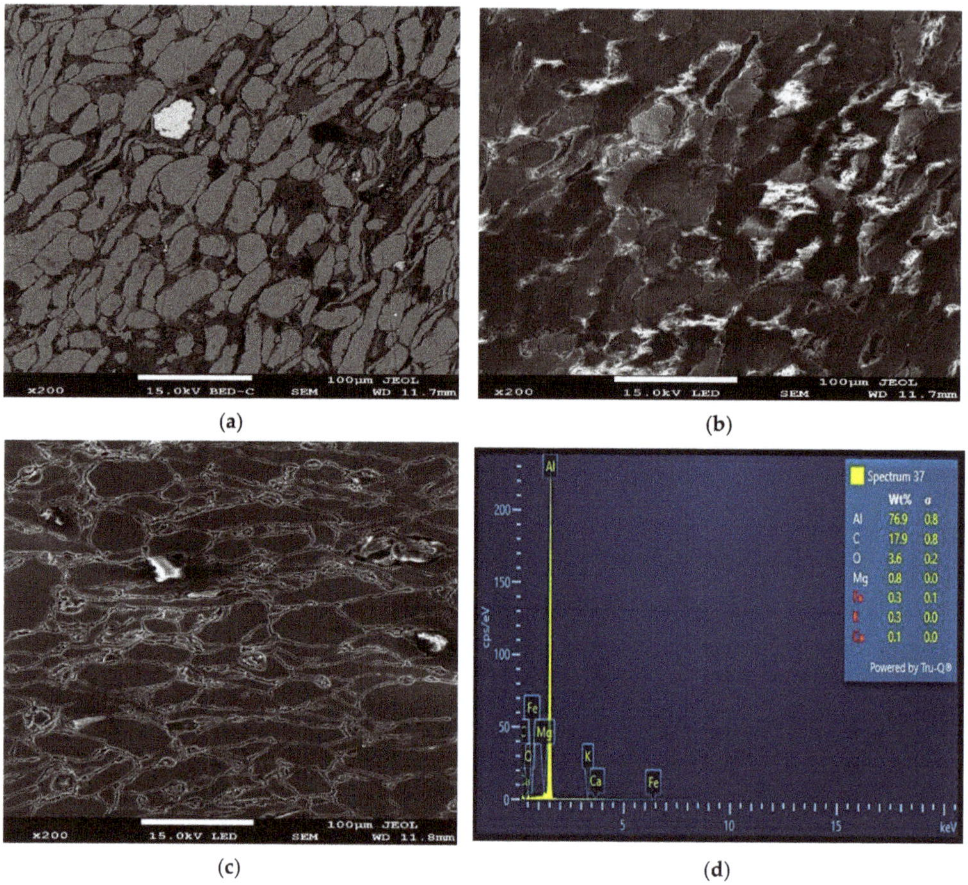

Figure 6. SEM images and SEM-EDS analysis of Al/Mg/PPA composites (**a**,**b**) A5 (**c**,**d**) A5.

The compaction pressure is another important factor in the powder metallurgy process. The initial consolidation of powder particles is facilitated by compaction pressure. In SPS, the sintering temperature and compaction pressure are applied simultaneously. A certain amount of cold welding happens when the sintering pressure rises, increasing powder particle consolidation through proper neck formation. More so, the inherent friction between the powder particles and the die walls is reduced through the sintering pressure. It is challenging to achieve complete densification because friction between powder particles alters density and decreases plastic flow. However, higher densification can be achieved when compaction pressure and sintering temperature are applied concurrently because this increases the mass transport of powder particles by plastic flow and diffusion. Furthermore, it can be seen from Tables 4 and 5 that the heating rate and holding time have a smaller impact on the RD of the composites when compared with the reinforcement material and sintering temperature. The holding time enables the composites to absorb enough thermal energy at the sintering temperature before quenching.

3.3. Influence of Matrix Reinforcement Characteristics and Microstructure on RD

The homogeneity of the powder particles is vital for higher densification, while powder particle agglomeration results in increased porosity. It is obvious from the optical and SEM images that the Al, Mg, and PPA particles are homogeneously distributed with little to no considerable agglomeration. This indicates that high-energy ball milling is an efficient method for achieving powder particle homogeneity [18,25].

More so, it can be deduced from Table 3 that as the weight fraction of PPA increases, the RD of the fabricated composites rises. This is a confirmation that the PPA reinforcement has a significant influence on the RD of the composites. From Table 3, sample 5 (Al/2Mg/PPA) has the highest porosity or the lowest RD, while sample 16 (Al/2Mg/15PPA) has the maximum RD or the lowest porosity. According to Equation (3), the composites' RD rises as the theoretical densities drop. As the weight fraction of PPA increases, the theoretical densities drop, as seen in Table 2. For example, the theoretical density of sample A2 (Al/2Mg/5PPA) drops from 2.6636 g/cm^3 to 2.6476 g/cm^3 as the weight fraction of PPA reinforcement increases from 5 wt% to 10 wt%, and so on. When the concentration of the PPA increased from 15% to 20%, the composite's RD decreased from 99.05% to 98.02%. The porosity of the composite may have increased slightly due to an increase in particle clustering caused by high PPA concentration.

Additionally, composites densify more when fine and mixed grains are used because they pack better and have fewer pores than coarse grains. From the grain size distribution shown in Figure 8b,d using Image J software, version 1.54, it is observed that the grain sizes are mixed and range from 20 μm to 140 μm. When compared with the as-received powder particles, it can be inferred that some of the coarse grains have been fractured into finer particles through the process of ball milling and compaction. Powders with diverse particle sizes, shown in Figure 8, will minimize porosity and increase densification more so than powders with uniform particle sizes. For instance, in contrast to a combination of 20 μm (matrix) and 20 μm (reinforcement) or 120 μm (matrix) and 120 μm (reinforcement), a combination of powder particle sizes of 20 μm (matrix) and 60 μm (reinforcement) or 25 μm (matrix) and 41 μm (reinforcement) is more likely to produce higher densification. Fine and mixed grain sizes pack better, resulting in higher densification.

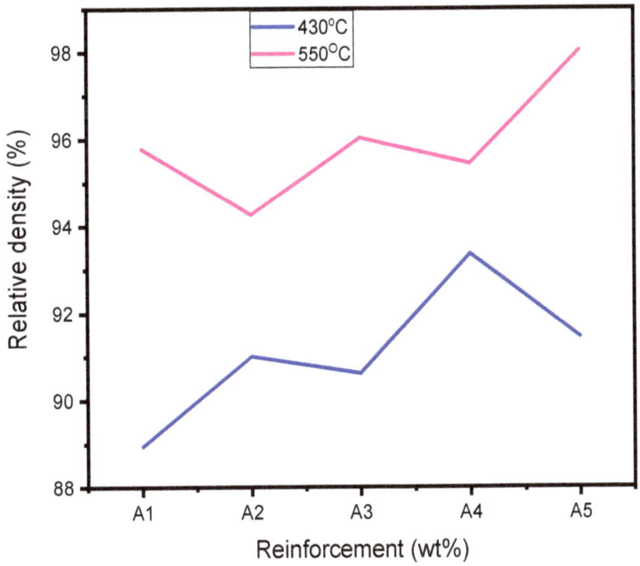

Figure 7. Influence of PPA reinforcement and sintering temperature on the RD of Al/Mg/PPA composites.

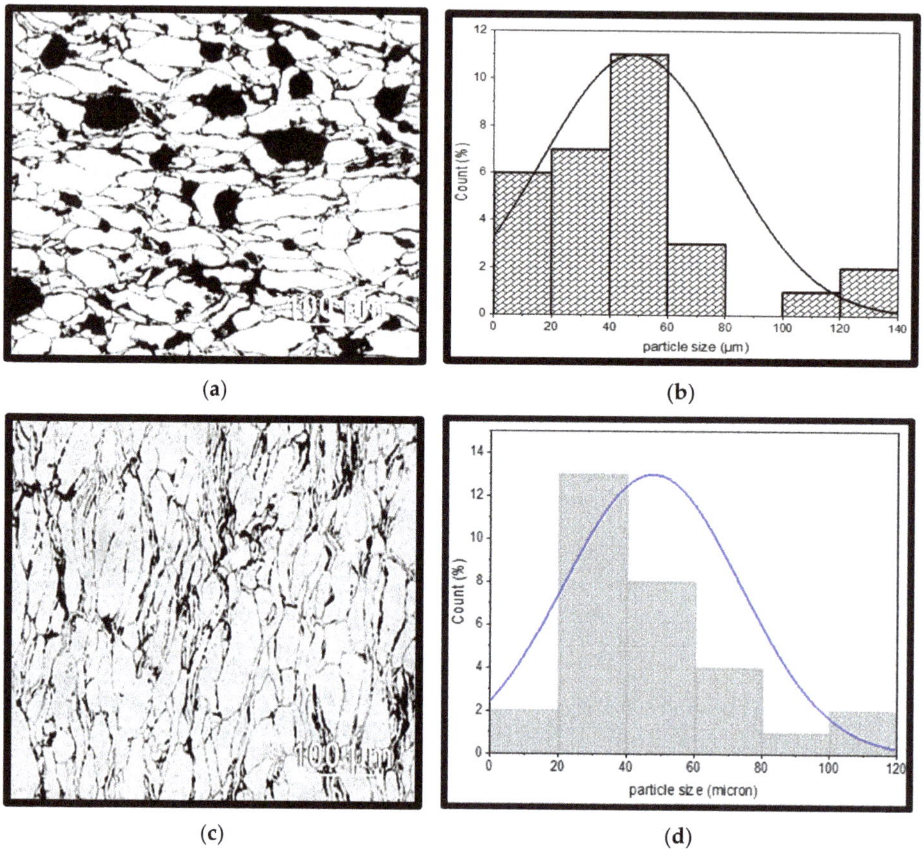

Figure 8. Optical micrographs and particle size distribution of sintered Al/Mg/PPA samples, (**a**,**b**) A3, (**c**,**d**) A1.

3.4. Micro-Hardness of Sintered Al-Mg-PPA Samples

As earlier stated, the hardness of the fabricated Al/Mg/PPA samples was determined using a Vickers' hardness testing machine equipped with a diamond pyramid indenter. The result of the Vickers hardness test is presented in Table 3. From the Table, the average micro-hardness of the sintered aluminum matrix is 41.2 HV, while the highest micro-hardness of the sintered composites is 77.0 HV. Figure 9a,b shows the indentation for the hardness of the composites during the hardness test. Factors such as reinforcement material characteristics and processing parameters can be attributed to the improved hardness of the Al/Mg/PPA composites [25,26].

3.5. Optimal Input Parameters for Micro-Hardness

The Taguchi optimization process is used to determine the optimal processing parameters that maximize the micro-hardness of the composites. Using Minitab 20 and the larger-the-better approach expressed in Equation (5), the optimization results are presented in Tables 6 and 7 and Figure 10.

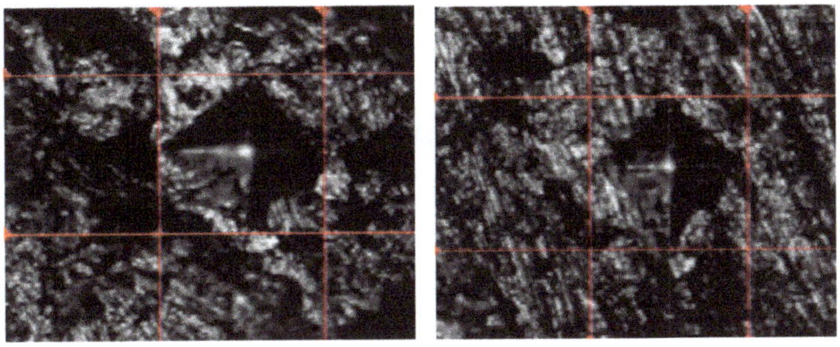

(**a**) (**b**)

Figure 9. Indentation images for the hardness of sintered Al/Mg/PPA samples (**a**) first, (**b**) second image.

Table 6. Signal-to-noise ratios of micro-hardness (larger is better).

Level	Temperature	Holding Time	Pressure	Material	Heating Rate
1	35.37	35.96	35.59	34.67	35.63
2	35.28	35.63	35.55	36.38	35.33
3	35.78	35.61	35.61	34.66	35.73
4	36.05	35.66	35.57	36.04	36.09
5	36.11	35.73	36.27	36.85	35.81
Delta	0.83	0.35	0.72	2.19	0.76
Rank	2	5	4	1	3

Table 7. ANOVA for micro-hardness.

Source	DF	Adj SS	Adj MS	F-Value	p-Value
Regression	5	643.95	128.789	2.63	0.057
Temperature	1	137.12	137.117	2.80	0.111
Holding time	1	5.51	5.511	0.11	0.741
Pressure	1	51.01	51.005	1.04	0.320
Material	1	406.70	406.695	8.31	0.010
Heating rate	1	43.62	43.618	0.89	0.357
Error	19	929.86	48.940		
Total	24	1573.81			

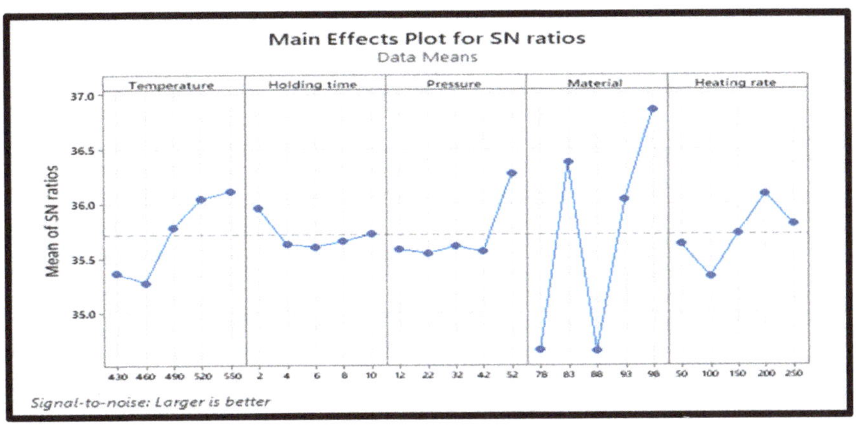

Figure 10. Main effects plot of signal-to-noise ratio for hardness.

The results presented in Tables 6 and 7 show that the material for the composites has the strongest influence on the hardness of the composites (rank 1), followed by the sintering temperature (rank 2), and so on; the holding time has the weakest influence on the hardness. The main effects plot for SN ratios shown in Figure 10 reveals that the optimal processing parameters for the best hardness of Al-Mg-PPA composites include a sintering temperature = 550 °C, holding time = 2 min, compaction pressure = 52 MPa, reinforcement concentration = 15 wt% PPA, and a heating rate = 200 °C. As evident from Table 7, the effect of material on the micro-hardness is statistically significant.

3.6. *Influence of Sintering Parameters and Microstructure on Micro-Hardness*

As illustrated in Figure 10, high sintering temperature and compaction pressure are required for the enhanced Al/Mg/PPA composites' hardness. The sintering temperature facilitates the consolidation of the composites' powder particles by causing interparticle welding to occur. This enhances interfacial bonding strength, increasing hardness. At higher sintering temperatures, greater consolidation of powder particles takes place, reducing porosity. However, at higher sintering temperatures, higher interfacial precipitation occurs.

The SEM-EDS analysis, XRD peaks, and SEM mapping illustrated in Figures 6d, 11, 12b,d and 13 show the presence of precipitates distributed within the grain boundaries. These precipitates are more in samples reinforced with PPA and, as a result, have larger areas of grain boundaries. These precipitates, which are chemically composed of elements such as Al, Mg, C, Si, K, O, and Fe, originated from the starting materials in this study. More so, aluminum easily reacts with oxygen to form oxide, which serves as impurities, hence the presence of oxygen in the samples, as shown in Figure 13. The presence of aluminum oxide in the samples may have influenced the result of the EDS analysis. The XRD pattern illustrated in Figure 11 reveals the presence of major and minor phases. Using Xpert HighScore 5.2 software for the peak analysis and from the literature, peaks such as (38.58°), (45.01°), (65.24°), (78.44°), and (82.54°) correspond to the Al, K_2O, Al_2O_3, $MgAl_2O_4$, and MgO phases, respectively. Considering all the elements and compounds present in the composite samples, it can be inferred that precipitates such as AlMgSi and Al_4C_3 are also present in the microstructure. Interfacial precipitates such as AlMgSi and $MgAl_2O_4$ may be responsible for the improved hardness of the composites. At higher sintering temperatures, Al_4C_3 precipitates increase. Al_4C_3 is inherently brittle. While a small concentration of it can improve hardness, large concentrations in composite samples can reduce hardness [27,28]. This explains why the composite samples reinforced with 20 wt% PPA consistently showed a decline in hardness. The SPS machine uses a very high heating rate and, as a result, may not require the use of a long holding time, hence an optimal holding time of 2 min, as illustrated in Figure 10. Furthermore, the sintering pressure transmutes the loose powder into compacts during sintering. Higher pressure is required for adequate compaction, and a well-compacted sample has reduced porosity.

3.7. *Influence of Reinforcement Characteristics and Microstructure on Micro-Hardness*

As evident in Tables 6 and 7, the reinforcement material has the greatest impact on the hardness of the Al/Mg/PPA samples. At an optimal reinforcement of 15 wt% PPA, the developed composites had the highest hardness. Beyond the 15 wt% PPA, the hardness of the composites consistently declined, as depicted in Figure 14. More so, it can be seen from Figure 10 that the highest peak for the material, which is the highest point above the dotted line, is 98 wt% Al. This represents the sintered samples without PPA reinforcement. On average, the sintered samples without the PPA reinforcement displayed the best value of hardness. This can be attributed to the stronger surface adhesion in Al/Mg samples than in Al/Mg/PPA samples. As stated earlier, the microstructure of the composites reinforced with PPA shows the presence of Al, Mg, C, Si, K, O, and Fe, as indicated by the SEM-EDS analysis in Figures 6d and 12b. Similarly, the EDS analysis of the samples without PPA reinforcement shown in Figure 12d indicates the presence of Al, Mg, and O phases. This suggests that precipitates such as MgO and $MgAl_2O_4$ are present in the microstructure, as confirmed by the XRD peaks in Figure 11. While substantial precipitation of Al_4C_3 deteriorates the microstructure and reduces the mechanical properties such as hardness, interfacial precipitation, such as that of $MgAl_2O_4$, increases the interfacial bonding strength and hardness [29,30].

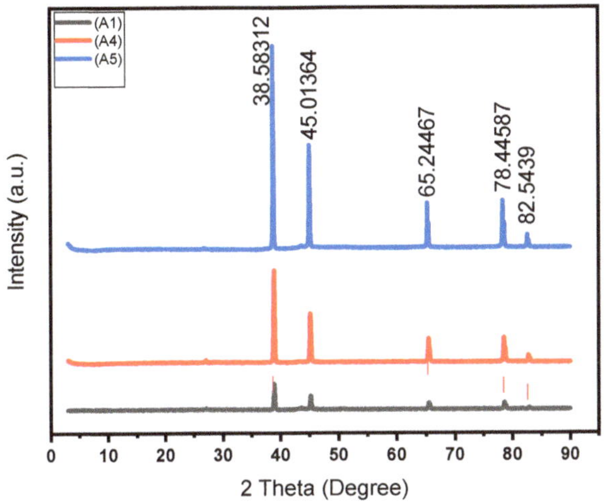

Figure 11. XRD pattern for the sintered Al/Mg/PPA samples.

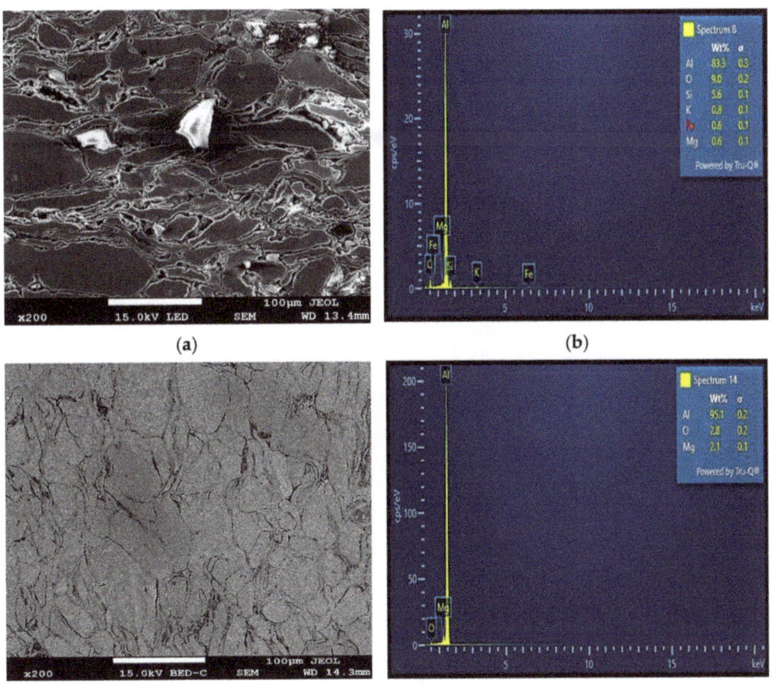

Figure 12. SEM micrographs and EDS analysis of Al/Mg/PPA samples (**a**,**b**) A2 and (**c**,**d**) A1.

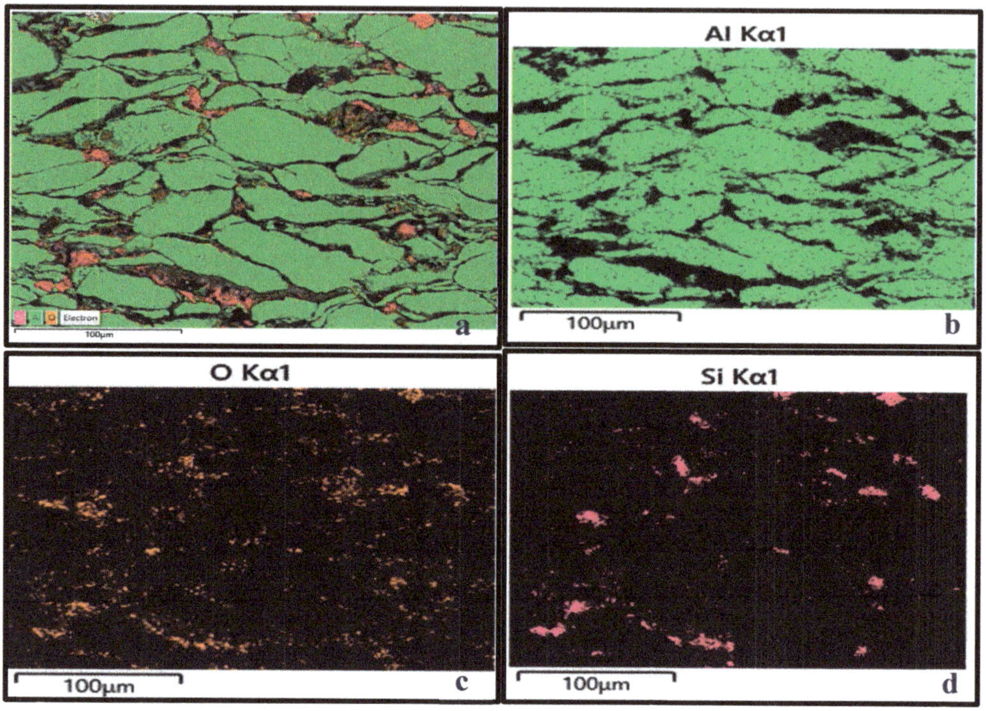

Figure 13. SEM mappings of Al/Mg/PPA composites (**a–d**) A2.

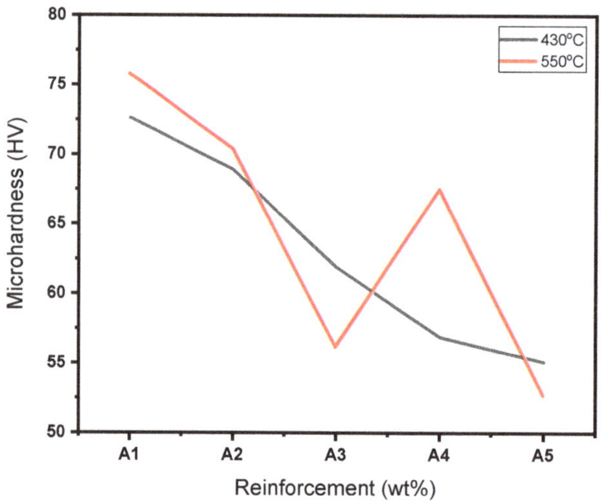

Figure 14. Influence of reinforcement concentration and sintering temperature on the hardness of Al/Mg/PPA composites.

Furthermore, adding harder particles to metal matrices raises hardness. The concentration of the harder particles in the samples determines the magnitude of hardness. For instance, Tosun and Kurt [31] discovered that the hardness of AMC reinforced with SiC was higher than the ones reinforced with Al_2O_3. This is so because SiC (2350 HV) is harder than Al_2O_3 (880 HV). Therefore, adding PPA and Mg particles to the Al matrix enhanced the hardness of the fabricated samples.

The degree of wettability between the matrix and reinforcement influences hardness. To improve the wettability in this study, 2 wt% Mg was introduced into the Al/PPA mixes. A maximum hardness of 77 HV was achieved at 15 wt% PPA concentration, as seen in Table 3. However, the composites synthesized with 20 wt% PPA (A5) consistently showed a decline in hardness, as illustrated in Figure 14. This can be explained by the weak interfacial bonding strength and poor surface adhesion. Therefore, it can be inferred that as the concentration of PPA increases, the degree of wettability and surface adhesion drops. In addition, the hardness of Al/Mg/PPA composites reinforced with 20 wt% PPA dropped because the PPA particles may not be as hard as the synthetic reinforcement particles such as B_4C and SiC. More so, the K_2O phase in the microstructure, as indicated by the XRD peaks, may be one of the weak phases that can impair hardness. Therefore, hardness reduces as the weight fraction of the weak phase increases in the composites.

The homogeneity of the powder particles, as indicated by the SEM and optical images, is due to the improved hardness of the composites. Particle agglomeration results in micro-cracks, weak interfacial bonding, and increased pores, which impair hardness. At higher concentrations of PPA, particle clustering is inevitable, thereby reducing hardness.

The refinement of the matrix and reinforcement particles significantly influenced hardness. Fine particles enhance hardness compared to coarse particles, and mixed particle sizes promote better packing than uniform-size particles [32,33]. The improved hardness of Al/Mg/PPA composites can be attributed to the fine and mixed particle sizes, as shown in Figure 8. During milling, the high-energy balls collide with the powder particles, thereby fracturing larger particles into fine ones. In addition, the plastic deformation that occurs during the ball milling process causes strain hardening and particle transmutation. Fine and diverse particle sizes result in better particle packing and alignment, making it tougher for dislocation motion to pass through. This ultimately enhances hardness.

4. Conclusions

This article reported an investigation of the densification and hardness properties of Al/Mg/PPA samples with reference to PPA reinforcement and SPS parameters. The results show that the enhanced densification and hardness of the synthesized composites are caused by factors such as processing parameters, reinforcement characteristics, and particle surface adhesion. As the sintering temperatures increased, the porosity of the sintered samples decreased. The RD was significantly impacted by the sintering temperature and material, whereas the matrix/reinforcement characteristics significantly impacted the hardness property. The best RD of Al/Mg/PPA composites was obtained at 550 °C sintering temperature, 4 min holding time, 42 MPa compaction pressure, 250 °C heating rate, and 15 wt% PPA reinforcement. The values of these parameters for the optimal hardness were 550 °C, 2 min, 52 MPa, 200 °C, and 15 wt%, respectively.

Author Contributions: Conceptualizing and designing experiments—O.O.E., F.K.T., P.A.O. and K.G.; Investigation and Software—O.O.E.; Analysis and optimization—O.O.E., F.K.T., P.A.O. and K.G.; Writing original draft—O.O.E.; Review, Editing, and Supervision—F.K.T., P.A.O. and K.G. All authors have read and agreed to the published version of the manuscript.

Funding: This research received no external funding.

Institutional Review Board Statement: Not applicable.

Informed Consent Statement: Not applicable.

Data Availability Statement: The original contributions presented in this study are included in the article. Further inquiries can be directed to the corresponding author.

Conflicts of Interest: The authors declare that they have no known competing financial interests or personal relationships that could have appeared to influence the work reported in this paper.

References

1. Kulkarni, P.P.; Siddeswarappa, B. A study on microstructure and mechanical behavior of AA6063 metal matrix composite reinforced with areca sheath ash (ASA) and rice husk ash (RHA). *Mater. Today Proc.* **2022**, *52*, 445–451. [CrossRef]
2. Dwivedi, S.P.; Srivastava, A.K.; Maurya, N.K.; Sahu, R. Microstructure and mechanical behavior of Al/SiC/agro-waste RhA hybrid metal matrix composite. *Rev. Des. Compos. Mater. Av.* **2020**, *30*, 43–47. [CrossRef]
3. Bodunrin, M.O.; Alaneme, K.K.; Chown, L.H. Aluminum matrix hybrid composites: A review of reinforcement philosophies; Mechanical, corrosion and tribological characteristics. *J. Mater. Res. Technol.* **2015**, *4*, 434–445. [CrossRef]
4. Bahrami, A.; Soltani, N.; Pech-Canul, M.I.; Gutiérrez, C.A. Development of metal-matrix composites from industrial/agricultural waste materials and their derivatives. *Crit. Rev. Environ. Sci. Technol.* **2016**, *46*, 143–207. [CrossRef]
5. Sonker, P.K.; Jackson Singh, T.; Kumar Verma, Y. Effect of reinforced diverse synthetic and agricultural-industrial waste materials on the microstructural and mechanical performance of hybrid metal matrix composite. *Mater. Today Proc.* **2023**, *106*, 94–98. [CrossRef]
6. Olusesi, O.S.; Udoye, N.E. Development and characterization of AA6061 aluminum alloy/clay and rice husk ash composite. *Manuf. Lett.* **2021**, *29*, 34–41. [CrossRef]
7. Phiri, R.; Rangappa, S.M.; Siengchin, S.; Marinkovic, D. Agro-waste natural fiber sample preparation techniques for bio-composites development: Methodological insights. *Facta Univ. Ser. Mech. Eng.* **2023**, *21*, 631. [CrossRef]
8. Alaneme, K.K.; Sanusi, K.O. Microstructural characteristics, mechanical and wear behavior of aluminum matrix hybrid composites reinforced with alumina, rice husk ash, and graphite. *Eng. Sci. Technol. Int. J.* **2015**, *18*, 416–422. [CrossRef]
9. Haigh, L.; de Wit, M.; von Daniels, C.; Colloricchio, A.; Hoogzaad, J. *The Circularity Gap Report 2021*; Circle Economy: Amsterdam, The Netherlands, 2021; pp. 1–71.
10. Abshalomu, Y.; Jyothi, Y.; Balamurugan, K.; Selvaraj, R. Effect of Varied Cashew Nut Ash Reinforcement in Aluminum Matrix Composite. *Adv. Mater. Sci. Eng.* **2023**, *2023*, 3383777. [CrossRef]
11. Prasad, D.S.; Shoba, C.; Ramanaiah, N. Investigations on mechanical properties of aluminum hybrid composites. *J. Mater. Res. Technol.* **2014**, *3*, 79–85. [CrossRef]
12. Ikele, U.S.; Alaneme, K.K.; Oyetunji, A. Mechanical behavior of stir cast aluminum matrix composites reinforced with silicon carbide and palm kernel shell ash. *Manuf. Rev.* **2022**, *9*, 12. [CrossRef]
13. Surappa, M.K. Aluminium matrix composites: Challenges and opportunities. *Sadhana Acad. Proc. Eng. Sci.* **2003**, *28*, 319–334. [CrossRef]
14. Saad, M.H.; Nazzal, M.A.; Darras, B.M. A general framework for sustainability assessment of manufacturing processes. *Ecol. Indic.* **2019**, *97*, 211–224. [CrossRef]
15. Ayodele, O.O.; Babalola, B.J.; Olubambi, P.A. Characterization, nanomechanical, and wear attributes of sintered Al–TiB2 composites. *J. Mater. Res. Technol.* **2023**, *24*, 4153–4167. [CrossRef]
16. Sharma, S.K.; Saxena, K.K.; Salem, K.H.; Mohammed, K.A.; Singh, R.; Prakash, C. Effects of various fabrication techniques on the mechanical characteristics of metal matrix composites: A review. *Adv. Mater. Process. Technol.* **2024**, *10*, 277–294. [CrossRef]
17. Soares, E.; Bouchonneau, N.; Alves, E.; Alves, K.; Araújo Filho, O.; Mesguich, D.; Chevallier, G.; Laurent, C.; Estournès, C. Microstructure and mechanical properties of AA7075 aluminum alloy fabricated by spark plasma sintering (SPS). *Materials* **2021**, *14*, 430. [CrossRef] [PubMed]
18. Lade, J.; Mohammed, K.A.; Singh, D.; Verma, R.P.; Math, P.; Saraswat, M.; Gupta, L.R. A critical review of fabrication routes and their effects on mechanical properties of AMMCs. *Mater. Today Proc.* **2023**. [CrossRef]
19. Edosa, O.O.; Tekweme, F.K.; Olubambi, P.A.; Gupta, K. Microstructural Analysis, Compressive Strength, and Wear Properties of Spark-Plasma-Sintered Al–Mg–PPA Composites. *Quantum Beam Sci.* **2024**, *8*, 32. [CrossRef]
20. Guan, H.D.; Li, C.J.; Gao, P.; Yi, J.H.; Bao, R.; Tao, J.M.; Fang, D.; Feng, Z.X. Fe-based metallic glass particles reinforced Al-7075 matrix composites prepared by spark plasma sintering. *Adv. Powder Technol.* **2020**, *31*, 3500–3506. [CrossRef]
21. Anand Sekhar, R.; Rakesh Pillai, R.; Mubarak Ali, M.; Shyam Kumar, C.N. Enhanced mechanical and wear properties of aluminium-based composites reinforced with a unique blend of granite particles and boron carbide for sustainable material recycling. *J. Alloys Compd.* **2023**, *963*, 171165. [CrossRef]
22. Rominiyi, A.L.; Shongwe, M.B.; Tshabalala, L.C.; Ogunmuyiwa, E.N.; Jeje, S.O.; Babalola, B.J.; Olubambi, P.A. Spark plasma sintering of Ti–Ni–TiCN composites: Microstructural characterization, densification and mechanical properties. *J. Alloys Compd.* **2020**, *848*, 156559. [CrossRef]
23. Akbarpour, M.R. Microstructure, Densification, and Compressive Properties of Al-CNT Metal Matrix Composites Fabricated by Flake Powder Metallurgy and Conventional Sintering Process. *Adv. Ceram. Prog.* **2022**, *8*, 27–33.

24. Rahimian, M.; Ehsani, N.; Parvin, N.; Baharvandi, H. The effect of particle size, sintering temperature and sintering time on the properties of Al-Al2O3 composites, made by powder metallurgy. *J. Mater. Process. Technol.* **2009**, *209*, 5387–5393. [CrossRef]
25. Ponhan, K.; Tassenberg, K.; Weston, D.; Nicholls, K.G.M.; Thornton, R. Effect of SiC nanoparticle content and milling time on the microstructural characteristics and properties of Mg-SiC nanocomposites synthesized with powder metallurgy incorporating high-energy ball milling. *Ceram. Int.* **2020**, *46*, 26956–26969. [CrossRef]
26. Surya, M.S.; Gugulothu, S.K. Fabrication, Mechanical and Wear Characterization of Silicon Carbide Reinforced Aluminium 7075 Metal Matrix Composite. *Silicon* **2022**, *14*, 2023–2032. [CrossRef]
27. Xiong, B.; Liu, K.; Yan, Q.; Xiong, W.; Wu, X. Microstructure and mechanical properties of graphene nanoplatelets reinforced Al matrix composites fabricated by spark plasma sintering. *J. Alloys Compd.* **2020**, *837*, 155495. [CrossRef]
28. Wu, X.; Zhang, W. A review on aluminum matrix composites' characteristics and applications for automotive sector. *Heliyon* **2024**, *10*, 38576. [CrossRef] [PubMed]
29. Seyed Pourmand, N.; Asgharzadeh, H. Aluminum Matrix Composites Reinforced with Graphene: A Review on Production, Microstructure, and Properties. *Crit. Rev. Solid State Mater. Sci.* **2020**, *45*, 289–337. [CrossRef]
30. Dudina, D.V.; Georgarakis, K.; Olevsky, E.A. Progress in aluminium and magnesium matrix composites obtained by spark plasma, microwave and induction sintering. *Int. Mater. Rev.* **2023**, *68*, 225–246. [CrossRef]
31. Tosun, G.; Kurt, M. The porosity, microstructure, and hardness of Al-Mg composites reinforced with micro particle SiC/Al2O3 produced using powder metallurgy. *Compos. Part B Eng.* **2019**, *174*, 106965. [CrossRef]
32. Manohar, G.; Pandey, K.M.; Maity, S.R. Effect of spark plasma sintering on microstructure and mechanical properties of AA7075/B4C/ZrC hybrid nanocomposite fabricated by powder metallurgy techniques. *Mater. Chem. Phys.* **2022**, *282*, 126000. [CrossRef]
33. Delavari, M.; Salarvand, A.; Rahi, A.; Shahri, F. The effect of powder metallurgy process parameters on mechanical properties of micro and nano-iron powder. *Int. J. Eng. Sci. Technol.* **2012**, *3*, 86–94. [CrossRef]

MDPI AG
Grosspeteranlage 5
4052 Basel
Switzerland
Tel.: +41 61 683 77 34

Materials Editorial Office
E-mail: materials@mdpi.com
www.mdpi.com/journal/materials

www.ingramcontent.com/pod-product-compliance
Lightning Source LLC
Chambersburg PA
CBHW061441210926
51850CB00008B/1573
9783725851973